Lecture Notes in Mathematics   2233

More information about this series at http://www.springer.com/series/304

Kenneth W. Johnson

Group Matrices, Group Determinants and Representation Theory

The Mathematical Legacy of Frobenius

 Springer

Kenneth W. Johnson
Department of Mathematics
Pennsylvania State University
Abington, PA, USA

ISSN 0075-8434   ISSN 1617-9692   (electronic)
Lecture Notes in Mathematics
ISBN 978-3-030-28299-8   ISBN 978-3-030-28300-1   (eBook)
https://doi.org/10.1007/978-3-030-28300-1

Mathematics Subject Classification (2010): Primary: 20-XX; Secondary: 05-XX

This Springer imprint is published by the registered company Springer Nature Switzerland AG.
The registered company address is: Gewerbestrasse 11, 6330 Cham, Switzerland

I have often pondered over the roles of knowledge or experience on the one hand and imagination or intuition on the other, in the process of discovery. I believe that there is a fundamental conflict between the two, and knowledge, by advocating caution, tends to inhibit the flight of imagination. Therefore, a certain naiveté, unburdened by conventional wisdom, can sometimes be a positive asset.

—Harish-Chandra

This work is dedicated to my wife, Susan, and our friend, Helen Eves, who provided support at a crucial time.

Preface

In 1896, after correspondence with Dedekind, Frobenius published two papers in which group character theory was introduced and was used to describe the factorization of the group determinant for a finite group. Subsequently, again, after correspondence with Dedekind, he showed that characters are traces of matrix representations. From this simple beginning, group representation theory has gradually expanded to pervade much of mathematics and many of its areas of application. The collection of articles in the book of Mackey [209] gives some indication of how hard it is to put boundaries on its influence.

Frobenius was a student of Weierstrass in Berlin and had been called back in 1893 to assume the position of Full Professor (Ordentliche Professor). This achievement was then a pinnacle in the world of mathematics. In 1911, the American magazine *Popular Science Monthly* lamented that no Americans could be named among the foremost mathematicians of the eighteenth and nineteenth centuries and listed as examples of the top mathematicians of the day Poincaré, Klein, Hilbert, Frobenius, Jordan, Picard, and Darboux (who represented Paris, Berlin, and Göttingen). However, during his tenure in Berlin, Frobenius must have experienced some disappointment in that the school of Berlin lost its preeminence in German mathematics to Göttingen whose ascendance began after the appointment of Hilbert in 1895. Gradually, the abstract approach began to dominate. The fashion in research drifted away from the way he looked at mathematics towards what would become the approach of Bourbaki, and his methods based on the extremely skillful use of matrices and determinants became overshadowed by those of the Göttingen school. In Germany and elsewhere, important mathematicians continued to use his methods, most notably Schur, Brandt, and Richard Brauer, but the impact of their successors seems to have been overshadowed by the decline of German mathematics and the rise of Bourbaki. Nonetheless, Frobenius was a great mathematician, and the techniques emanating from him and his pupil Schur are intrinsic to many areas of modern mathematics, especially in the interface between combinatorics and algebra, and perhaps because it is often difficult for modern mathematicians to follow their methods, their work still contains parts which have not entered the modern canon. Partly because of the startling impact of physics on mathematics in recent years, it is now acknowledged that the ideas on how mathematics progresses are subject to constant change and that there are no absolutes (e.g., see [50]).

Frobenius was regarded as an exponent of a more rigorous way of doing mathematics, especially in his approach to linear algebra, but after the abstract methods of E. Noether and her collaborators became accepted, his techniques using matrices and determinants seem to have been looked upon as old-fashioned. Whereas Frobenius made comments on the mathematics of Klein and Lie as being unrigorous and fanciful, Noether regarded even the use of matrices as being unnecessary.

When the author began asking questions in the mid-1980s about the work of Frobenius on group matrices and group determinants, it proved hard to find mathematicians who were aware of all the techniques in his original group determinant papers. There were descriptions in historical journals, most notably those of Hawkins, but it appeared that simple questions such as whether the group determinant contained sufficient information to determine a group had never been formulated. Eventually, Hoehnke, who was a student of Brandt and who was expert in the use of several of Frobenius' tools but who had been somewhat isolated in East Germany, heard of the questions. He quickly indicated relevant techniques, and this stimulated new results which have shed light on the theory of representations of groups and algebras. In this book, an attempt is made to set out exactly what Frobenius did, and to highlight the places where his techniques have appeared in modern work, often in disguised form. The constructive approach to the theory of algebras represented by the work of Hoehnke and the school of A. Bergmann may be said to be a continuation of Frobenius' ideas and has provided tools to answer questions arising in the theory. This has so far been available only in journal articles, mostly in German, and an account of the relevant parts is given here. The following discussion indicates some other reasons why the book has been written.

In 1963, at a high point in the theory of finite groups, Richard Brauer wrote in the survey paper [30]:

> It has been written said by E. T. Bell that 'Wherever groups disclosed themselves or could be introduced, simplicity crystallized out of comparative chaos.' This may often be true, but, strangely enough, it does not apply to group theory itself, not even when we restrict ourselves to groups of finite order. We are reminded of the educators who want to educate the world and cannot handle our own children. A tremendous effort has been made by mathematicians for more than a century to clear up the chaos in group theory. Still, we cannot answer some of the simplest questions.

Brauer then sets out a series of unsolved questions in the theory of group characters and representations, many of which remain without satisfactory answers.

The quotes from the prefaces to the two editions of Burnside's book [39] are also revealing.

From the first edition (1897):

> Cayley's dictum that "a group is defined by means of the laws of combination of its symbols" would imply that, in dealing purely with the theory of groups, no more concrete mode of representation should be used than is absolutely necessary. It may then be asked why, in a book which professes to leave all applications aside, a considerable space is

devoted to substitution groups; while other particular modes of representation, such as groups of linear transformations, are not even referred to. My answer to this question is that while, in the present state of our knowledge, many results in the pure theory are arrived at most readily by dealing with the properties of substitution groups, it would be difficult to find a result that could be directly obtained by consideration of groups of linear transformations.

In modern parlance, a substitution group is a permutation group.
From the second edition (1911):

In particular, the theory of groups of linear substitutions has been the subject of numerous and important investigations by several writers.... In fact it is now more true to say that for further advances in the abstract theory one must look largely to the representation of a group as a group of linear substitutions.

Over 50 years after Brauer's comments and over a century after those of Burnside, there remains a murkiness, at least to the author, in that whereas statements of questions and results in finite group theory can be formulated very precisely, it is hard to understand why some of the proofs which have been obtained so far are inaccessible to the average mathematician. The difficulty of understanding the Classification Theorem overshadows the area, and several very good mathematicians have put forward the view that a new way of looking at group theory is needed to provide an accessible proof of it. The question of why representation theory appears to be necessary to group theory as well as in its many applications to other areas also remains elusive. This leads to a further reason for writing this book: to take a different look in a naive way at the basics of group representation theory which hopefully may address some of the above concerns.

The naive aspect of the approach here may be helpful in other ways. In order to bring students to the frontiers of group representation theory in as reasonably short a time as possible, most texts take advantage of the condensation which modern abstract notation enables, so that statements which mathematicians such as Cayley, Klein, or Poincaré made at length and perhaps not very precisely can often be rewritten in a few lines. This is entirely appropriate in many cases, especially in statements about finite groups.

However, the new areas of mathematics which have been suggested by modern physics, or the way in which representation theory has been used in differential geometry, make it difficult to come away with the impression that the essential concepts can be expressed in a clear-cut way. The differential geometer S. S. Chern made the following statement ([59], concluding remarks). "Modern differential geometry is a young subject. Not counting the strong impetus it received from relativity and topology, its developments have been continuous. I am glad that we do not know what it is and, unlike many other mathematical disciplines, I hope it will not be axiomatized. With its contact with other domains in and outside of mathematics and with its spirit of relating local to global, it will remain a fertile area for years to come." It seems that however hard mathematicians strive to make the subject clear and precise, it has the habit of escaping from the bounds imposed. The following example which is described in the book Emergence of the Theory of Lie Groups: An Essay in the History of Mathematics 1869–1926 by Hawkins

[138] may be a paradigm. The context was the interaction between mathematics and the "new" quantum theory in the early 1920s. Hermann Weyl had reluctantly used tensor analysis which at that time involved a plethora of indices. On pp. 450–455 of the book, an exchange between Study and Weyl is described, which began when Study implicitly chided Weyl for neglecting the tools of the algebraic theory of invariants and which led to Weyl quickly absorbing ideas in invariant theory to simplify his current work at the frontiers. It is somewhat ironic that, according to Hawkins, Frobenius held much of the work on invariants up to ridicule.

The homogeneous polynomials which were the focus of Frobenius' investigations, unlike matrix representations, are independent of a choice of basis, and where there is no apparent homomorphism into a linear group, they offer possibilities of extension to "group-like" objects such as quantum groups. Furthermore, the rapidly improving symbolic manipulation packages now available can enable us to calculate examples without needing the expertise of Frobenius.

The study of the homology and cohomology of finite groups has important applications to other areas, but the impact within finite group theory seems to be peripheral, in the sense that there are few results for which cohomological or homological tools are necessary for the proofs. The k-characters of Frobenius have some similarity with cocycles, and the corresponding polynomials have similar algebraic properties to Chern classes, and it seems worthwhile to examine these tools as perhaps an improvement on group cohomology.

It is easy to present a further reason for the dissemination of the material here, in that group matrices have appeared spontaneously in several areas of current interest, among which are Fourier analysis, control theory, random walks on groups, wavelets, and group rings. The formulae defining k-characters in Frobenius's description of group determinant factors have also appeared in diverse areas such as the invariant theory of matrices [239], Helling's characterization of group characters [141], Wiles and Taylor's papers leading up to the proof of Fermat's theorem [275], work by Buchstaber and Rees on symmetric spaces [36], and papers on higher cumulants in probability [249]. The connections with Frobenius have usually been made a posteriori, and it is hoped that this book can avoid some of the repetition involved, as well as to bring his tools to the attention of a wider mathematical public.

The motivation to examine the original papers of Frobenius was an attempt to extend character theory to loops and quasigroups, which may be regarded as non-associative analogues of groups. This led to the connections between the work of Frobenius and the theory of association schemes and related objects. The intuition carried over from group theory, especially using the viewpoint of Frobenius, has produced new results and ideas related to association schemes and group characters, and a chapter of this book develops this. Although this chapter has some overlap with existing accounts, it differs significantly from them.

In his letters to Dedekind [137], Frobenius provides a humorous analogy between mathematical research and buying horses: "One should never look at the horse that one really wants to buy." Perhaps, the results that this author is furtively examining out of the corner of his eye are the extensions of the tools arising from group determinant work in finite group theory to infinite harmonic analysis in all its forms, including relevant parts of mathematical physics.

An attempt has been made to develop the theory as much as possible from the point of view of Frobenius, in order to present the tools he used in as simple a form as possible, as well as to preserve his insight, although in some places the exposition is simplified by using subsequent work. Many of the topics presented have not previously appeared in book form.

Frobenius was the product of the very formal Prussian milieu, which is described by Siegel in his "Erinnerungen an Frobenius" in [110], p. iv. (This is summarized in Appendix B.) It is likely that Frobenius would be displeased at the informal manner in which parts of this book have been written. The author would like to apologize to those who are similarly offended.

This book originated as a joint project with S. Strunkov in Moscow, but the vagaries of life in the ex-Soviet Union made the collaboration too difficult. The book has benefitted from discussions with him especially during joint stays at Oberwolfach, the Euler Institute in St. Petersburg and at Charles University in Prague.

Since there are several books which have overlap with the material presented here, there follows brief descriptions of the contents of some of these, together with an indication of their relevance to this volume.

1. *Pioneers of Representation Theory*: Frobenius, Burnside, Schur, and Brauer, by Curtis [68]. This book is historical in emphasis and does not claim to expound on current developments of the group determinant work of Frobenius. It gives an account of his 1896 papers and his subsequent development of representation theory without going into the k-characters. As might be expected, given that Curtis has also co-written one of the bibles of group representation theory, his book is very useful. His description of Schur's work is followed closely here.

2. *The Scope and History of Commutative and Noncommutative Harmonic Analysis* by Mackey [209]. This is a compendium of articles written for different audiences, and as Mackey indicates, his project expanded considerably from its initial conception. It is invaluable as it sets out harmonic analysis and its relationship to group representation theory in a very wide context. Infinite groups are perhaps the main concern, but much is relevant to the theory of finite groups and good explanations of how extensions of the theory from the finite to the infinite case are natural. It mentions Frobenius' group determinant work, but quickly reverts to the more conventional approach to representations.

3. *Emergence of the Theory of Lie Groups: An Essay in the History of Mathematics 1869–1926* by Hawkins [138]. This contains a historical account of the work of Lie, Killing, Cartan, and Weyl up to the 1920s. It focuses on what Lie was trying to achieve and how the ideas which he initiated were carried on by his main successors. Although it does not deal much with finite groups and their representations, in the opinion of this author, it indicates many directions in the theory of Lie groups and Lie algebras which could benefit from an examination of the tools presented in this book.

The next two books introduce an important area of applications.

4. *Group Representations in Probability and Statistics* by Diaconis [82]. Probabilistic concepts are inherent right at the foundations of Frobenius's group
   determinant work. This book, which is a set of lecture notes, gives many
   facts about group representation theory and association schemes on the way
   to applications in probability theory. It has several results on representations
   which are hard to find in other sources. It is wide-ranging and is written from
   the applied point of view, in that tools are developed insofar as they are useful
   for applications. It does not explicitly mention the connection to Frobenius' work
   but uses the group matrix.

5. *Harmonic Analysis on Finite Groups* by Ceccherini-Silberstein et al. [54]. This
   may be regarded as a continuation of part of the book by Diaconis, filling in some
   of the areas. Chapters 4 and 7 below have some overlap with this book, but the
   approach here is significantly different.

   Another book which discusses Fourier analysis on finite groups is [276].

6. *Algebraic Combinatorics I: Association Schemes* by Bannai and Ito [10]. This set
   of lecture notes contained the first discussion of the theory of association schemes
   in book form and is a valuable source. Unfortunately, the promised successor has
   not yet appeared. Their point of view treats association schemes as a means of
   abstracting the mathematics behind group character theory, although they bring
   in other sources as well. There are now other books which have perhaps a wider
   combinatorial flavor such as *Algebraic Combinatorics*, by C. D. Godsil. Again,
   there is overlap with Chap. 4 but the exposition here is essentially different. The
   lecture notes [297] also have an interesting viewpoint.

7. *Circulant Matrices* by Davis [72]. The book is written at an elementary level
   giving a compendium of properties of circulants and their applications. From the
   point of view here, circulants are group matrices of cyclic groups. Chapter 2 of
   this book extends the discussion of some of the questions which Davis examined
   for circulants to arbitrary group matrices, and in Chap. 7 a generalization of his
   discussion of Fourier transforms appears.

8. *An Introduction to Quasigroups and Their Representations* by Smith [263].
   This book is an intensive study of how representation theoretic concepts can
   be extended to the theory of quasigroups. The book contains many interesting
   insights into the tools of group representation theory, since their extensions to
   quasigroups often bring out the essence of the techniques. Parts of Chap. 4 may
   be used as an introduction to this book.

   The most recent addition to the literature on Frobenius is the following.

9. *The Mathematics of Frobenius in Context: A Journey Through 18th to 20th
   Century Mathematics* by Hawkins [139]. Hawkins has contributed essential
   articles in historical journals on many of the contributions of Frobenius, and this
   book is invaluable in that it provides a thorough account of the various parts
   of his work. In particular, it indicates areas of mathematics where the modern
   "canonical" presentation may essentialy be found in Frobenius' work and, also,
   those where the approach of Frobenius, initially superceded, was subsequently
   found to be fundamentally important in modern applications. It describes

Frobenius' relationship with the mathematics of Weierstrass and his place in German mathematics. The influence of Dedekind, not only with regard to the group determinant but also related to Frobenius' work on complexes of elements, is related. Hawkins explains how Frobenius' work on theta functions led directly to the techniques which he used in his papers on the group determinant. The book also gives interesting insights into his personal characteristics. The book provides a good explanation of the continued influence of Frobenius and his successor Schur on current mathematics.

It is not easy to have a clear view of Frobenius as a person. Some discusion of Frobenius' career and his interactions within the world of mathematics is given in Appendix B.

Acknowledgments Thanks are due to the following institutions which supported the author during the writing of this book:

- Brigham Young University
- Charles University, Prague
- Iowa State University
- New College of Florida, Sarasota
- Rutgers University
- The Euler Institute, St. Petersburg
- The Mathematisches Forschungsinstitut Oberwolfach
- The Ohio State University
- The Pennsylvania State University, University Park
- The University of Denver
- The University of Düsseldorf, Germany
- The University of Potsdam, Germany

The book would not have been written without the help and encouragement of many colleagues. The following is an incomplete list of those who gave help and of various kinds.

- A. Bergmann
- P. Cameron
- P. Diaconis
- T. Hawkins
- H-J. Hoehnke
- S. P. Humphries
- T. Hurley
- E. Poimenidou
- D. St.P. Richards
- S. Sahi
- Surinder K. Sehgal
- J. D. H. Smith
- S. Strunkov
- P. Vojtechovsky
- S. Waldron

In addition, R. Taylor and P. Deligne gave brief but invaluable consultations, the first on work involving "pseudocharacters" and the second on Poincaré's differential equation work.

Abington, PA, USA   Kenneth W. Johnson

How to Read the Book

Parts of the book are designed to be read by a very general audience of mathematicians and others in different subject areas who use group representation theory. There are other parts which require technical expertise of various kinds.

Chapter 1 is designed for a general audience, which could include good undergraduates. Chapter 2 is designed first to give an account of group matrices in a naive way, similar to that in the book by Davis on circulants. This could be read by those who use group matrices in electrical engineering. The work on supermatrices and Hopf algebras needs some familiarity with these and could be omitted at first reading.

Chapter 3 contains technical work arising from the use of norm forms in the theory of algebras and could be skipped through on a first reading.

Chapter 4 brings in algebraic combinatorics and the group theory associated to S-rings. The prerequisites are moderate, but some character theory of groups is assumed. There are also parts which are of interest to those in probability theory.

Chapter 5 can be read most easily by those who have a good background in group character theory. It also demands some category theory.

Chapter 6 describes links with classical representation theory of S_n and $GL(n, \mathbb{C})$ as set out in the book by Fulton and Harris [112] (and references therein). Those interested in these important connections are advised to have this at their side when reading the chapter.

Chapter 7 is designed for probablists and those group theorists who are interested in the connection with probability theory and harmonic analysis in the wider sense. The books by Diaconis [85] (and references therein) and Ceccerini et al. [54] (and references therein) can be used to fill in background material.

Chapter 8 is primarily designed for those who are interested in the two areas which are described. It illustrates how the basic combinatorics behind k-characters provides more insight into interesting areas of geometry and probability. However, it is hoped that the general mathematical reader finds it accessible.

In Chap. 9, some of the first part is technical and is for the group theorist with a bent towards algebraic combinatorics. The work on supercharacters of upper triangular groups should interest those in probability theory and the theory of Hopf algebras, but the general reader could browse through since it provides an example of how the ideas in Chap. 4 have had applications.

Chapter 10 is a mixture of topics. The general reader should browse it and select topics of interest to them. Specialists in the various areas may find that results in the chapter can help them avoid repetition of existing results.

Contents

List of Diagrams

Chapter 1
Multiplicative Forms on Algebras and the Group Determinant

Abstract The ideas in the initial papers by Frobenius on characters and group determinants are set out and put into context. It is indicated how the theory goes back to the search for "sums of squares identities", the construction of "hypercomplex numbers" and the investigation of quadratic forms. The underlying objects, the group matrix, and its determinant, the group determinant, are introduced. It is shown that group matrices can be constructed as block circulants. The first construction by Frobenius of group characters for noncommutative groups is explained. This led to his construction of the irreducible factors of the group determinant. The k-characters, used to construct the irreducible factor corresponding to an irreducible character, are defined. Resulting developments are then discussed, with an indication of how the ideas of Frobenius were taken up by other mathematicians and how his approach and its continuation in the theory of norm forms on algebras have been useful. A summary of the various ways in which the work has impacted current areas is included.

1.1 Introduction

In this chapter the basic ideas in the initial papers by Frobenius on characters and group determinants are set out and put into context. It is indicated how the theory goes back to the search for "sums of squares identities", the construction of "hypercomplex numbers" and the investigation of quadratic forms. The underlying object, the group matrix, is introduced. It is shown that group matrices can be constructed as block circulants (a more extensive discussion is given in Chap. 2). The other basic object, the group determinant, is the determinant of the group matrix. The first construction by Frobenius of group characters was from the class algebra, independent of matrix representations. Although this was published before his paper introducing group determinants, it seems to have been motivated by the problem of factoring the group determinant. A description is given of how Frobenius sets out the correspondence between irreducible factors of the group determinant and irreducible characters of the group. The k-characters, which he used to construct

© Springer Nature Switzerland AG 2019
K. W. Johnson, *Group Matrices, Group Determinants
and Representation Theory*, Lecture Notes in Mathematics 2233,
https://doi.org/10.1007/978-3-030-28300-1_1

the irreducible factor corresponding to a character, are defined. Proofs are given of the main results of Frobenius on group determinants, using simplifications which follow from Schur's approach to representation theory. Some of the subsequent developments are then given, indicating how the ideas of Frobenius were taken up by other mathematicians and how his approach and its continuation in the theory of norm forms on algebras have led to results for groups, for example the theorem that the 3-characters of a group determine the group. The chapter concludes by indicating further developments and work in progress, and the rest of the book expands on this.

1.2   Sums of Squares and Norm Forms on Algebras

Sources for the following are [68, 96] and [209].

The roots of the theory of norm forms on algebras may be said to be the "sums of squares identities". The familiar multiplicative property of the absolute value in the complex numbers:

$$|z_1 z_2| = |z_1|.|z_2| \qquad (1.2.1)$$

leads to the following identity, which seems to have been known to Diophantus

$$(a_1 a_2 - b_1 b_2)^2 + (a_1 b_2 + b_1 a_2)^2 = (a_1^2 + b_1^2)(a_2^2 + b_2^2). \qquad (1.2.2)$$

This is obtained by setting $z_1 = a_1 + ib_1$, $z_2 = a_2 + ib_2$ with $a_1, b_1, a_2, b_2 \in \mathbb{R}$ in (1.2.1). Significant effort was spent in determining the values of k for which there are similar identities which express a product of two sets of k squares as a sum of k squares (see [93] for an account of the work). Euler produced the identity for four squares in a letter to Goldbach in 1748, and the identity for eight squares was given in 1818 by Degen. The well-documented "discovery" of the quaternions $\mathbb{H}$ by Hamilton on October 16, 1843 while walking with his wife along the Royal Canal in Dublin, and that of the Octonions $\mathbb{K}$ in December 1843 by Graves, provided explanations for the existence of these identities. However according to Taussky [274], the multiplication rule for the quaternions appeared in the letter of Euler mentioned above and Gauss circa 1822–1823 also had the rule. The octonions were subsequently found independently by Cayley. Many of the historical details are given in [8]. More recently Pfister [231] showed that under weaker assumptions identities are available for 2^n squares for any n (a brief description of this appears below).

Identities among squares led to the examination of systems of **hypercomplex numbers**. The model for the construction of a system of hypercomplex numbers is the construction of $\mathbb{C}$ from $\mathbb{R}$. In general a set of "numbers" is obtained by adding "imaginary units" to the real numbers so that each number can be represented in the form

$$a_1 + a_2 i + a_3 j + \ldots + a_n l$$

where the a_i lie in $\mathbb{R}$, as in the case of Hamilton's construction of $\mathbb{H}$. This theory has now become absorbed into the theory of algebras.

A formal definition follows:

Definition 1.1 Consider the set $\mathbb{U}$ of expressions of the form

$$a_0 + a_1 \mathbf{i}_1 + a_2 \mathbf{i}_2 + \ldots + a_n \mathbf{i}_n, \tag{1.2.3}$$

where n is a fixed positive integer, $a_0, a_1, \ldots, a_n$ are real numbers and $\mathbf{i}_1, \mathbf{i}_2, \ldots, \mathbf{i}_n$ are symbols, called "imaginary units". Further suppose that

$$a_0 + a_1 \mathbf{i}_1 + a_2 \mathbf{i}_2 + \ldots + a_n \mathbf{i}_n = b_0 + b_1 \mathbf{i}_1 + b_2 \mathbf{i}_2 + \ldots + b_n \mathbf{i}_n,$$

if and only if

$$a_0 = b_0, \quad a_1 = b_1, \quad \ldots, \quad a_n = b_n.$$

Let $\bar{a} = (a_0 + a_1 \mathbf{i}_1 + a_2 \mathbf{i}_2 + \ldots + a_n \mathbf{i}_n)$ and $\bar{b} = (b_0 + b_1 \mathbf{i}_1 + b_2 \mathbf{i}_2 + \ldots + b_n \mathbf{i}_n)$. Define addition and subtraction of elements of $\mathbb{U}$ by

$$\bar{a} + \bar{b} = (a_0 + b_0) + (a_1 + b_1) \mathbf{i}_1 + \ldots + (a_n + b_n) \mathbf{i}_n.$$

$$\bar{a} - \bar{b} = (a_0 - b_0) + (a_1 - b_1) \mathbf{i}_1 + \ldots + (a_n - b_n) \mathbf{i}_n.$$

For each pair (α, β) in $\{1, \ldots, n\}$ there is given the product

$$\mathbf{i}_\alpha \mathbf{i}_\beta = c_{\alpha\beta}^1 \mathbf{i}_1 + c_{\alpha\beta}^2 \mathbf{i}_2 + \ldots + c_{\alpha\beta}^n \mathbf{i}_n,$$

where the $c_{\alpha\beta}^k$ are real numbers. Thus $\mathbf{i}_\alpha \mathbf{i}_\beta$ is an element of $\mathbb{U}$. The product of $\bar{a}$ and $\bar{b}$ is then defined by using the distributive law, for example

$$(a_0 + a_1 \mathbf{i}_1)(b_0 + b_1 \mathbf{i}_1) = a_0(b_0 + b_1 \mathbf{i}_1) + a_1 \mathbf{i}_1(b_0 + b_1 \mathbf{i}_1)$$
$$= a_0 b_0 + a_0 b_1 \mathbf{i}_1 + a_1 b_0 \mathbf{i}_1 + a_1 b_1 (\mathbf{i}_1)^2$$

and the second expression is an element of $\mathbb{U}$ of the form (1.2.3) on replacing $(\mathbf{i}_1)^2$ by

$$c_{11}^1 \mathbf{i}_1 + c_{11}^2 \mathbf{i}_2 + \ldots + c_{11}^n \mathbf{i}_n$$

and using the addition formula. Then with the operations above $\mathbb{U}$ is defined to be a **system of hypercomplex numbers**.

1.2.1   Diophantine Equations and the Composition of Quadratic Forms

The theory of Diophantine equations is usually understood as the process of finding integer solutions of polynomial equations. Linear Diophantine equations are not difficult to address, but quadratic equations in two variables become much more of a challenge. The general problem is to find all integer solutions of

$$Ax^2 + Bxy + Cy^2 + Dx + Ey + F = 0 \tag{1.2.4}$$

where A, B, C, D, E, F are given integers. The linear terms in (1.2.4) can be eliminated by standard transformations and it is equivalent to solve equations of the form

$$Ax^2 + Bxy + Cy^2 = -F$$

which in turn is equivalent to the problem of deciding whether an integer $n \, (= -F)$ can be represented by the binary quadratic form

$$Q(x, y) = Ax^2 + Bxy + Cy^2,$$

and if so in how many ways.

   The first significant progress on the general case of this problem appears to be due to Lagrange. For any matrix of integers

$$\begin{bmatrix} a & b \\ c & d \end{bmatrix}$$

of determinant ± 1, the substitution $x = ax' + by'$, $y = cx' + dy'$ transforms $Q(x, y)$ into

$$Q'(x', y') = A'x'^2 + B'x'y' + C'y'^2$$

which represents the same set of integers as $Q(x, y)$. This action induces an equivalence relation on the set of binary quadratic forms and implies that it is only necessary to consider the problem for one member of each equivalence class. The **discriminant** of the form $Q(x, y)$ is $D = B^2 - 4AC$ and is an invariant of the equivalence class, but inequivalent forms may have the same discriminant. In trying to relate the problem to the prime factorization of n, Lagrange discovered that

(a) for a given discriminant D there is only a finite number of equivalence classes of forms with discriminant D and
(b) when there is more than one class of forms with the same discriminant it is convenient to consider them all together.

If $Q(x, y) = p_1 p_2$ where p_1 and p_2 are distinct primes and x and y are integers it does not necessarily follow that either of the equations $Q(x, y) = p_1$ or $Q(x, y) = p_2$ has a solution in integers. It can only be deduced that there exist forms $Q'(x, y) = A'x^2 + B'xy + C'y^2$ and $Q''(x, y) = A''x^2 + B''xy + C''y^2$ each with discriminant D such that $Q'(x, y) = p_1$ and $Q''(x, y) = p_2$ both have integer solutions. The further progression of Lagrange's ideas is summarized in [209, p. 11].

The problem was taken up by Gauss. He developed a theory of composition for the equivalence classes of binary quadratic forms of a fixed discriminant over $\mathbb{Z}$. He restricted the forms to be of the following type

$$Q(u) = Au_1^2 + 2Bu_1u_2 + Cu_2^2,$$

so that here the discriminant is $B^2 - AC$. His definition of the composition of classes involves complicated technical results but a consequence is that the composition $C_1 \circ C_2$ of a pair of classes always contains a product $Q_1' Q_2'$ where Q_1' and Q_2' lie in C_1 and C_2 respectively. The following example illustrates.

Take $Q(u) = u_1^2 + 5u_2^2$ and $Q'(u) = 2u_1^2 + 2u_1u_2 + 3u_2^2$, which both have discriminant -5. It may be shown that the forms are inequivalent, and using an identity observed by Lagrange

$$(2u_1^2 + 2u_1u_2 + 3u_2^2)(2v_1^2 + 2v_1v_2 + 3v_2^2) = x^2 + 5y^2 \tag{1.2.5}$$

where

$$x = 2u_1v_1 + u_1v_2 + u_2v_1 - 2u_2v_2, \quad y = u_1v_2 + u_2v_1 + u_2v_2. \tag{1.2.6}$$

It follows that the composition $Q' \circ Q' = Q$ and it can also be shown that $Q \circ Q = Q$ and $Q \circ Q' = Q'$.

Accounts of the composition of classes of forms appears in [291, Ch. 14] and [69]. See also [138, Section 9.1]. The set of equivalence classes of forms of a given discriminant under composition forms a group, the **class group**, and the above example illustrates that the class group corresponding to discriminant -5 has two elements and therefore is the cyclic group of order 2 generated by Q'. If $\mathbb{Q}$ denotes the rational numbers the **class number** of $\mathbb{Q}(\sqrt{D})$ is the order of this class group and a high point of Dirichlet's work in number theory was the proof of a formula of Jacobi for this class number (see [202]).

A more conceptual way of describing the composition of quadratic forms recently became available in the work of Bhargava [18] which was preliminary to his results leading to the award of a Fields medal in 2014. This proceeds as follows. He considers a quadratic form

$$Ax^2 + Bxy + Cy^2.$$

with A, B, C integers where B is not assumed to be even. A form is **primitive** if A, B, C have no common factor. Let $\mathbb{F}_q$ denote the finite field of order q. He introduced a three-tensor U in $V(\mathbb{F}_2) \otimes V(\mathbb{F}_2) \otimes V(\mathbb{F}_2)$ (called a **cube**) with entries as indicated in the diagram.

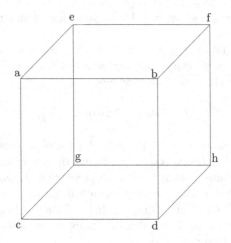

This has the following decompositions into "slices":

$$\{M_1 = \begin{bmatrix} a & b \\ c & d \end{bmatrix}, \ N_1 = \begin{bmatrix} e & f \\ g & h \end{bmatrix}\},$$

$$\{M_2 = \begin{bmatrix} a & c \\ e & g \end{bmatrix}, \ N_2 = \begin{bmatrix} b & d \\ f & h \end{bmatrix}\},$$

$$\{M_3 = \begin{bmatrix} a & e \\ b & f \end{bmatrix}, \ N_3 = \begin{bmatrix} c & g \\ d & h \end{bmatrix}\}.$$

Quadratic forms $Q_i, i = 1 \ldots 3$ are defined by

$$Q_i(x, y) = -Det(xM_i - yN_i).$$

Then each of the three forms Q_i have discriminant

$$a^2h^2 + b^2g^2 + c^2f^2 + d^2e^2 - 2(abgh + cdef + acfh + bdeg + aedh + bfcg)$$
$$+ 4(adfg + bceh)$$

which is the hyperdeterminant of U (see [119, p. 2]). It is interesting to note that hyperdeterminants date back to Cayley.

Bhargava's results leading to the group operation on classes of quadratic forms follow.

Let D be any integer congruent to 0 or 1 (mod 4), and let $Q_{id,D}$ be any primitive binary quadratic form of discriminant D such that there is a cube U_0 with

$$Q_1^{U_0} = Q_2^{U_0} = Q_3^{U_0} = Q_{id,D} .$$

Then there exists a unique group law on the set of $SL_2(\mathbb{Z})$-equivalence classes of primitive binary quadratic forms of discriminant D such that:

(a) $[Q_{id,D}]$ is the additive identity;
(b) For any cube U of discriminant D such that Q_1^U, Q_2^U, Q_3^U are primitive,

$$[Q_1^U] + [Q_2^U] + [Q_3^U] = [Q_{id,D}].$$

Conversely, given Q_1, Q_2, Q_3 with

$$[Q_1] + [Q_2] + [Q_3] = [Q_{id,D}]$$

there exists a cube U of discriminant D, unique up to Γ-equivalence, such that

$$Q_1^U = Q_1, \ Q_2^U = Q_2, \text{ and } Q_3^U = Q_3.$$

Here Γ is $SL_2(\mathbb{Z}) \times SL_2(\mathbb{Z}) \times SL_2(\mathbb{Z})$ acting on $2 \times 2 \times 2$ matrices. For a full account of this see [18].

In number theory the work has been extended by replacing $\mathbb{Q}(\sqrt{D})$ by an arbitrary field, where quadratic forms are replaced by "norm-type" forms of degree equal to the degree of the number field, with a composition similar to that for quadratic forms (see for example [288]).

1.2.2   The Relation to the Theory of Algebras

The theory of hypercomplex numbers was formalized as the theory of algebras and the sums of squares problem came to be closely related to this theory.

1.2.2.1   Definitions

Definition 1.2 An **algebra** A over a field K is a vector space over K with a bilinear product xy for any $x, y \in A$ satisfying

(a)

$$x(y+z) = xy + xz$$
$$(x+y)z = xz + yz$$

for all $x, y, z \in A$, and

(b)

$$(\lambda x)y = x(\lambda y) = \lambda(xy)$$

for $\lambda \in K$ and $x, y \in A$.

A is **associative** if

$$(xy)z = x(yz)$$

for all $x, y, z \in A$.

A is **unital** if there exists an element $1 \in A$ such that

$$1x = x1 = x$$

for all $x \in A$.

A is **alternative** if

$$(xy)y = x(yy)$$

and

$$(yy)x = y(yx)$$

for all $x, y \in A$.

Definition 1.3 An algebra A is **normed** if for each x, y in A there is defined a bilinear scalar product $(x, y) \in K$ and if $|x|$ is defined as $|x| = \sqrt{(x, x)}$, then

$$|xy| = |x||y|.$$

If A is an arbitrary algebra, it is a **division algebra** if it contains a non-zero element and if for any pair of elements $a, b \in A$ with b non-zero there exists precisely one element $x \in A$ such that $a = bx$ and there exists precisely one element $y \in A$ such that $a = yb$.

Definition 1.4 The **Jacobson radical** $J(R)$ of a ring R, is the intersection of all maximal right ideals of the ring.

Definition 1.5 An algebra A is **simple** if it contains no non-trivial two-sided ideals and the set $\{ab | a, b \in A\} \neq \{0\}$. A is **semisimple** if it has trivial Jacobson radical. If the algebra is finite dimensional this is equivalent to A being a Cartesian product of simple subalgebras.

Definition 1.6 An algebra A over a field K is **separable** if for every field extension L/K the algebra $A \otimes_K L$ is semisimple.

The quaternions form a normed non-commutative algebra and the octonions form a normed non-associative alternative algebra.

Definition 1.7 Given a finite group G and a field K the **group algebra** KG is the set of "formal" sums

$$\sum_{g \in G} a_g g, a_g \in K.$$

Addition is defined by

$$(\sum_{g \in G} a_g g) + (\sum_{g \in G} b_g g) = \sum_{g \in G} (a_g + b_g) g,$$

and the element of KG with $a_g = 1$ and $a_h = 0$ for $h \neq g$ is identified with g. The elements of G then form a basis of KG as a vector space, and the element $\sum_{g \in G} a_g g$ may be regarded as an actual sum of basis elements. The multiplication on KG is obtained by the group multiplication on the basis vectors $\{g\}_{g \in G}$ and extending linearly to KG.

1.2.2.2 Outline of the Results Concerning Sums of Squares and Algebras

Theorem 1.1 (Hurwitz) *Every normed unital algebra over* $\mathbb{R}$ *is one of* $\mathbb{R}$, $\mathbb{C}$, $\mathbb{H}$ *or* $\mathbb{K}$.

The assumption of the existence of an identity element is needed for this result. If this is dropped then there are other algebras such as the "para-Hurwitz" and "pseudo-octonion" algebras described in [228], but the dimensions of such algebras are restricted to $1, 2, 4$ and 8.

Theorem 1.2 (Frobenius) *Every associative finite-dimensional division algebra over* $\mathbb{R}$ *is isomorphic to* $\mathbb{R}$, $\mathbb{C}$, *or* $\mathbb{H}$.

This has the following extension obtained by Kurosh: Every finite-dimensional alternative division algebra over $\mathbb{R}$ is isomorphic to $\mathbb{R}$, $\mathbb{C}$, $\mathbb{H}$ or $\mathbb{K}$.

Theorem 1.3 (Kervaire, Milnor–Bott [24, 191]) *The possible dimensions of any real division algebra are limited to* 1, 2, 4, 8.

The proof of this theorem uses topological reasoning on a seven-dimensional sphere. A purely algebraic proof is unknown. (A complex algebra is never a division algebra unless it has dimension 1.)

A further result is: every identity of the form

$$(x_1^2 + x_2^2 + \ldots + x_n^2)(y_1^2 + y_2^2 + \ldots + y_n^2) = (\phi_1^2 + \phi_2^2 + \ldots + \phi_n^2) \qquad (1.2.7)$$

where the ϕ_i are bilinear forms in $\{x_i, y_i\}$ corresponds to an n-dimensional normed algebra over $\mathbb{R}$, and hence $n = 1, 2, 4$ or 8 (See [93] for an elementary proof of this).

For a more complete discussion of the above results the reader is referred to [64, 186] and [228]. In [228] interesting connections are made with physics.

If the condition above that the ϕ_i are bilinear forms in $\{x_i, y_i\}$ is weakened to the condition that ϕ_i are linear forms in $\{x_i\}$ then it was shown by Pfister in [231] and [232] that if (1.2.7) is satisfied over any field then

(a) any such form must be of degree 2^n, and
(b) such forms occur for every value of n. See also [64, p. 78].

The well-known Cayley-Dickson "doubling" procedure may also be mentioned. This constructs an algebra of dimension 2^{n+1} from an algebra of dimension 2^n. Starting with $\mathbb{R}$ this produces successively $\mathbb{C}$, $\mathbb{H}$, and $\mathbb{K}$ but for $n > 3$ a division algebra is not produced. There have been several constructions of algebras of degree 16 over $\mathbb{R}$ in which the various properties have been weakened. These have been given the name of **sedenions** (see [64] and [262]).

Quadratic forms intervene as follows. The above theorem of Hurwitz can be expressed as a theorem on the classification of composition algebras, i.e. of pairs (A, Q), where A is a finite-dimensional algebra over a field K and $Q : A \to K$ is a non-singular quadratic form for which

$$Q(x \cdot y) = Q(x) \cdot Q(y),$$

see [198]. In the case where a composition algebra A is of dimension 2 over K this implies that either $A \cong K \times K$ or that A may be constructed as follows. Each element of A is of the form $c + d\omega$ where c, d lie in K and ω is associated to the quadratic form

$$Q(x) = x_1^2 + bx_1x_2 + cx_2^2,$$

which is irreducible over K with non-zero discriminant, so that $\omega^2 = -b\omega - c$. If $u = u_1 + u_2\omega$ and $v = v_1 + v_2\omega$, then the multiplication in A is uniquely determined as

$$u \cdot v = u_1v_1 - cu_2v_2 + (u_1v_2 + u_2v_1 - bu_2v_2)\omega.$$

A discussion of algebras of dimension 4 was initiated by Brandt [26, 27] and led to quaternion algebras, which form a special case of central simple algebras. He showed that the multiplication in such an algebra is uniquely determined by the "generic norm" up to an involutory antiautomorphism. This leads to the more general concept of a composition algebra of arbitrary degree. The generalization to the theory of norm-type forms on algebras will be discussed in Chap. 3.

1.3   Group Matrices and the Group Determinant

The basic objects which led to Frobenius' construction of group representation theory were given in [106] and [107]. They will now be presented. It will be seen that the group algebra of a finite group has a collection of "multiplicative norms" which are closely connected to group representations.

1.3.1   Group Characters Before Frobenius

Prior to the paper [106] characters had been defined for abelian groups. In the introduction to [106] Frobenius states that in the proof of the theorem that each linear function of a single variable represents infinitely many prime numbers provided that its coefficients are relatively prime integers, Dedekind used for the first time certain systems of roots of unity. He also states that these also appear in the closely related question of the number of ideal classes in a cyclotomic field (referring to the remarks of Dedekind in [96]), as well as in the generalization of the above result to quadratic forms and in the investigation of their division into equivalence classes then called genera. The characteristic properties arise from an arbitrary function $\chi(n)$ on the positive integers which can take on a finite number of values and such that

$$\chi(m)\chi(n) = \chi(mn).$$

He then introduces the abstract definition, as follows. Let the elements of an abelian group G be $\{g_1, g_2, \ldots, g_n\}$ and let $\chi_r(g_i)$, $i = 1, \ldots, n$, $r = 1, \ldots, n$ be a set of functions whose values are roots of unity such that for all i, j, r

$$\chi_r(g_i)\chi_r(g_j) = \chi_r(g_i g_j)$$

then the χ_r are called characters of the group G (he indicates that the term character is due to Gauss). He states that Gauss understood a character to be a relation between the numbers represented by a quadratic form and the set of odd prime divisors p of its determinant together with 4 or 8.

The composition of classes of quadratic forms corresponds to the composition of the characters, i.e.

$$(\chi_i \circ \chi_j)(g) = \chi_i(g)\chi_j(g).$$

He notes that under this operation the group of characters of an abelian group G is isomorphic to G. He quotes Weber [289, 290] as a source of results on the relation of characters to subgroups of an abelian group.

There does not seem to be any awareness before [106] that characters could be defined for nonabelian groups.

1.3.2   The Results on Characters Which Appear in [106] and [107]

Characters are defined for an arbitrary finite group G. This extended the definition from abelian groups (see above). Initially the term character is used by Frobenius only for an irreducible character. A character χ is defined via the solution of equations in the class algebra of G. If the conjugacy classes of G are $\{C_i\}_{i=1}^r$, and $\overline{C}_i = \sum_{g \in C_i} g, i = 1, \ldots, r$, this algebra is generated by the $\overline{C}_i$. The algorithm which Frobenius used constructs the irreducible characters of G as functions from G to $\mathbb{C}$. The following results appear.

(1) The number of irreducible characters is equal to the number of conjugacy classes of the group.
(2) A character is constant on a conjugacy class.
(3) The value of a character is an algebraic integer.
(4) A table for the irreducible characters is produced.
(5) The orthogonality relations for characters are set out.
(6) The degree of a character χ is defined to be $\chi(e)$. The sum of the squares of the degrees of the irreducible characters is equal to $|G|$.
(7) The expression for the number of ways an element g of a group G can be expressed as a commutator is set out as

$$|G| \sum_{i=1}^{r} \frac{\chi_i(g)}{\chi_i(e)}.$$

where the irreducible characters are $\{\chi_1, \chi_2, \ldots, \chi_r\}$.

He defined a "relative character" of a group G with respect to a group H which contains G. This will be explained below.

Most of the results above appear in the early stages of an exposition of character theory (see [162]), but (7) usually appears later in the development.

He illustrated his methods by calculating the irreducible characters of various groups. In particular he calculated the irreducible characters of $PSL(2, p)$, the projective special linear group over the field with characteristic p. (This is a computational tour de force.)

1.3.3 Group Determinant Factors

Initially he did not associate a representation to a character. He exhibited a 1:1 correspondence between irreducible characters and irreducible factors of the group determinant. This implies that the number of irreducible factors of the group determinant is equal to the number of conjugacy classes of the group. He showed the following.

1. A convolution property characterizes factors of the group determinant.
2. The multiplicity of an irreducible factor of the group determinant is equal to its degree (as a polynomial).
3. Linear factors correspond to elements of the factor group G/G' where G' is the derived subgroup of G.

In addition he defined "k-characters" for a group and used these to obtain a formula for the irreducible group determinant factor corresponding to an irreducible character.

The results on characters are incidental in the description of the factorization of the group determinant and the first account which sets out character theory in a modern way was that of Schur in [252] (see below).

1.3.4 Group Matrices

As mentioned above, the motivation for the definition of group characters appears in [107] where the group matrix and group determinant are introduced.

Let G be a finite group of order n with a listing of elements $\{g_1 = e, g_2, \ldots, g_n\}$ and let $\{x_{g_1}, x_{g_2}, \ldots, x_{g_n}\}$ be a set of independent commuting variables indexed by the elements of G.

Definition 1.8 The **(full) group matrix** X_G is the matrix whose rows and columns are indexed by the elements of G and whose (g, h)th entry is $x_{gh^{-1}}$.

The matrix X_G may be obtained from a Cayley table of G (more specifically the table whose (i, j)th entry is $g_i g_j$) by replacing each element g by x_g and interchanging columns so that x_e appears in the diagonal. Any matrix obtained by specializing the variables x_{g_i} is also called a group matrix. If G (and the ordering) is fixed then group matrices are determined by their first row (or column), and provide an example of **patterned matrices** (see [83, 292]). Other well known examples are Hankel matrices and Toeplitz matrices. If A is such a matrix the patterns of the powers A^i will in general change. It will be shown that all (full) group matrices X_G are examples of matrices where the pattern of A^i is the same as that of A. The connection with graphs is explored in [292].

The definition of a linear representation appeared later.

Definition 1.9 Given a finite group G a **representation** of G is a homomorphism $\rho : G \to GL(n, K)$.

Here $GL(n, K)$ denotes the group of invertible linear transformations of the vector space of dimension n over a field K. Equivalently, if the elements of G are $\{g_1, g_2, \ldots, g_n\}$, a representation ρ assigns a set of invertible matrices $\{\rho(g_i)\}_{i=1,\ldots,n}$ with entries in K, such that

$$\rho(g_i)\rho(g_j) = \rho(g_i g_j).$$

The relevance of the group matrix to representation theory is a follows. Let $\mathcal{S}_n$ denote the symmetric group on n objects. Given a finite group G of order n, the **right regular representation** of G is the homomorphism $\pi : G \to \mathcal{S}_n$ where $\pi(g)$ is defined by $h^{\pi(g)} = hg$ for all h, g in G.

Definition 1.10 Given a permutation σ on a set $\{a_1, a_2, \ldots, a_n\}$, the **permutation matrix** $T(\sigma)$ is the $(0, 1)$ matrix whose (i, j)th entry is 1 if and only if $\sigma(a_i) = a_j$.

If the elements of G are listed as $\{g_1, g_2, \ldots, g_n\}$ an explicit permutation matrix representation is obtained by taking the basis

$$\{v_1, v_2, \ldots, v_n\} = \{g_1^{-1}, g_2^{-1}, \ldots, g_n^{-1}\}$$

and then $\pi(g) : v_i \to v_j$ where $g_j^{-1} = g_{i^{-1}}g$. Thus the (i, j)th entry of the corresponding matrix $T(\pi(g))$ is 1 if and only if $g_j^{-1} = g_{i^{-1}}g$, and hence in the matrix $T(\pi(g))x_g$ the entry x_g appears in the (i, j)th position if and only if $g = g_i g_j^{-1}$. This leads to

Proposition 1.1 *The group matrix X_G may be written as*

$$X_G = \sum_{g \in G} T(\pi(g))x_g,$$

where $T(\pi(g))$ is a permutation matrix for the right regular representation of G.

The term group matrix was later extended as follows to any matrix of the form

$$\sum_{g \in G} A(g)x_g$$

where the set of $m \times m$ matrices $\{A(g)\}_{g \in G}$ form a representation of G. These will appear below and will be examined more closely in Chap. 2, where further generalizations are discussed.

In most modern accounts of representation theory (see for example [69, 112, 162]), it is developed by using algebras and modules, but in order to explain the insights of Frobenius his development of the theory in terms of the group matrix and its determinant is given.

Throughout the book, if an ordering of group G is given as $\{e = g_1, g_2, \ldots, g_n\}$ together with an associated set of commuting variables $\{x_{g_1}, x_{g_2}, \ldots, x_{g_n}\}$ the abbreviation $x_i = x_{g_i}$ will often be used when describing the group matrix. Sometimes x_i will be replaced by i.

Example 1.1 Let G be C_n, the cyclic group of order n, ordered by the powers of a generator and listed as $\{1, 2, \ldots, n\}$. The group matrix is

$$X_G = \begin{bmatrix} x_1 & x_n & x_{n-1} & \ldots & x_2 \\ x_2 & x_1 & x_n & \ldots & x_3 \\ x_3 & x_2 & x_1 & \ldots & x_4 \\ & & \ldots & & \\ x_n & x_{n-1} & x_{n-2} & \ldots & x_1 \end{bmatrix} \qquad (1.3.1)$$

and is an example of a *circulant*.

A **circulant** matrix is a square matrix in which any row distinct from the first is obtained from the previous row by translating each entry one place to the right and wrapping around. In fact a circulant of order n could be defined to be any (full) group matrix associated to C_n ordered as above. In the following $C(a_1, a_2, \ldots, a_n)$ will denote the circulant with first column $(a_1, a_2, \ldots, a_n)$.

Example 1.2 The circulant $C(1, 2, 3)$ is the matrix

$$\begin{bmatrix} 1 & 3 & 2 \\ 2 & 1 & 3 \\ 3 & 2 & 1 \end{bmatrix}.$$

If P is the matrix

$$\begin{bmatrix} 1 & 1 & 1 & \ldots & 1 \\ 1 & \rho & \rho^2 & \ldots & \rho^{n-1} \\ 1 & \rho^2 & \rho^4 & \ldots & \rho^{2(n-1)} \\ & & \ldots & & \\ 1 & \rho^{n-1} & \rho^{2n-2} & \ldots & \rho^{(n-1)^2} \end{bmatrix} \qquad (1.3.2)$$

with $\rho = e^{\frac{2\pi i}{n}}$ then if $G = C_n$, ordered as above, $P X_G P^{-1}$ is the diagonal matrix with entries

$$(x_1 + x_2 + \ldots + x_n), (x_1 + \rho x_2 + \ldots + \rho^{n-1} x_n), (x_1 + \rho^2 x_2 + \ldots + \rho^{2(n-1)} x_n),$$

$$\ldots, (x_1 + \rho^{n-1} x_2 + \ldots + \rho^{(n-1)^2} x_n). \tag{1.3.3}$$

This diagonalization is associated with the finite Fourier transform and will be discussed later in Chap. 7. The book [72] has collected much information on circulants and their uses. For P above, $\frac{1}{\sqrt{n}} P$ is called the **Fourier matrix** F_n.

For arbitrary G with a fixed ordering, the matrix X_G has very strong symmetry properties. As mentioned above, its entries are determined by the first row. The following theorem of Frobenius was basic to his results. Let Y_G and Z_G be obtained from X_G by replacing each variable x_g by y_g and z_g respectively.

Theorem 1.4 *Let the variables* $\{x_g\}_{g \in G}$, $\{y_g\}_{g \in G}$, $\{z_g\}_{g \in G}$ *be related by*

$$z_g = \sum_{hk=g} x_h y_k \text{ for all } g \in G. \tag{1.3.4}$$

Then the matrices X_G *and* Y_G *have the property that*

$$X_G Y_G = Z_G. \tag{1.3.5}$$

Proof This is immediate from matrix multiplication. The (g, h)th entry of $X_G Y_G$ is

$$[x_{gg_1^{-1}}, x_{gg_2^{-1}}, \ldots, x_{gg_n^{-1}}].[y_{g_1h^{-1}}, x_{g_2h^{-1}}, \ldots, x_{g_nh^{-1}}] = \sum_{i=1}^{n} x_{gg_i^{-1}} y_{g_i h^{-1}}$$

$$= \sum_{kl=gh^{-1}} x_k y_l.$$

$\square$

From Theorem 1.4 it follows that any power of X_G (including its inverse) has the same pattern, i.e. it is obtained by replacing each x_g by a function ϕ_g which in the case of positive powers is a polynomial in the $\{x_g\}$ and in the case of negative powers is a rational function in the $\{x_g\}$.

Example 1.3 Let $G = C_3$ with generator a. Then $X_G = C(x_e, x_a, x_{a^2})$ and

$$X_G^2 = C(\phi_e, \phi_a, \phi_{a^2})$$

where

$$\phi_e = x_e^2 + 2x_{a^2} x_a, \quad \phi_a = x_{a^2}^2 + 2x_e x_a, \quad \phi_{a^2} = x_a^2 + 2x_{a^2} x_e$$

and

$$X_G^{-1} = C(\psi_e, \psi_a, \psi_{a^2})$$

where $\psi_e = (x_e^2 - x_{a^2}x_a)/\Delta$, $\psi_a = (x_{a^2}^2 - x_e x_a)/\Delta$ and $\psi_{a^2} = (x_a^2 - x_{a^2}x_e)/\Delta$, with

$$\Delta = \det(X_G) = x_e^3 + x_a^3 + x_{a^2}^3 - 3x_e x_a x_{a^2}.$$

A consequence of Theorem 1.4 is that any pair of group matrices for C_n which are in the form (1.3.1) commute. This is set out in the following Lemma.

Lemma 1.1 *If U and V are any $n \times n$ circulants with entries which are in a commutative ring then $UV = VU$.*

Proof Let the entries in U and V be $\{u_i\}_{i=1}^n$ and $\{v_i\}_{i=1}^n$ respectively. Then by Theorem 1.4

$$UV = C(z_1, z_2, \ldots, z_n), \text{ and } VU = C(w_1, w_2, \ldots, w_n)$$

where for each i

$$z_i = z_{g_i} = \sum_{hk=g_i} u_h v_k, \text{ and } w_i = w_{g_i} = \sum_{kh=g_i} v_k u_h$$

with multiplication in the cyclic group of order n. It is clear that $z_i = w_i$ and the Lemma follows.   □

1.3.4.1 The Connection with Probability

Given a function $f : G \to \mathbb{C}$, consider $X_G(f)$, the specialization of X_G obtained by inserting $f(g)$ for x_g, for all $g \in G$. It is sometimes convenient to regard f as a probability on G, when the further conditions $f(g) \in \mathbb{R}$, $f(g) \geq 0$, and $\sum_{g \in G} f(g) = 1$ are imposed. If functions f_1 and f_2 are given then the **convolution** $f_1 * f_2$ is defined by

$$f_1 * f_2 = \sum_{h \in G} f_1(gh^{-1}) f_2(h), \tag{1.3.6}$$

and it is readily seen that if $\{x_g\}$, $\{y_g\}$ correspond respectively to f_1 and f_2 and if $\{x_g\}$, $\{y_g\}$ and $\{z_g\}$ are related as in (1.3.4), then $\{z_g\}$ corresponds to $f_1 * f_2$, or alternatively

$$X_G(f_1)X_G(f_2) = X_G(f_1 * f_2).$$

This connection will be explored more fully later. Thus this vital operation in probability, systems analysis and other areas appears at the foundations of group representation theory.

Definition 1.11 If G is a finite group and sets of variables $\underline{x} = \{x_g\}_{g\in G}$, $\underline{y} = \{y_g\}_{g\in G}$, $\underline{z} = \{z_g\}_{g\in G}$ are given which satisfy (1.3.4) then $\underline{z}$ will be called the **convolution** of $\underline{x} * \underline{y}$ of $\underline{x}$ and $\underline{y}$.

1.3.5 Right Division in Groups and Group Matrices of Small Groups

1.3.5.1 The Ward Quasigroup Corresponding to a Group

The group matrix encodes the operation of right division in a group. Let G be a group and define the operation $(*)$ by

$$x * y = xy^{-1}$$

The operation $(*)$ is not associative, and although e is a right identity it is not a left identity.

A **quasigroup** Q is a set with binary operation $\circ$ such that if in the equation $x \circ y = z$ any two of the elements are known then the third is uniquely determined. If Q is finite this is equivalent to the unbordered multiplication table of Q being a latin square. If Q has an identity element then Q is a **loop**. The set of elements of G under $(*)$ give a quasigroup which is denoted by $(Q, *)$. It is clear that Q is closely connected to X_G. This right division operation has appeared in several works, the first apparently being [287]. The reader is referred to [182] for references.

The quasigroup Q is easily seen to satisfy the identity

$$(x * y) * (z * y) = x * z, \tag{1.3.7}$$

but it is less obvious that if Q is a quasigroup satisfying (1.3.7) a group G can reconstructed from Q. Quasigroups $(Q, *)$ satisfying (1.3.7) are known as **Ward quasigroups**, although other names seem to have been used.

The following proposition seems to have been part of the folklore for some time.

Proposition 1.2 *If $G = (G, \cdot)$ is a group and $Q = (Q, *)$ is defined by $x * y = x \cdot y^{-1}$ then Q is a quasigroup satisfying the identity (1.3.7). Conversely assume that Q is a quasigroup satisfying the identity (1.3.7). Then:*

(a) There is a unique element $e \in Q$ such that $x * x = e$ for every $x \in Q$.
(b) This element e is the right neutral element of Q.
(c) The map $\widehat{\ } : Q \to Q$ defined by $\widehat{x} = e * x$ satisfies $\widehat{\widehat{x}} = x$, $\widehat{x * y} = y * x$.

(d) *If the groupoid $(G, \cdot)$ is constructed from Q by $x \cdot y = x * \widehat{y}$, then G is a group with identity element e, and the inverse of x in G is $\widehat{x}$.*

(e) *If the Ward quasigroup Q arises from a group H, then $(G, \cdot)$ is isomorphic to H.*

The proof is left as an exercise to the reader (see [182]). If G is a group, let $Wa(G)$ denote the unique Ward quasigroup associated to right division in G.

1.3.5.2 Multiplication Tables

Let G be a finite group of order n, and let $Q = Wa(G)$, and let S be a cyclic subgroup of G of order m with generator s. Then S is a subquasigroup of Q and the elements of S can be listed as $e, s, s^2, \ldots, s^{m-1}$, where the powers are calculated in G. Let $\widehat{x} = e * x \ (= x^{-1}$, the inverse of x in G).

Let $r = n/m$. Assume that $a_1 = e, a_2, \ldots, a_r$ form a set of representatives of the left cosets $\{gS; \ g \in G\}$ of S in G. Define an ordering of G as follows: order the elements of the coset $a_i S$ as $a_i, a_i s, \ldots, a_i s^{m-1}$. Then list the elements of G as $S = a_1 S, a_2 S, \ldots, a_r S$. Let M be the multiplication table of Q obtained by labelling both rows and columns by this ordering. This table is comprised of $m \times m$ blocks with the (i, j)th block consisting of the elements in $a_i S a_j^{-1}$.

Example 1.4 Let $G = Q_8$, the quaternion group. Let $S = \{e, i, -1, -i\}$. Then a suitable ordering of G with respect to S is

$$\{e, i, -1, -i, j, -k, -j, k\}$$

and the multiplication table for $(G, *)$ is

	e	i	-1	$-i$	j	$-k$	$-j$	k
e	e	$-i$	-1	i	$-j$	k	j	$-k$
i	i	e	$-i$	-1	$-k$	$-j$	k	j
-1	-1	i	e	$-i$	j	$-k$	$-j$	k
$-i$	$-i$	-1	i	e	k	j	$-k$	$-j$
j	j	k	$-j$	$-k$	e	i	-1	$-i$
$-k$	$-k$	j	k	$-j$	$-i$	e	i	-1
$-j$	$-j$	$-k$	j	k	-1	$-i$	e	i
k	k	$-j$	$-k$	j	i	-1	$-i$	e

(1.3.8)

Proposition 1.3 *Let M be the multiplication table of Q as described above. Then*

(i) *$M = (m_{ij})$ consists of r^2 circulant matrices C_{ij}, each of size m;*

(ii) *if any pair of rows of M are taken, then the product under $(*)$ of the entries in the same column is constant, i.e., $m_{ij} * m_{kj} = m_{il} * m_{kl}$ for every i, j, k, l;*

(iii) *if the jth column of M is labelled by $q \in Q$, then $m_{1j} = q^{-1}$;*

(iv) *all the diagonal elements of M are equal to e;*

(v) *if the entry in the (i, j)th position of M is q, then the entry in the (j, i)th position of M is $\widehat{q} = q^{-1}$.*

Proof A circulant of order m is determined by the following property: an entry in the (i, j)th position is equal to the entry in the $(i + 1, j + 1)$th position, where $i + 1$ and $j + 1$ are reduced modulo m. In the block C_{ij}, if the (k, l)th entry is $x * y$ then the $(k + 1, l + 1)$th entry is $(xs) * (ys) = (x * s^{-1}) * (y * s^{-1}) = x * y$, where again $k + 1$ and $l + 1$ are reduced modulo m. Thus every block C_{ij} is a circulant matrix, and (i) is shown. Assume that the jth column is labelled by q. Then

$$m_{ij} * m_{kj} = (m_{i1} * q) * (m_{k1} * q) = m_{i1} * m_{k1}$$

by direct application of (1.3.7) which shows (ii). Moreover, $m_{1j} = e * q = \widehat{q}$ $(= q^{-1})$, which shows (iii). By Proposition 1.2, $x * x = e$ for every $x \in Q$, and this implies (iv). Finally, since from Proposition 1.2, $(x * y) = \widehat{(y * x)}$ and (v) follows.   □

The properties described in Proposition 1.3 are illustrated in Example 1.4.

Remark 1.1 If the table for a Ward quasigroup is constructed with any ordering of the elements, then condition (ii) is satisfied, and conversely if any quasigroup table satisfies (ii) then the quasigroup is a Ward quasigroup.

The symmetric group on n symbols will be denoted by $\mathcal{S}_n$.

Example 1.5 Let G be a group with six elements, and suppose that it is known only that there is a cyclic subgroup S of order 3 in G, and that the elements not in S are involutions. Let $\{e = 1, 2, 3\}$ denote the elements of S. Let M be the unbordered multiplication table of $Q = Wa(G)$. Then the first row of M is $[1, 3, 2, 4, 5, 6]$, and the first column is $[1, 2, 3, 4, 5, 6]^t$. Proposition 1.3 implies that, with respect to the ordering, M is of the form

B_{11}	B_{12}
B_{21}	B_{22}

where $B_{11} = C(1, 2, 3)$, $B_{12} = C(4, 6, 5)$, and $B_{21} = C(4, 5, 6)$. The remaining entries on the diagonal are e, and hence M is partially determined as

	1	2	3	4	5	6
1	1	3	2	4	5	6
2	2	1	3	6	4	5
3	3	2	1	5	6	4
4	4	6	5	1		
5	5	4	6		1	
6	6	5	4			1

Condition (ii) of Proposition 1.3 for rows 1 and 3 and forces $4 * 5 = 1 * 3 = 2$, and the complete table is determined as

	1	2	3	4	5	6
1	1	3	2	4	5	6
2	2	1	3	6	4	5
3	3	2	1	5	6	4
4	4	6	5	1	2	3
5	5	4	6	3	1	2
6	6	5	4	2	3	1

Since there is only one group of order six with three involutions, the Ward quasigroup for $\mathcal{S}_3$ has been constructed.

1.3.6 Symmetrical Forms of the Group Matrix

If G is a finite group, with a given ordering and with associated Ward quasigroup Q, the group matrix X_G with respect to the ordering may be obtained from the multiplication table of Q by replacing each element g by the variable x_g. The following lemma is a restatement of the results in Proposition 1.3.

Lemma 1.2 *For any cyclic subgroup S of order r of a finite group G of order $n = rm$, there is an ordering of G such that the group matrix X_G may be written as a block matrix, each of whose blocks is a circulant of size r. Moreover, for any fixed pair (i, j) of rows of X_G, if x_g is the entry in the (i, k)th position and x_h is the entry in the (j, k)th position, the product gh^{-1} is independent of k. For arbitrary i, k, if x_g is the element in the (i, k)th position then $x_{g^{-1}}$ is the element in the (k, i)th position.*

To be more explicit, for every cyclic subgroup S of G with the ordering by left cosets as described above, X_G is an $m \times m$ block matrix $(B_{ij})_{m \times m}$, where each B_{ij} is a circulant of the form $C(x_{g_{k_1}}, \dots, x_{g_{k_r}})$ where $g_{k_1}, \dots, g_{k_r}$ are elements of the set $a_i S a_j^{-1}$. This special group matrix will be denoted by $X_G(S)$.

Example 1.6 From Example 1.5, the group matrix X_G, $G = \mathcal{S}_3$ is

$$\begin{bmatrix} C(x_1, x_2, x_3) & C(x_4, x_6, x_5) \\ C(x_4, x_5, x_6) & C(x_1, x_3, x_2) \end{bmatrix}.$$

Example 1.7 If D_n is the dihedral group of order $2n$ its group matrix can be put in the form

$$\begin{bmatrix} C(x_1, x_2, \dots, x_n) & C(x_{n+1}, x_{2n}, x_{2n-1}, \dots, x_{n+2}) \\ C(x_{n+1}, x_{n+2}, .., x_{2n}) & C(x_1, x_n, x_{n-1}, \dots x_2) \end{bmatrix}$$

The proof is similar to the argument given for S_3, S being the unique cyclic subgroup of order n.

Example 1.8 The group matrix for Q_8 can be put in the form

$$\begin{bmatrix} C(x_1, x_2, x_3, x_4) & C(x_7, x_6, x_5, x_8) \\ C(x_5, x_6, x_7, x_8) & C(x_1, x_4, x_3, x_2) \end{bmatrix}$$

This is obtained from the table (1.3.8) with the ith element of the group replaced by x_i.

Let Fr_n denote the Frobenius group of order n.

Example 1.9 Let $G = Fr_{21}$. G has a normal cyclic subgroup S of order 7. An ordering of G exists with respect to the cosets of S similar to those above such that, abbreviating x_{g_i} as x_i, the group matrix is

$$\begin{bmatrix} B_{11} & B_{12} & B_{13} \\ B_{21} & B_{22} & B_{23} \\ B_{31} & B_{32} & B_{33} \end{bmatrix},$$

where

- $B_{11} = C(x_1, x_2, x_3, x_4, x_5, x_6, x_7)$,
- $B_{12} = C(x_{15}, x_{19}, x_{16}, x_{20}, x_{17}, x_{21}, x_{18})$,
- $B_{13} = C(x_8, x_{10}, x_{12}, x_{14}, x_9, x_{11}, x_{13})$,
- $B_{21} = C(x_8, x_9, x_{10}, x_{11}, x_{12}, x_{13}, x_{14})$,
- $B_{22} = C(x_1, x_5, x_2, x_6, x_3, x_7, x_4)$,
- $B_{23} = C(x_{15}, x_{17}, x_{19}, x_{21}, x_{16}, x_{18}, x_{20})$,
- $B_{31} = C(x_{15}, x_{16}, x_{17}, x_{18}, x_{19}, x_{20}, x_{21})$,
- $B_{32} = C(x_8, x_{12}, x_9, x_{13}, x_{10}, x_{14}, x_{11})$,
- $B_{33} = C(x_1, x_3, x_5, x_7, x_2, x_4, x_6)$.

This group is a split extension and in Chap. 2 there is given a systematic way to obtain the group matrices in this case.

1.3.7   The Group Determinant

The object which gave rise to Frobenius' initial work can now be defined.

Definition 1.12 The **group determinant** $\Theta(G)$ is $\det(X_G)$.

This will often be denoted by Θ if no ambiguity occurs. In the following, if $\underline{x} = (x_{g_1}, x_{g_2}, \ldots, x_{g_n})$, the notation $\Theta(\underline{x}) := \det(X_G)$ will be used. It is clear that Θ is a polynomial of degree n in n variables. It was the question of Dedekind of how this polynomial factorizes which prompted Frobenius' work. Then Θ has the important property

$$\Theta(\underline{x})\Theta(\underline{y}) = \Theta(\underline{x} * \underline{y}) \tag{1.3.9}$$

(take the determinant of both sides in (1.3.5)).

Note that $\Theta(\underline{x})$ always has the **trivial factor** $x_{g_1} + x_{g_2} + \ldots + x_{g_n}$. This is seen by adding all rows of X_G to the first, after which all the entries in the first row become this sum.

Example 1.10 Let G be $C_3 = \{e, a, a^2\}$ and abbreviate the elements (x_e, x_a, x_{a^2}) by (x_1, x_2, x_3). Then,

$$\Theta(G) = x_1^3 + x_2^3 + x_3^3 - 3x_1x_2x_3.$$

This factors as

$$(x_1 + x_2 + x_3)(x_1 + \omega x_2 + \omega^2 x_3)(x_1 + \omega^2 x_2 + \omega x_3).$$

where $\omega = e^{\frac{2\pi i}{3}}$. The product of the two non-trivial factors is

$$\phi = x_1^2 + x_2^2 + x_3^2 - x_1x_2 - x_1x_3 - x_2x_3. \tag{1.3.10}$$

It will be proved below that for any factor ϕ of the group determinant,

$$\phi(\underline{x})\phi(\underline{y}) = \phi(\underline{x} * \underline{y}),$$

and therefore the ϕ in (1.3.10) is a "composable" quadratic form. Also since $\Theta(\underline{x})\Theta(\underline{y}) = \Theta(\underline{x} * \underline{y})$ this gives perhaps what is the best approximation to a "three cubes" identity.

Example 1.11 Let G be S_3. A specific ordering which produces the table in Example 1.5 is $\{e, (123), (132), (12), (23), (13)\}$ which will be abbreviated by $1, \ldots, 6$ in the order given. Then X_G is given in Example 1.6 and

$$\Theta = \det\left(\begin{bmatrix} C(x_1, x_2, x_3) & C(x_4, x_6, x_5) \\ C(x_4, x_5, x_6) & C(x_1, x_3, x_2) \end{bmatrix}\right).$$

It is an exercise in the elementary theory of determinants to show that the factors of Θ are

$$\phi_1 = (x_1 + x_2 + x_3 + x_4 + x_5 + x_6),$$

$$\phi_2 = (x_1 + x_2 + x_3 - x_4 - x_5 - x_6)$$

and

$$\phi_3 = (x_1^2 + x_2^2 + x_3^2 - x_1x_2 - x_2x_3 - x_3x_1 - x_4^2 - x_5^2 - x_6^2 + x_4x_5 + x_5x_6 + x_6x_4)$$

where ϕ_3 appears with multiplicity 2. A consequence of Frobenius's results is that ϕ_3 is irreducible (over $\mathbb{C}$).

1.3.7.1   Changing the Ordering

For a group G of order n two different orderings $\{g_1, \ldots, g_n\}$ and $\{g_{\sigma(1)}, \ldots, g_{\sigma(n)}\}$ with σ in $\mathcal{S}_n$ produce group matrices X_G' and X_G'', such that there is a permutation matrix T with $T^{-1} X_G' T = X_G''$.

Definition 1.13 Let ϕ_1 and ϕ_2 be polynomials in the variables $\{x_1, \ldots, x_n\}$. The equivalence relation E is defined by

$$\phi_1 E \phi_2 \text{ if and only if } \phi_1(x_1, \ldots, x_n) = \pm \phi_2(x_{\sigma(1)}, \ldots, x_{\sigma(n)})$$

for some permutation σ in $\mathcal{S}_n$.

It follows directly that any pair of group determinants of a group G corresponding to different orderings are E-equivalent.

1.4   The Original Motivation for the Examination of the Group Determinant

The group determinant was brought to the attention of Frobenius by a diffident suggestion from Dedekind to investigate the way in which Θ factorizes (over $\mathbb{C}$ or some extension of $\mathbb{C}$). The correspondence between Dedekind and Frobenius is discussed in [137, 138] and [139]. In number theory, if K is a normal extension field of the field L with Galois group $G = \{\pi_1, \pi_2, \ldots, \pi_n\}$ and $\{\omega_1, \ldots, \omega_n\}$ is a basis of K over L the **discriminant** Δ of this system is defined to be D^2 where

$$D = \det \begin{bmatrix} \omega_1\pi_1 & \omega_2\pi_1 & \cdots & \omega_n\pi_1 \\ \omega_1\pi_2 & \omega_2\pi_2 & \cdots & \omega_n\pi_2 \\ & & \cdots & \\ \omega_1\pi_n & \omega_2\pi_n & \cdots & \omega_n\pi_n \end{bmatrix}.$$

Often the basis which is used consists of the images of a single element ω,

$$\{\omega\pi_1, \omega\pi_2, \ldots, \omega\pi_n\},$$

and in this case

$$D = \det \begin{bmatrix} \omega\pi_1\pi_1 & \omega\pi_2\pi_1 & \dots & \omega\pi_n\pi_1 \\ \omega\pi_1\pi_2 & \omega\pi_2\pi_2 & \dots & \omega\pi_n\pi_2 \\ & \dots & & \\ \omega\pi_1\pi_n & \omega\pi_2\pi_n & \dots & \omega\pi_n\pi_n \end{bmatrix}.$$

Thus the determinant of (x_{gh}) is suggested, and it is E-equivalent to the group determinant of the Galois group.

In the case where G is abelian it is not difficult to deduce that Θ is a product of linear factors with coefficients which are roots of unity (this follows for cyclic groups from (1.3.3) and can be easily extended to arbitrary abelian groups). Apparently the hope of Dedekind was that, in the case of nonabelian G, new systems of hypercomplex numbers would be discovered which could be used to split Θ into linear factors. He had already calculated that the group determinant of Q_8, which he regarded as the group of units $\{\pm 1, \pm i, \pm j, \pm k\}$ in $\mathbb{H}$ of order 8, factors into four linear factors and the square of the quadratic factor

$$\psi = \alpha^2 + \beta^2 + \gamma^2 + \delta^2, \tag{1.4.1}$$

where $\alpha = x_1 - x_2$, $\beta = x_3 - x_4$, $\gamma = x_5 - x_6$ and $\delta = x_7 - x_8$ for the ordering $\{1, -1, i, -i, j, -j, k, -k\}$ of the group. While ψ is irreducible over $\mathbb{C}$, if the skewfield of coefficients is extended to $\mathbb{H}$ it can be factored as

$$q\bar{q} = (\alpha + i\beta + j\gamma + k\delta)(\alpha - i\beta - j\gamma - k\delta) \tag{1.4.2}$$

which relates back to the results on sums of squares in Sect. 1.2, and therefore the group determinant of Q_8 splits into linear factors over $\mathbb{H}$.

In the case of the Octonions the analogous object formed from the units is the non-associative Octonion loop O_{16}. If the standard basis $\{e_i\}_{i=1}^8$ is taken for $\mathbb{K}$ the elements of O_{16} are the 16 units $\{\pm e_i\}_{i=1}^8$. This loop satisfies the Moufang property (see [8]). A polynomial derived from the Cayley table of O_{16}, the loop determinant, can be defined and this polynomial factors into 8 linear factors and the fourth power of a quadratic factor of the form

$$\psi = \alpha^2 + \beta^2 + \gamma^2 + \delta^2 + \lambda^2 + \mu^2 + \nu^2 + \xi^2,$$

where, similarly to the quaternion case, the $\alpha, \beta, \dots$ are all of the form $x_{e_i} - x_{-e_i}$. Then ψ factorizes into linear factors over $\mathbb{K}$ as $u.\bar{u}$ in an analogous manner to the quaternionic case (see for example [186, p. 47]). Perhaps this is another way of recognizing the special place of $\mathbb{K}$ among nonassociative algebras. This example will be examined in Chap. 2.

The initial direction envisaged by Dedekind became almost forgotten after Frobenius defined character theory for non-commutative groups in order to present his solution of the factorization problem. In a remarkably short period Frobenius established most of the basic properties of characters and applied them to his solution of the factorization problem.

1.5 Frobenius' Definition of Group Characters

The definition of group characters which appeared in Frobenius' first paper is as follows. Let $\{C_i\}_{i=1}^r$ be the conjugacy classes of G. For each i let $\overline{C_i}$ be the element of $\mathbb{Z}G$ defined by $\overline{C_i} = \sum_{g \in C_i} g$. Then

$$\overline{C_i}.\overline{C_j} = \sum a_{ij}^k \overline{C_k}$$

where the **structure constants** a_{ij}^k are non-negative integers. Actually, Frobenius defined the structure constants as solutions of certain sets of equations, but his definition is equivalent to that above. It is an elementary exercise to check that the $\overline{C_i}$ generate a commutative subalgebra $\mathbf{S}$ of $\mathbb{C}G$, the **class algebra** which is in fact the center of $\mathbb{Z}G$. The matrices B_j defined by $B_j(i, k) = \{a_{ij}^k\}$ give a representation of $\mathbf{S}$ (this will be explained more fully in Chap. 4). They form a normal set and by standard theory they have a common set of eigenvectors $\{v_1, v_2, \ldots, v_r\}$. Let $\lambda(i, j)$ be the eigenvalue of B_j with respect to v_i. Frobenius defined the **irreducible character** $\chi_i : G \to \mathbb{C}$ to be the function which has the value

$$\frac{m_i \lambda(i, j)}{|C_j|}, \tag{1.5.1}$$

on an element of C_j, where the m_i are defined in a rather mysterious manner, which is only later clarified by Frobenius' discussion of the group determinant. In fact m_i is the degree of the corresponding representation. The **character table** of G is the table whose rows are indexed by the irreducible characters χ_i of G and columns by the conjugacy classes of G and whose (i, j)th entry is $\chi_i(g)$, where $g \in C_j$. In modern usage, for any set $\{n_i\}$ of non-negative integers (not all zero) the linear combination $\sum n_i \chi_i$ is called a **character** of G. A more detailed account of Frobenius' first paper appears in [68]. The following example illustrates his method of calculating group characters.

Example 1.12 Let $G = S_3$. Using the ordering in Example 1.6 the conjugacy classes are $C_1 = \{1\}$, $C_2 = \{2, 3\}$ and $C_3 = \{4, 5, 6\}$. The structure constants are given by

$$\overline{C_1}.\overline{C_i} = \overline{C_i}, \overline{C_2}^2 = 2\overline{C_1} + \overline{C_2}, \overline{C_2}.\overline{C_3} = 2\overline{C_3}, \overline{C_3}^2 = 3\overline{C_1} + 3\overline{C_2}.$$

Then

$$B_1 = I, \; B_2 = \begin{bmatrix} 0 & 1 & 0 \\ 2 & 1 & 0 \\ 0 & 0 & 2 \end{bmatrix}, \; B_3 = \begin{bmatrix} 0 & 0 & 1 \\ 0 & 0 & 2 \\ 3 & 3 & 0 \end{bmatrix},$$

the matrix of eigenvalues is

$$\begin{bmatrix} 1 & 2 & 3 \\ 1 & 2 & -3 \\ 1 & -1 & 0 \end{bmatrix}$$

and the character table of $\mathcal{S}_3$ is

Class order	1	2	3
χ_1	1	1	1
χ_2	1	1	-1
χ_3	2	-1	0

(1.5.2)

It is left as an exercise to the reader to calculate in this way the character tables of D_4 and Q_8, the dihedral and quaternion groups of order 8. They coincide as

Class order	1	1	2	2	2
χ_1	1	1	1	1	1
χ_2	1	1	1	-1	-1
χ_3	1	1	-1	1	-1
χ_4	1	1	-1	-1	1
χ_5	2	-2	0	0	0

(1.5.3)

Frobenius also discussed the case where G is normal in a larger group H. The orbits $\{U_j\}$ of H acting by conjugacy on G are then used instead of the usual conjugacy classes to generate a subalgebra of $\mathbf{S}$. In Chap. 9 this will be described as a **fusion** of $\mathbf{S}$. The characters are related to eigenvalues of generators in a similar way to that given above. The example he gives in [106] is the following. Let $G = \mathcal{A}_4$, the alternating group on four symbols. Its ordinary character table is

class order	1	3	4	4
χ_1	1	1	1	1
χ_2	1	1	ω	ω^2
χ_3	1	1	ω^2	ω
χ_4	3	-1	0	0

(1.5.4)

If the overgroup S_4 is taken as H the classes represented by the last two columns fuse into a single orbit and the last two linear characters fuse. Frobenius computes the relative character table as

$$
\begin{array}{|ccc|}
\hline
1 & 1 & 1 \\
2 & 2 & -1 \\
3 & -1 & 0 \\
\hline
\end{array}
\tag{1.5.5}
$$

This would be usually written nowadays as either the "$\mathcal{P}$-matrix" of an association scheme

$$
\begin{bmatrix}
1 & 3 & 8 \\
1 & 3 & -4 \\
1 & -1 & 0
\end{bmatrix}
\tag{1.5.6}
$$

or as the "group-normalized" table

Class order	1	3	8
χ_1	1	1	1
χ_2	$\sqrt{2}$	$\sqrt{2}$	$-1/\sqrt{2}$
χ_3	3	-1	0

$$\tag{1.5.7}$$

which exhibits the usual orthogonality conditions but which violates several of the conditions for a group character table, for example the values of a character on e are not necessarily integers and the entries are not all algebraic integers.

In [135] a theory is put forward that the first paper of Frobenius did not introduce the group determinant because he had not yet proved his "main theorem" which was that the multiplicity of a group determinant factor is equal to the degree. There Hawkins discusses the rivalry between Frobenius and Burnside and suggests that the first paper was prepared to provide evidence of Frobenius' priority in defining group characters. However this may be, it is remarkable that Frobenius pointed the way in his first paper to a large area of mathematics which now includes the theory of association schemes, Hecke Algebras, S-rings and Gelfand pairs which will be discussed in Chap. 4.

1.6 Group Characters and the Group Determinant

In Frobenius' second paper [107] the character theory of a group G was tied to its group determinant as follows. If, for each conjugacy class C_i of G, the variables x_g, $g \in C_i$, in X_G are replaced by a single variable x_{C_i}, then the **reduced group matrix** X_R is obtained. The **reduced group determinant** Θ_R is $\det(X_R)$. Frobenius showed that Θ_R factorizes into linear factors of the form

$$\sum_{i=1}^{r} \lambda(i, j)x_{C_j}, \tag{1.6.1}$$

where the $\lambda(i, j)$ are the eigenvalues of the matrices B_i as described above. He then associated to each character of a group a factor of the group determinant.

Frobenius' result may be explained as follows. As in Proposition 1.1 X_G may be written as $\sum_{g \in G} \pi(g)x_g$. Then on replacing each variable x_g by x_{C_i}, X_R becomes

$$\bullet \qquad X_R = \sum_{i=1}^{r} A_i x_{C_i}$$

where each A_i is a $(0, 1)$ matrix. The matrices A_i form a commutative algebra under usual matrix addition and multiplication which is isomorphic to the class algebra **S** described above, and the eigenvalues of A_j are the same as those of B_j. It follows that the A_j can be simultaneously diagonalized i.e. there is a non-singular matrix P such that $P^{-1}X_R P$ is diagonal with entries

$$\sum_{j=1}^{r} \lambda(i, j)x_{C_j}. \tag{1.6.2}$$

Note that this implies that the Fourier transform of the reduced group matrix is $T^{-1}X_R T$ (see Chap. 7).

Thus the factors of Θ_R are precisely as described in (1.6.1). The result of Frobenius that such a factor appears with multiplicity m_i^2 where m_i is the degree of the corresponding character χ_i (which he regarded as a high point in the theory) will be proved later.

In the subsequent paper [107] after another suggestion of Dedekind, Frobenius introduced the representation theory of groups by matrices and it became clear that characters were traces of representing matrices. It was gradually realized that Frobenius had discovered a major tool which, after many generalizations, has become central to much of modern mathematics and physics. The early development is set out in [68, 135–138] and [139].

1.7   Frobenius' Basic Results and Their Proofs

It is difficult to follow Frobenius' intricate arguments which lead to his results on the group determinant factors. The paper by Dieudonné [95] includes a justification for some of the techniques which he and Schur used. In the following some departures are made from his original approach because the theory of group representation by matrices, initially not available to him, can be used to clarify some places where his work is difficult to put into a modern framework.

For any homogeneous polynomial $\varphi(x_1, x_2, \ldots, x_n)$, $\varphi(\underline{x})$, $\varphi(\underline{y})$, $\varphi(\underline{z})$ will be written to denote respectively the polynomials in the set of variables $\underline{x} = \{x_1, x_2, \ldots, x_n\}$, $\underline{y} = \{y_1, y_2, \ldots, y_n\}$, $\underline{z} = \{z_1, z_2, \ldots, z_n\}$. Let ε denote the specialization $\{1, 0, 0, \ldots, 0\}$.

Proposition 1.4 *If φ is any factor of Θ and $\underline{x}, \underline{y}, \underline{z}$ are sets of variables indexed by the elements of G such that $\underline{x} * \underline{y} = \underline{z}$, then*

$$\varphi(\underline{x})\varphi(\underline{y}) = \varphi(\underline{z}). \tag{1.7.1}$$

Conversely, if φ is any irreducible homogeneous polynomial which satisfies (1.7.1) then φ is a factor of Θ.

Proof Consider a factor φ of Θ of degree r and assume that the coefficient of x_e^r in φ is 1. Now suppose that $\underline{x}, \underline{y}, \underline{z}$ are sets of variables with $\underline{x} * \underline{y} = \underline{z}$ as in the statement of the theorem. From the decomposition of

$$\Theta(\underline{x})\Theta(\underline{y}) = \Theta(\underline{z})$$

into irreducible factors, it follows that since $\varphi(\underline{z})$ divides $\Theta(\underline{z})$ it also divides the left-hand product and thus it must be the product of a polynomial $\Lambda(\underline{x})$ independent of the $\{y_i\}$ and a polynomial $\Psi(\underline{y})$ independent of the $\{x_i\}$, i.e.

$$\varphi(\underline{z}) = \Lambda(\underline{x})\Psi(\underline{y}). \tag{1.7.2}$$

Now if in (1.7.2), $\underline{y}$ is replaced by ε, considered as set indexed by the ordering of G, it follows that $\underline{x} * \varepsilon = \underline{x}$ and thus

$$\varphi(\underline{x}) = \Lambda(\underline{x})\alpha \text{ where } \alpha = \varphi(\varepsilon) \in \mathbb{C}$$

(the above statement and those following rely on unique factorization in the algebra of multivariate polynomials, see [139]).

Similarly, if β is obtained by setting $\underline{x} = \varepsilon$ in $\Lambda(\underline{x})$ it follows that $\varphi(\underline{y}) = \beta\Psi(\underline{y})$. If both of the substitutions $\underline{x} = \varepsilon$ and $\underline{y} = \varepsilon$ are inserted into (1.7.2), $\underline{z}$ also becomes ε and therefore $\varphi(\underline{z})$ becomes 1 which implies that $\alpha\beta = 1$. Thus $\varphi(\underline{x})\varphi(\underline{y}) = \varphi(\underline{z})$.

Conversely, suppose that φ is an irreducible polynomial which satisfies (1.7.1). Let X_G as usual be the group matrix in terms of the variables $\{x_{g_i}\}$. Then, after abbreviating x_{g_i} to x_i, the adjoint $adj(X_G) = Y_G$ is also a group matrix with entries $\{x_i^*\}$ which are polynomials in the $\{x_i\}$ and

$$X_G Y_G = \det(X_G)I = \Theta(\underline{x})I. \tag{1.7.3}$$

Now the corresponding $\{z_i = (x * y)_i\}$ can be read off from the right-hand side of (1.7.3), namely $z_1 = \Theta(\underline{x})$, $z_i = 0$ for $i \neq 1$. It follows that for this specialization of $\underline{z}$, $\varphi(\underline{z}) = \Theta(\underline{x})^r$. But also, on inserting these specializations into (1.7.1)

$$\Theta(\underline{x})^r = \varphi(\underline{z}) = \varphi(\underline{x})\varphi(\underline{y}).$$

Since $\varphi(\underline{x})$ is a polynomial in the $\{x_i\}$, it follows from the irreducibility of $\varphi(\underline{x})$ that it must be a factor of $\Theta(\underline{x})$. □

Corollary 1.1 *If ψ is any product of irreducible factors of Θ and $\underline{x}$, $\underline{y}$, $\underline{z}$ are sets of variables indexed by the elements of G such that $\underline{x} * \underline{y} = \underline{z}$ then*

$$\psi(\underline{x})\psi(\underline{y}) = \psi(\underline{z}).$$

In order to make use of the simplifications which Schur introduced in [252] an account of his work is given below. It is based on that given in [68, Chap. 4]. It is assumed that the underlying field is $\mathbb{C}$.

1.7.1 Schur's Approach to Representation Theory

A representation is a function $\rho : G \to GL(n, \mathbb{C})$. Then the **degree** of ρ, $\deg(\rho) = n$. For each such representation Schur chooses matrices $A(g) = \rho(g)$ and constructs the group matrix

$$X_A = \sum A(g)x_g.$$

He then assumes that $\det(X_A) \neq 0$, which is equivalent to $\det(A(e)) \neq 0$, i.e. $A(e) = I$. Group matrices X_A and X_B are **equivalent** if $X_B = P^{-1}X_A P$ for an invertible constant matrix P. The definition of equivalence for representations is a direct translation:

Definition 1.14 Given representations ρ_1 and $\rho_2 : G \to GL(n, F)$ for some field F, ρ_1 and ρ_2 are **equivalent** if there is an invertible constant matrix P such that

$$P^{-1}\rho_1(g)P = \rho_2(g) \text{ for all } g \text{ in } G.$$

Schur's definition of a reducible representation is in terms of group matrices. Any group matrix X is **reducible** if it is equivalent to a matrix of the form

$$\begin{bmatrix} X_1 & 0 \\ U & X_2 \end{bmatrix} \text{ or equivalently } \begin{bmatrix} X_1 & V \\ 0 & X_2 \end{bmatrix}$$

in which case X_1 and X_2 are also group matrices associated to representations. The following lemma is usually referred to as Schur's Lemma.

Lemma 1.3

(a) *Let X and X' be irreducible group matrices of respective degrees f and f'. Let P be a constant $f \times f'$ matrix such that*

$$XP = PX'.$$

Then either $P = 0$ or X and X' are equivalent and P is an invertible $f \times f$ matrix.

(b) *Let X be an irreducible group matrix of degree f. Then each constant matrix P such that $XP = PX$ has the form aI_f where $a \in \mathbb{C}$ and I_f is the $f \times f$ identity matrix.*

Proof

(a) If P is any matrix it may be reduced by row and column operations to a matrix of the form

$$Q = \begin{bmatrix} I_m & 0 \\ 0 & 0 \end{bmatrix}, \tag{1.7.4}$$

or equivalently there exist nonsingular matrices A and B such that

$$APB = Q. \tag{1.7.5}$$

Now X and X' be as in the statement of the lemma and let P be a non-zero constant matrix, such that $XP = PX'$, and let A, B, Q be as in (1.7.4) and (1.7.5). Set $X_1 = AXA^{-1}$ and $X'_1 = B^{-1}X'B$. Since

$$AXPB = APX'B$$

it follows that

$$AXA^{-1}APB = APBB^{-1}X'B,$$

and hence $X_1 Q = QX'_1$. If the matrices X and X' are represented in block form as

$$X_1 = \begin{bmatrix} X_{mm} & Y \\ Z & W \end{bmatrix}, \quad X'_1 = \begin{bmatrix} X'_{mm} & Y' \\ Z' & W' \end{bmatrix}$$

with X_{mm} and X'_{mm} both of size $m \times m$, it follows that $Z = 0 = Y'$. Therefore if either $m < f$ or $m < f'$ then the matrix X (resp) X' is reducible. Thus $m = f = f'$ and P is invertible with $P^{-1}XP = X'$.

(b) Let $PX = XP$ and suppose that α is an eigenvalue of P, i.e.

$$\det(P - \alpha I_f) = 0.$$

Now $X(P - \alpha I_f) = (P - \alpha I_f)X$ and by part (a) using the fact that $P - \alpha I_f$ is singular it must be the zero matrix, i.e. $P = \alpha I_f$.

$\square$

Schur's lemma was used to obtain the following set of orthogonality relations for the coefficients of irreducible representations. The definition of orthogonality given is a slight variation on the usual way in which orthogonality is defined, in order to be consistent with the notation used in [68]. Let $\delta_{il} = 1$ if $i = l$, 0 otherwise.

Theorem 1.5 *Let $g \mapsto A(g) = (a_{ij}(g))$ be an irreducible representation of degree f of G. Then for all $i, j, k, l \in \{1, \ldots, f\}$,*

$$\sum_{g \in G} a_{ij}(g^{-1})a_{kl}(g) = \frac{|G|}{f}\delta_{il}\delta_{jk}. \tag{1.7.6}$$

If $g \mapsto B(g) = (b_{ij}(g))$ is a second irreducible representation not equivalent to A then

$$\sum_{g \in G} a_{ij}(g^{-1})b_{kl}(g) = 0 \tag{1.7.7}$$

for all i, j, k, l.

Proof For the first relation consider the $f \times f$ matrix $U = (u_{ij})$. Let

$$V = \sum_{g \in G} A(g)^{-1} U A(g) \tag{1.7.8}$$

It is seen directly that for any $h \in G$, using the fact that $g \mapsto A(g)$ is a representation,

$$A(h)^{-1}VA(h) = \sum_{g \in G} A(h^{-1}g^{-1})UA(gh).$$

But the right-hand expression is

$$\sum_{k \in G} A(k)^{-1}UA(k) = V,$$

i.e. $A(h)^{-1}VA(h) = V$. Therefore $VA(h) = A(h)V$ for all $h \in G$. From part (b) of Schur's Lemma, $V = vI_f$ for some $v \in \mathbb{C}$. But from (1.7.8) the entries in V are linear combinations of the entries of U, and thus

$$v = \sum_{\lambda,\mu} c_{\lambda\mu} u_{\lambda\mu}$$

for $c_{\lambda\mu} \in \mathbb{C}$. More precisely, from (1.7.8)

$$V = \sum_{g \in G} \sum_{j,k} a_{ij}(g^{-1}) u_{jk} a_{kl}(g) = \delta_{il} \sum_{\lambda,\mu} c_{\lambda\mu} u_{\lambda\mu} \qquad (1.7.9)$$

for all choices of i and l.

Now U is arbitrary. If u_{jk} is taken to be 1 for fixed j, k and 0 otherwise, it follows that

$$\sum_{g \in G} \sum_{j,k} a_{ij}(g^{-1}) a_{kl}(g) = \delta_{il} c_{jk}$$

and if c_{jk} is evaluated by taking $i = l$ and summing (1.7.9) on i, a consequence is

$$\sum_{g \in G} \sum_{j,k} a_{ij}(g^{-1}) a_{kl}(g) = f c_{jk}$$

and since $A(g^{-1})A(g) = I_f$ it follows that $|G|\delta_{jk} = f c_{jk}$.

The second relation is proved in a similar way. Suppose the representations $g \mapsto A(g)$ and $g \mapsto B(g)$ are as in the statement of the theorem. Then let U be an arbitrary $f \times f'$ matrix, and let

$$V = \sum_{g \in G} A(g^{-1}) U B(g).$$

With an argument similar to that above it is shown that $A(h)V = V B(h)$ for all $h \in G$ and again this implies that $V = 0$. Then it follows as above that

$$\sum_{g \in G} \sum_{j,k} a_{ij}(g^{-1}) u_{jk} b_{kl}(g) = 0$$

for all i, l and all choices of the elements u_{jk}. The second orthogonality relation follows by specializing the u_{jk} as above.   □

Thus the functions $a_{kl}^{(i)}$ form a set of $|G|$ orthogonal functions on G. The corollary below is a direct consequence.

Corollary 1.2 *Given the finite set of mutually inequivalent irreducible representations of G, $\{g \mapsto A^k(g)\}$, $k = 1 \ldots r$, where the $A^k(g) = (a_{ij}^k(g))$ are square matrices of size f_k, the coefficients a_{ij}^k, regarded as functions from G to $\mathbb{C}$, form a linearly independent set.*

The following sets out the well known character orthogonality relations.

The **inner product** $\langle , \rangle$ of two class functions ψ_1 and ψ_2 on a group is defined by

$$\langle \psi_1, \psi_2 \rangle = \frac{1}{|G|} \sum_{g \in G} \psi_1(g)\psi_2(g^{-1}).$$

This is consistent with the orthogonality in (1.7.6) and (1.7.7).

Let $C(g)$ denote the conjugacy class of g in G.

Corollary 1.3 *If G is a finite group and $\{\chi_1, \chi_2, \ldots, \chi_r\}$ are the set of irreducible characters of G as defined above then*

(i)

$$\langle \chi_i, \chi_j \rangle = \delta_{ij}.$$

(ii) For fixed g and h,

$$\sum_{i=1}^{r} \chi_i(g)\chi_i(h^{-1}) = 0$$

if g is not conjugate to h and equal to $\frac{|G|}{|C(g)|}$ otherwise.

Frobenius' second description of a character as a trace is as follows. Let $Tr(M)$ be the trace of a matrix M. Then corresponding to the matrix representation $g \mapsto A(g)$ the character $\widehat{\chi}$ is defined by $\widehat{\chi}(g) = Tr(A(g))$.

Definition 1.15 For any finite group G, let $Irr(G)$ be the set of traces of irreducible representations of G.

Corollary 1.4 *If $Irr(G) = \{\widehat{\chi}_i\}_{i=1}^{r}$ with $\widehat{\chi}_i(e) = m_i$ then*

$$\langle \widehat{\chi}_i, \widehat{\chi}_j \rangle = \delta_{ij}. \tag{1.7.10}$$

Proof Let $\widehat{\chi}_i$ correspond to the representation $g \mapsto A^{(i)}(g) = (a_{kl}^{(i)}(g))$, $i = 1, \ldots, r$. Then $\widehat{\chi}_i(g) = \sum_{k=1}^{m_i} a_{kk}^{(i)}(g)$. And

$$\langle \widehat{\chi}_i, \widehat{\chi}_j \rangle = \frac{1}{|G|} \sum_{g \in G} \sum_{k=1}^{m_i} a_{kk}^{(i)}(g) \sum_{l=1}^{m_j} a_{ll}^{(j)}(g^{-1})$$

$$= \frac{1}{|G|} \sum_{k=1}^{m_i} \sum_{l=1}^{m_j} (\sum_{g \in G} a_{kk}^{(i)}(g) a_{ll}^{(j)}(g^{-1})).$$

Using (1.7.7) if $i \neq j$ the inner sum is 0. If $i = j$ it is $\delta_{kl} \frac{|G|}{m_i}$. Therefore

$$\langle \widehat{\chi}_i, \widehat{\chi}_i \rangle = \frac{1}{|G|} m_i \frac{|G|}{m_i} = 1.$$

□

Corollary 1.5

$$\sum_{\widehat{\chi}_i \in Irr(G)} \widehat{\chi}_i(g)\widehat{\chi}_i(h^{-1}) = 0$$

if g is not conjugate to h and equal to $\frac{|G|}{|C(g)|}$ otherwise.

Proof Let $\Psi(G)$ be the character table of G regarded as an $r \times r$ matrix. Suppose $\{C_i\}_{i=1}^r$ are the conjugacy classes and that $\{g_i\}_{i=1}^r$ are representatives so that $g_i \in C_i$. Let D be the $r \times r$ diagonal matrix with entry $|C_i|$ in the ith position in the diagonal, then (1.7.10) is equivalent to

$$|G|\delta_{ij} = \sum_{g \in G} \widehat{\chi}_i(g)\widehat{\chi}_j(g^{-1}) = \sum_{k=1}^r |C_k|\widehat{\chi}_i(g_k)\widehat{\chi}_j(g_k^{-1}). \qquad (1.7.11)$$

There are r^2 equations of the form 1.7.11 and these are equivalent to the matrix equation

$$|G|I = \Psi(G)D\overline{\Psi(G)}^T.$$

This in turn is equivalent to $\frac{1}{|G|}D\overline{\Psi(G)}^T = \Psi(G)$ and hence to

$$|G|I = D\overline{\Psi(G)}^T \Psi(G). \qquad (1.7.12)$$

The system (1.7.12) may be written as

$$|G|\delta_{ij} = \sum_{k=1}^r |C_i|\overline{\widehat{\chi}_k(g_i^{-1})}\widehat{\chi}_k(g_j).$$

This gives

$$\sum_{k=1}^r \overline{\widehat{\chi}_k(g_i^{-1})}\widehat{\chi}_k(g_j) = \frac{|G|}{|C_i|}\delta_{ij}$$

which is equivalent to the statement of the Corollary.

□

The following is the version of Maschke's Theorem stated by Schur.

Theorem 1.6 *Let X be a group matrix for a finite group G, which is equivalent to a group matrix of the form*

$$\begin{bmatrix} X_1 & 0 \\ U & X_2 \end{bmatrix}.$$

Then it is also equivalent to a group matrix of the form

$$Z = \begin{bmatrix} X_1 & 0 \\ 0 & X_2 \end{bmatrix}.$$

See [68] for Schur's proof. Shorter proofs are available (for example see [162, p. 4]).

Now suppose that $\widehat{\pi}$ is the regular character of G, the trace of the right regular representation, i.e. $\widehat{\pi}(e) = |G|$, $\widehat{\pi}(g) = 0$ if $g \neq e$, it follows that $\langle \widehat{\chi_i}, \widehat{\pi} \rangle = \frac{1}{|G|} m_i |G| = m_i$. This is equivalent to

$$\pi = \sum m_i \widehat{\chi_i}. \tag{1.7.13}$$

On evaluating both sides of (1.7.13) at e the result $\sum m_i^2 = |G|$ is obtained.

Example 1.13 Let $G = S_3$, ordered as in Example 1.5. There are two representations of degree 1. If a function f on the elements of G is given by the vector of values $(f(g_1), f(g_2), \ldots, f(g_n))$, then these representations become two of the functions in Corollary 1.2, namely $(1, 1, 1, 1, 1, 1)$ and $(1, 1, 1, -1, -1, -1)$. There is one remaining irreducible representation, and a corresponding irreducible group matrix is

$$\begin{bmatrix} x_1 + \omega x_2 + \omega^2 x_3 & x_4 + \omega x_5 + \omega^2 x_6 \\ x_4 + \omega^2 x_5 + \omega x_6 & x_1 + \omega^2 x_2 + \omega x_3 \end{bmatrix},$$

where $\omega = e^{(2\pi i/3)}$. This gives rise to the following four coefficient functions

$$(1, \omega, \omega^2, 0, 0, 0), (0, 0, 0, 1, \omega, \omega^2), (0, 0, 0, 1, \omega^2, \omega), (1, \omega^2, \omega, 0, 0, 0).$$

Note that the last four functions are dependent on the basis chosen to describe the representation of degree 2. It will be seen later that the coefficient functions may be used to transform a group matrix X_G into a block diagonal matrix whose blocks correspond to irreducible group matrices.

Corollary 1.6 *If $g \mapsto A^{(k)}(g)$ is an irreducible representation of G with corresponding group matrix $X^{(k)}$, then $\phi = \det(X^{(k)})$ is an irreducible polynomial.*

Proof For any k, if $X^{(k)}$ is the group matrix corresponding to $A^{(k)}$ then

$$X^{(k)} = \sum_{g \in G} A^{(k)}(g)x_g = (u_{ij}^k)$$

where

$$u_{ij}^{(k)} = \sum_{g_\mu \in G} a_{ij}^{(k)}(g_\mu)x_{g_\mu}, \qquad (1.7.14)$$

Let $\{A^{(k)}\}_{k=1}^r$ be the set of irreducible representations of G. As k runs from 1 to r (1.7.14) can be regarded as a set of $|G|$ linear equations expressing the $u_{ij}^{(k)}$ in terms of the $|G|$ variables x_{g_μ}. From the orthogonality relations (1.7.6) and (1.7.7) on the $a_{ij}^{(k)}$ it follows that the matrix of coefficients is non-singular and hence each x_{g_μ} may be expressed in terms of the u_{ij}:

$$x_{g_\mu} = \sum_{k=1}^r \sum_{i,j=1}^{f_k} b_{ij}^{(k)} u_{ij}^{(k)}, \qquad (1.7.15)$$

so that any algebraic relation between the $\{u_{ij}^{(k)}\}$ would translate to an algebraic relation between the variables $\{x_{g_\mu}\}$, which are assumed to be algebraically independent. It follows that the $u_{ij}^{(k)}$ form an algebraically independent set. Now it is well known that if the elements of a matrix M are independent variables then $\det(M)$ is an irreducible polynomial in these variables, [22, pp. 176–177], which implies that, for any k, $\det(X^{(k)})$ is an irreducible polynomial in the $\{u_{ij}^{(k)}\}$. It is necessary to show that $\det(X^{(k)})$ remains irreducible as a polynomial in the $\{x_{g_\mu}\}$ when the $u_{ij}^{(k)}$ are replaced by the expressions in (1.7.14). Suppose that $\det(X^{(k)})$ factors as a polynomial in the $\{x_g\}$ as $\phi_1\phi_2$, with $\deg(\phi_1) = s$, $\deg(\phi_2) = t$ with $s, t \geq 1$. If the x_{g_μ} in ϕ_1 and ϕ_2 are replaced using (1.7.15) a non-trivial factorization of $\det(X)$ as a polynomial in the $\{u_{ij}^t\}$ is obtained, which gives a contradiction. $\square$

It follows from Maschke's theorem that the group matrix X is similar to the block diagonal matrix

$$Z = diag(X_1, X_2, \ldots, X_s),$$

where the X_i are irreducible group matrices. Therefore $\Theta(G) = \varphi_1\varphi_2 \ldots \varphi_s$ where $\varphi_i = \det(X_i)$.

Corollary 1.7 *Each irreducible factor of Θ is of the form $\det(X)$, where X is the group matrix corresponding to an irreducible representation of G.*

Let $\underline{x} = (x_{g_1}, x_{g_2}, \ldots, x_{g_n})$. Any irreducible factor φ of Θ of degree m can be expanded in powers of $x_1 = x_{g_1}$:

$$\varphi(\underline{x}) = x_1^m + t_1 x_1^{m-1} + t_2 x_1^{m-2} + \ldots + t_m.$$

And this defines the polynomials $\{t_i\}_{i=1}^m$ where $\deg(t_i) = i$. Now

$$t_1 = \sum_{g \in G - \{e\}} v(g) x_g$$

where $v(g) \in \mathbb{C}$. Frobenius extended v to a function from G to $\mathbb{C}$ by defining $v(e) = m$ and called v the character corresponding to φ. Since it can be assumed that $\varphi = \det(X)$ where $X = \sum A(g) x_g$, the characteristic equation for X, $\det(\lambda I - X)$, expands as

$$\lambda^m - \lambda^{m-1} Tr(X) + \ldots$$

and in effect $Tr(X) = \sum_{g \in G} Tr(A(g)) x_g$. Consequently, $v(g) = Tr(A(g))$ for all $g \in G$, and v coincides with the character $\widehat{\chi}$ defined above. It is clear that $\widehat{\chi}(gh) = \widehat{\chi}(hg)$, since $Tr(A(gh)) = Tr(A(g)A(h)) = Tr(A(h)A(g)) = Tr(A(hg))$.

Lemma 1.4 *If φ is an irreducible factor Θ and the character $\widehat{\chi}$ is defined as above, then $\widehat{\chi}$ is an irreducible character as defined in Sect. 1.5.*

Proof Suppose that $\widehat{\chi}$ is obtained from the irreducible factor φ of degree f. Then $\varphi = \det(X_{\widehat{\chi}})$ where $X_{\widehat{\chi}} = \sum_{g \in G} A(g) x_g$ is a group matrix corresponding to the irreducible representation $g \mapsto A(g)$. Let $X_{\widehat{\chi}}^{red} = \sum_i \sum_{g \in C_i} A(g) x_{C_i}$ and $\varphi_{\widehat{\chi}}^{red} = \det(X_i^{red})$. Now

$$X_{\widehat{\chi}}^{red} X_{\widehat{\chi}} = \sum_i \sum_{h \in C_i} A(h) x_{C_i} \sum_{g \in G} A(g) x_g = \sum_i \sum_{g \in G} \sum_{h \in C_i} A(h) A(g) x_g x_{C_i}$$

$$= \sum_i \sum_{g \in G} \sum_{h \in C_i} A(hg) x_g x_{C_i}.$$

Similarly,

$$X_{\widehat{\chi}} X_{\widehat{\chi}}^{red} = \sum_i \sum_{g \in G} \sum_{h \in C_i} A(gh) x_g x_{C_i}.$$

Now for fixed g,

$$\sum_{h \in C_i} A(hg) = \sum_{h \in C_i} A((ghg^{-1})g) = \sum_{h \in C_i} A(gh)$$

(this statement is essentially equivalent to the proof that the class sums are in the center of the group algebra). It follows that $X_{\widehat{\chi}}^{red} X_{\widehat{\chi}} = X_{\widehat{\chi}} X_{\widehat{\chi}}^{red}$. By Schur's Lemma, it follows that $X_{\widehat{\chi}}^{red} = \eta I$ so that $\varphi_{\widehat{\chi}}^{red} = \eta^f$. Since $\varphi_{\widehat{\chi}}^{red}$ is a factor of $\Theta_R = \det(X$ it follows that $\eta = \sum_{i=1}^{r} \lambda(i, j) x_{C_i}$, and $Tr(X_{\widehat{\chi}}^{red}) = f\eta$. But $Tr(X_{\widehat{\chi}}^{red})$ may be obtained from $X_{\widehat{\chi}}$ by replacing each x_g by x_{C_i} where $g \in C_i$. Therefore

$$Tr(X_{\widehat{\chi}}^{red}) = \sum_{g \in G} Tr(A(g)) x_{C_i} = \sum_{g \in G} \widehat{\chi}(g) x_{C_i} = \sum_{i=1}^{r} \widehat{\chi}(g)|C_i| x_{C_i},$$

where the fact that $\widehat{\chi}$ is constant on conjugacy classes has been used. It follows that $f\lambda(i, j) = \widehat{\chi}(g)|C_i|$, or equivalently $\widehat{\chi}(g) = \frac{f\lambda(i,j)}{|C_i|}$, i.e $\widehat{\chi}$ is a character as defined by the formula (1.5.1). The Lemma is proved.   □

Frobenius' two definitions of a character have now been reconciled, and the use either definition will be used in subsequent work. In particular, the term character will be used for the trace of an arbitrary representation.

Corollary 1.8 *If φ is an irreducible factor of $\Theta(G)$ corresponding to the character χ of degree f then φ^{red} factors as η^f where $\eta = \sum_{i=1}^{r} \lambda(i, j) x_{C_i}$, $\lambda(i, j) = \chi(g)|C_i|/f$ and g is any element of C_i.*

Corollary 1.9 *If the distinct linear factors of Θ_R are $\eta_1, \eta_2, \ldots, \eta_r$ corresponding to irreducible characters $\chi_1, \chi_2, \ldots, \chi_r$ of degrees $m_1, m_2, \ldots, m_r$ then $\Theta_R = \prod_{i=1}^{r} \eta_i^{m_i^2}$.*

Proof The character orthogonality relations imply that $\eta_i \neq \eta_j$ if $\chi_i \neq \chi_j$. Now if π is the regular character of G, by (1.7.13) $\pi = \sum_{i=1}^{r} m_i \chi_i$ and therefore X_G is similar to a block diagonal matrix which for each i in $\{1, \ldots, r\}$ has an entry of the form

$$\text{Diag}(X_{\chi_i}, X_{\chi_i}, \ldots, X_{\chi_i}) \ (m_i \text{ copies}).$$

The contribution of this entry to Θ_R is $\eta_i^{m_i^2}$ and the Corollary follows.   □

A consequence of the above corollaries is that an irreducible factor φ of $\Theta(G)$ of degree m appears in $\Theta(G)$ with multiplicity m, a result which Frobenius regarded a high point in his theory.

1.8 k-Characters and the Construction of Irreducible Factors

Frobenius introduced "k-characters" in order to describe the construction of an irreducible factor of the group determinant from an irreducible group character. In order to use his definition in a wider context, the following generalization due to Buchstaber and Rees is used. Let A be an algebra over a field K (unless otherwise stated algebras will be assumed to be associative).

Definition 1.16 The function $f : A \to K$ is **trace-like** if $f(xy) = f(yx)$ for all x, y in K.

In particular a character is trace-like.

Definition 1.17 If f is a trace-like function, define $\delta^{(r)}(f) : A^r \to K$ for $r = 1, 2, \ldots$ inductively by

$$\delta^{(1)}(f)(a) = f(a),$$

$$
\begin{aligned}
\delta^{(k)}(f)(a_1, a_2, \cdots, a_k) = {} & f(a_1)\delta^{(k-1)}(f)(a_2, a_3, \cdots, a_k) \\
& - \delta^{(k-1)}(f)(a_1 a_2, a_3, \cdots, a_k) \\
& - \delta^{(k-1)}(f)(a_2, a_1 a_3, \cdots, a_k) \\
& - \cdots - \delta^{(k-1)}(f)(a_2, a_3, \cdots, a_1 a_k).
\end{aligned}
\tag{1.8.1}
$$

Definition 1.18 The k-**character** $\chi^{(k)} : G^k \to \mathbb{C}$ is defined for $k = 1, 2, \cdots$ by

$$\chi^{(k)}(g_{i_1}, g_{i_2}, \cdots, g_{i_k}) = \delta^{(k)}\chi(g_{i_1}, g_{i_2}, \cdots, g_{i_k}).
\tag{1.8.2}$$

In other words $\chi^{(k)}$ is $\delta^{(k)}\chi$ applied to k-tuples of the basis $\{g_1 = e, g_2, \ldots, g_n\}$ of $\mathbb{C}G$. For $k = 1, 2, 3$ there are the following explicit formulae:

$$\chi^{(1)}(g) = \chi(g),$$

$$\chi^{(2)}(g, h) = \chi(g)\chi(h) - \chi(gh)$$

and

$$\chi^{(3)}(g, h, k) = \chi(g)\chi(h)\chi(k) - \chi(g)\chi(hk) - \chi(h)\chi(gk) - \chi(k)\chi(gh)$$
$$+ \chi(ghk) + \chi(gkh).$$

An equivalent way of defining the operation $\delta^{(k)}(f)$ is contained in the following proposition.

Proposition 1.5 *Let f be a trace-like function on A. For any cycle $\sigma = (j_1, j_2, \ldots, j_t)$ of S_k define*

$$f_\sigma(a_{i_1}, a_{i_2}, \cdots, a_{i_k}) = f(a_{i_{j_1}} a_{i_{j_2}} \cdots a_{i_{j_t}}).$$

For any $\tau \in S_k$ expressed as a product of disjoint cycles, $\tau = \sigma_1 \sigma_2 \ldots \sigma_s$ (including cycles of length 1) define

$$f_\tau(a_{i_1}, a_{i_2}, \cdots, a_{i_k}) = f_{\sigma_1}(a_{i_1}, a_{i_2}, \cdots, a_{i_k}) f_{\sigma_2}(a_{i_1}, a_{i_2}, \cdots, a_{i_k})$$

$$\ldots f_{\sigma_s}(a_{i_1}, a_{i_2}, \cdots, a_{i_k}).$$

Then if $sgn(\tau)$ denotes the sign of τ

$$\delta^{(k)} f(a_{i_1}, a_{i_2}, \cdots, a_{i_k}) = \sum_{\tau \in S_k} sgn(\tau) f_\tau(a_{i_1}, a_{i_2}, \cdots, a_{i_k}). \tag{1.8.3}$$

Proof Suppose that the expressions for $\delta^{(k-1)} f$ in (1.8.1) and (1.8.3) agree. Note that this implies that $\delta^{(k-1)} f(a_{i_1}, a_{i_2}, \cdots, a_{i_{k-1}})$ is invariant under any permutation of the arguments. Consider the term

$$\delta^{(k-1)} f(a_{i_2}, a_{i_3}, \cdots, a_{i_1} a_{i_r}, \ldots, a_{i_k})$$

in (1.8.1). By assumption this is equal to

$$\sum_{\tau \in S_{k-1}} sgn(\tau) f_\tau(a_{i_1}, a_{i_2}, a_{i_1} a_{i_r}, \ldots, a_{i_k})$$

$$= \sum_{\tau \in S_{k-1}} sgn(\tau) f_\tau(a_{i_1} a_{i_r}, a_{i_1}, a_{i_2}, \ldots, a_{i_{r-1}}, a_{i_{r+1}}, \ldots, a_{i_k}).$$

Define the subset Λ_r of S_k to be the set of elements which take i_1 to i_r. Each element τ in Λ_r where $r > 1$ when expressed canonically as a product of disjoint cycles has a cycle σ of the form $(i_1, i_r, i_{t_1}, \ldots, i_{t_s})$. Associate to τ the permutation $\widehat{\tau}$ in the copy of S_{k-1} on $\{i_2, i_3, \ldots, i_k\}$ which has cycles identical to those of τ except that σ is replaced by $(i_r, i_{t_1}, \ldots, i_{t_s})$. This establishes a bijection between Λ_r and S_{k-1} such that $sgn(\tau) = -sgn(\widehat{\tau})$. Then if $r > 1$ the term $f_\tau(a_{i_1}, a_{i_2}, \cdots, a_{i_k})$ is equal to

$$f_{\widehat{\tau}}(a_{i_1} a_{i_r}, a_{i_2}, \ldots, a_{i_{r-1}}, a_{i_{r+1}}, \ldots, a_{i_k})$$

$$= f_{\widehat{\tau}}(a_{i_2}, a_{i_3}, \ldots, a_{i_{r-1}}, a_{i_1} a_{i_r}, a_{i_{r+1}}, \ldots, a_{i_k})$$

and hence

$$\sum_{\tau \in \Lambda_r} sgn(\tau) f_\tau(a_{i_1}, a_{i_2}, \cdots, a_{i_k}) = - \sum_{\hat{\tau} \in S_{k-1}} sgn(\hat{\tau}) f_{\hat{\tau}}(a_{i_1}, a_{i_2}, a_{i_1} a_{i_r}, \ldots, a_{i_k})$$

$$= -f^{(k-1)}(a_{i_2}, a_{i_3}, \cdots, a_{i_1} a_{i_r}, \ldots, a_{i_k}).$$

In the case $r = 1$ a bijection between Λ_1 and S_{k-1} is defined by $\hat{\tau}(i_r) = \tau(i_r)$ for $r \geq 2$ where this case $sgn(\tau) = sgn(\hat{\tau})$. Then $f(a_{i_1}) \delta^{(k-1)} f(a_{i_2}, a_{i_3}, \cdots, a_{i_k})$ is easily seen to be equal to

$$f(a_{i_1}) \sum_{\hat{\tau} \in S_{k-1}} sgn(\hat{\tau}) f_{\hat{\tau}}(a_{i_2}, a_{i_3}, \ldots, a_{i_r}, \ldots, a_{i_k})$$

$$= \sum_{\tau \in \Lambda_r} sgn(\tau) f_\tau(a_{i_1}, a_{i_2}, \cdots, a_{i_k}).$$

It has been shown that if (1.8.3) holds for $k - 1$ it holds for k and since it holds for $k = 1$ trivially the proposition is proved.   □

Remark 1.2   It is a consequence of the recursive definition of $\chi^{(k)}$ that if

$$\chi^{(k)}(g_{i_1}, g_{i_2}, \cdots, g_{i_k}) = 0$$

for all k-tuples $(g_{i_1}, g_{i_2}, \cdots, g_{i_k}) \in G^k$ then $\chi^{(r)}(g_{i_1}, g_{i_2}, \cdots, g_{i_r}) = 0$ for all $r \geq k$. Moreover from (1.8.3) $\chi^{(k)}(g_{i_1}, g_{i_2}, \cdots, g_{i_k})$ is symmetric in the sense that

$$\chi^{(k)}(g_{i_1}, g_{i_2}, \cdots, g_{i_k}) = \chi^{(k)}(g_{\sigma(i_1)}, g_{\sigma(i_2)}, \cdots, g_{\sigma(i_r)}))$$

for any element $\sigma \in S_k$ and for arbitrary k.

1.8.1   The Connection with the Cayley-Hamilton Theorem and Symmetric Functions

If $X \in M_n(R)$ is an $n \times n$ matrix over the commutative ring R its characteristic polynomial is defined by the formula $\gamma_X(z) = det(zI - X)$. The Cayley-Hamilton theorem states that

$$\gamma_X(X) = 0.$$

In characteristic 0 the polynomial $\gamma_X(z)$ can be computed using the elements $Tr(X^i), i = 1, 2, \ldots, n$. For example when $n = 2$

$$\chi_X(z) = z^2 - Tr(X)z + \frac{1}{2}(Tr(X)^2 - Tr(X^2)). \tag{1.8.4}$$

The method is based on the recursive algorithm expressing the elementary symmetric functions $s_k(x_1, x_2, \ldots, x_n)$

$$s_k = \sum_{1 \leq i_1 < i_2 < \ldots < i_n \leq n} x_{i_1} x_{i_2} \ldots x_{i_k} \text{ for } k = 1, 2, \ldots n,$$

in terms of the Newton functions

$$p_k = \sum_{i=1}^{n} x_i^{k}.$$

The relationship may be expresed in terms of generating functions. If

$$F(z) = \prod_{i=1}^{\infty} (1 + x_i z) = \sum_{i=1}^{\infty} s_k z^k$$

is the generating function for the s_k and

$$P(z) = \sum_{i=1}^{\infty} p_k z^k$$

is the generating function for the p_k then

$$\frac{d}{dz} \log(F(z)) = \frac{F'(z)}{F(z)} = P(-z).$$

Consider the characteristic polynomial $\gamma_{X_G}(z)$ of a group matrix X_G. On replacing z^k by $Tr(X^k)$ this gives rise to the expression for the polynomials $s_i(\underline{x})$ in terms of k-characters.

Newton's formula connecting the p_i with the s_j is

$$p_i - s_1(\underline{x})p_{i-1} + \ldots + (-1)^i i s_i(\underline{x}) = 0. \qquad (1.8.5)$$

If $\text{Char}(K) \nmid m$ an explicit formula for s_i in terms of the p_j is as follows

$$s_i = \sum_{i_1 + 2i_2 + \ldots + m i_m = i} (-1)^{i_1 + i_2 + i_3 + \ldots} \frac{p_1^{i_1} p_2^{i_2} \cdots p_m^{i_m} \cdot}{1^{i_1} 2^{i_2} 3^{i_3} \ldots m^{i_m} i_1! \cdots i_m!}$$

and Waring's formula, independent of characteristic, gives explicitly

$$p_i = i \cdot \sum_{i_1 + 2i_2 + \ldots + m i_m = i} (-1)^{i_2 + i_4 + i_6 + \ldots} \frac{(i_1 + \ldots + i_m - 1)!}{i_1! \cdots i_m!} s_1^{i_1} \ldots s_m^{i_m}.$$

The **trace identity** for arbitrary $m \times m$ matrices $A_1, A_2, \cdots, A_{m+1}$ is

$$\delta^{(m+1)} f(A_1, A_2, \cdots, A_{m+1}) = \sum_{\tau \in \mathcal{S}_{m+1}} sgn(\tau) f_\tau(A_1, A_2, \cdots, A_{m+1}) = 0,$$

$$(1.8.6)$$

where f is the trace and $\delta^{(k)} f$ has been defined in (1.8.3). One way of obtaining this is to start from the expressions of the form (1.8.4) arising from Newton's identity. For example in the 2×2 case (1.8.4) translates to the formula

$$A^2 - Tr(A)A + \det(A)I = 0$$

which on multilinearization (see Chap. 3) becomes

$$A_1 A_2 + A_2 A_1 - Tr(A_1)A_2 - Tr(A_2)A_1 + Tr(A_1)Tr(A_2) - Tr(A_1 A_2) = 0.$$

$$(1.8.7)$$

On multiplication of (1.8.7) on the right by A_3,

$$A_1 A_2 A_3 + A_2 A_1 A_3 - Tr(A_1)A_2 A_3 - Tr(A_2)A_1 A_3 + Tr(A_1)Tr(A_2)A_3$$
$$- Tr(A_1 A_2)A_3 = 0$$

is obtained and on taking traces,

$$Tr(A_1 A_2 A_3) + Tr(A_2 A_1 A_3) - Tr(A_1)Tr(A_2 A_3) - Tr(A_2)Tr(A_1 A_3)$$
$$+ Tr(A_1)Tr(A_2)Tr(A_3) - Tr(A_1 A_2)Tr(A_3) = 0$$

which is equivalent to

$$\delta^{(3)} f(A_1, A_2, A_3) = 0.$$

The result (1.8.6) implies that if χ is a character of a group G with $\chi(e) = m$ then $\chi^{(m+1)} = \delta^{(m+1)}\chi = 0$.

The description of Frobenius' construction of a factor from a character χ follows.

Theorem 1.7 *Suppose that the factor ψ_χ of Θ corresponds to the character χ of G and that m is the degree of χ. Then ψ_χ is given by*

$$\psi_\chi = \frac{1}{m!} \sum \chi^{(m)}(g_{i_1}, g_{i_2}, \cdots, g_{i_m}) x_{g_1} x_{g_2} \cdots x_{g_m}$$

where the summation runs over all m-tuples of elements of G.

The proof of this is deferred to Chap. 3, Theorem 3.6, where it is shown that the analogous result holds in general for norm-type forms.

Corollary 1.10 *Two irreducible factors φ_1 and φ_2 of Θ are E-equivalent if and only if the associated characters χ_1 and χ_2 are equal.*

Proof If $\varphi_1 = \varphi_2$ it is clear that $\chi_1 = \chi_2$. Now suppose that χ_1 and χ_2 are equal. Then φ_1 and φ_2 can be constructed from the algorithm in Theorem 1.7 and must be equal.   □

1.9   Historical Remarks

1.9.1   Early Group Determinant Work After Frobenius

The initial expositions of the group determinant work of Frobenius, for example [90] and [291], emphasize the difficulty of following his arguments which display great technical skill in manipulating determinants and complex formulae. Poincaré mentioned some of his work peripherally in the context of differential equations in [234], but it appears that even among prominent mathematicians the proofs were regarded as formidable. Alternative approaches to representation theory by Burnside and Frobenius' pupil Schur, some aspects of which are given above, made representation theory much more accessible, especially after the Noether school showed that the main results were a consequence of results in the theory of modules over algebras, and Frobenius' group determinant theory seems to have become largely ignored. At this time finite group theory was a respectable research area but was overshadowed by the dazzling school of Göttingen where the focus was in other directions, often pursuing mathematics related to the important developments in physics.

   L. E. Dickson was one other author who looked specifically at group matrices and group determinants. In a series of papers [89, 91, 92] he discussed the group determinant over fields of finite characteristic, and also gave some facts about arbitrary group matrices. An account of his work will appear in Chap. 2.

   Poincaré, in the paper [234] on differential equations, used a group determinant to characterize those equations with rational solutions, as well as making the suggestion that the representation theory work of Frobenius and that of Elie Cartan on representations of Lie groups and Lie algebras could "mutually clarify" ("mutuellement éclairer"). This is reinforced in his book on probability (see Chap. 7). Before Frobenius's work appeared, Poincaré in [233] had recognized the importance of hypercomplex numbers and had suggested that the characteristic polynomial of an element could shed light on the structure of a system. He later introduced a set of "fundamental invariants", essentially the coefficients of the characteristic equation of each member of a generating set, in connection with the monodromy group of a linear differential equation. It appears to be an interesting project to relate the work of Cartan and Killing on Lie groups and Lie algebras to that of Frobenius (especially the 2-form introduced by Killing).

These ideas are closely connected with k-characters and norm forms. Mention should also be made of a paper of Speiser [266] which relates the group determinant, discriminants of number fields and the Klein "Normformproblem". The group determinant for abelian groups has appeared in various places in number theory (see for instance [200]), and for those interested in this direction the paper [79] may be of interest.

1.9.1.1 Modern Reexamination of the Work of Frobenius

The catalyst to reexamine Frobenius' papers was an attempt to generalize character theory and representation theory to loops and quasigroups by Johnson and Smith [175] see also [263]. A homomorphism from a non-associative structure into a linear group loses the non-associativity, and a matrix representation theory in this context would appear to have limited use. It is possible to generalize Frobenius' first definition of characters given above using the theory of association schemes, but the character theory which emerges in general loses more of the structure theory than group characters (for example "almost all" quasigroups have trivial character tables). It seemed to be interesting to look at the further tools of Frobenius, especially those which are independent of matrix representations. An attempt was begun to understand his ideas and generalize them. Gradually it became apparent that even in the group case there were aspects of Frobenius' work which had not been absorbed into the modern canon, but which were relevant to current research.

A basic question was to determine whether the group determinant contains all the information on a group. This question was posed in the paper [168] where some of Frobenius' results were summarized. A later paper [169] showed that if a group G has a faithful irreducible representation of degree 2, then the group is determined from the knowledge of the corresponding complex 2-character, and pointed out orthogonality between "extended k-characters" (See also [171] and Chap. 6). A breakthrough occurred in 1989 when Hoehnke after hearing of the problem posed in [168] indicated that he thought that a result of Frobenius (Theorem 3.1, Chap. 3) could be used to show that groups with the same determinant are isomorphic. After some of his observations were sent to Formanek, the latter and Sibley proved that the group determinant determines the group in [100]. It then emerged that work of Hoehnke on norm-type forms (see Chap. 3) could be applied to show that with restrictions on characteristic the $1-$, $2-$ and 3-characters of the regular representation or equivalently $1-$, $2-$ and 3-characters of the set of irreducible representations determine the group. After an announcement of this result (published in [147], see also [148]) was communicated to him, Mansfield produced a direct proof of the Formanek-Sibley result, and his argument with minor modification also proves the 3-character result with restrictions on characteristic. Another character theoretic proof of this result was produced by Sibley (unpublished) and independently by Roggenkamp–Kimmerle in [194]. Further, in [149] previous invariants of groups due to Gallagher [114] and Roitman [244] were related to k-characters. A parallel treatment of norm forms is given in the papers of the school of Bergmann [15–17] and the 3-character result may also be derived from this work.

1.10 The Initial Results on the Information in k-Characters

In the following the assumptions on the underlying field K are relaxed. K is any field such that $\text{Char}(K) \nmid 2n$, where $|G| = n$.

Theorem 1.8 (Formanek-Sibley) *If groups G and H have E-equivalent group determinants then they are isomorphic.*

Theorem 1.9 (Hoehnke-Johnson) *For any group G, Θ_G is determined by any of the following.*

  (i) *The 1-, 2-, and 3-characters of any representation which contains at least one copy of every irreducible representation.*
 (ii) *The 1-, 2-, and 3-characters of the regular representation.*
(iii) *The 1-, 2- and 3-characters of all the irreducible representations.*

   Theorem 1.9 was initially proved using the norm form work of Hoehnke which will be developed in Chap. 3. Since then a more direct proof of (ii) has been discovered, and (iii) follows directly from this. It is less easy to produce a direct proof of (i), so the proof of this is deferred to Chap. 3, Theorem 3.6.
   The following calculations on the 2-character and 3-character of the regular character π of a group G of order n, appeared in [174]. It is clear that $\pi(g) = 0$ for $g \neq e$, and $\pi(e) = n$.
   Let g, h, m be arbitrary elements of G. Then

$$\text{If } gh \neq e \text{ then } \pi^2(g, h) = \pi(g)\pi(h) - \pi(gh) = 0.$$

$$\pi^2(e, e) = n^2 - n$$

$$\pi^2(g, g^{-1}) = -n \text{ if } g \neq e.$$

Now consider the formula

$$\pi^3(g, h, m) = \pi(g)\pi(h)\pi(m) - \pi(g)\pi(hm) - \pi(h)\pi(gm) - \pi(m)\pi(gh)$$
$$+ \pi(ghm) + \pi(gmh).$$

It follows that $\pi^3(g, h, m) = 0$ except for the following cases

(1) $g = h = m = e$. Then $\pi^3(e, e, e) = n^3 - 3n^2 + 2n$.
(2) $g = e, m = h^{-1} \neq e$. Then $\pi^3(e, h, h^{-1}) = -n^2 + 2n$.

(3) $g, h \neq e, gh \neq e, m = (gh)^{-1}$. Then

$$\pi^3(g, h, (gh)^{-1}) = n \text{ if } gh \neq hg$$
$$= 2n \text{ if } gh = hg.$$

It may be seen that all groups of the same order have the same regular character, and a pair (G, H) of groups with the same order and the same number of involutions have the same regular 2-character, via any bijection $\sigma : G \to H$ such that $\sigma(e) = e$, $\sigma(g^{-1}) = (\sigma(g))^{-1}$ for all $g \neq e$.

Theorem 1.10 *The* 1-, 2- *and* 3-*characters of the regular representation* π *of a group* G *determine all triples* $\{g, h, m\}$ *such that at least one of* $ghm = e$ *or* $gmh = e$ *hold.*

Proof The element e can be determined from the 1-character of π and from the 2-character of π the inverse g^{-1} for any $g \in G$ can be determined. Suppose that $\pi^3(g, h, m)$ is known for all triples (g, h, m) in G. From the results above $\pi^3(g, h, m) = 0$ if $ghm \neq e$. Assume that $\pi^3(g, h, m) \neq 0$. Since $\pi^3(g, h, m)$ is invariant under all the 3! orderings of the triple, one of the cases (1)–(3) above is applicable. Using the restriction on characteristic it may be remarked that in all the cases the values of $\pi^3(g, h, m)$ are distinct. If $gh = hg$ then gh may be deduced. Now suppose that $gh \neq hg$. If $\pi^3(g, h, m) = n$ this implies that either ghm or hgm is e, but not both, or equivalently either $gh = m^{-1}$ or $hg = m^{-1}$. In either case, there must be a further triple such that $\pi^3(g, h, m_1) = n$ and thus $gh \in \{m^{-1}, m_1^{-1}\}$. □

Theorem 1.11 *A finite group* G *is determined by the regular* 1-, 2- *and* 3-*characters.*

Proof The proof follows directly from the following lemma, which appeared in a slightly different form as an exercise in [25, Section 4 Ex. 26]. The proof given is due to Mansfield [212]. □

Lemma 1.5 *Suppose that in a finite group* G *there is given the set* $\{gh, hg\}$ *for every pair* $\{g, h\}$ *in* G. *Then* G *is determined up to isomorphism.*

Proof Under the assumption of the lemma, note that if $xy = yx$ then xy is determined. In particular, e and x^{-1}, for each x, are determined. Now let there be given fixed elements g, h in G such that $gh \neq hg$ and that gh is known. Suppose that x and y are arbitrary elements. Then it will be shown that the following hold.

(1) ghx can be calculated.
(2) hx can be calculated.
(3) if $hx \neq xh$, then xy can be calculated for any y.
(4) if $[g, x] = [g, y] = [h, x] = [h, y] = e$ then xy can be calculated.

To prove (1), consider $S_1 = \{(gh)m, m(gh)\}$ for any m. Since $\{hm, mh\}$ is known, $S_2 = \{g(hm), (hm)g, g(mh), (mh)g\}$ is also known. Now since $ghm \in S_1 \cap S_2$ and, if $|S_1 \cap S_2| = 1$, (1) follows. Therefore it can be assumed that $ghm \neq mgh$ and that one of

$$(a)\ mgh = gmh, \quad (b)\ mgh = hmg, \quad (c)\ mgh = mhg$$

hold. Case (c) can be eliminated since this implies directly that $gh = hg$. Assume case (a). Then it follows that mgh lies in both $S_3 = \{(gh)m, m(gh)\}$ and $S_4 = \{(mg)h, h(mg)\}$. If $|S_3 \cap S_4| = 1$ this implies that mgh is known, otherwise $ghm = hmg$. Now $hmg = hgm$ and hence $ghm = hgm$ and again this implies $gh = hg$.

There remains case (b). Suppose that $x = mgh = hmg$. Then $xh^{-1} = h^{-1}x = mg$. If $y = ghm$ does not satisfy $yh^{-1} = h^{-1}y \in \{mg, gm\}$, each element of S_1 can be tested for this property and thus y is this element. Now assume $y = h^{-1}(ghm) = (ghm)h^{-1} \in \{mg, gm\}$. If $h^{-1}(ghm) = mg$ then $hg = gh$, and if $(ghm)h^{-1} = mg$ then $ghm = gmh$, both possibilities having already been excluded. Thus (1) has been proved.

It follows directly that (2) holds by the substitution $g = g^{-1}$, $h = gh$ in (1).

To address (3) if $hx \neq xh$ then by (2) it can be assumed that hx is known. Then xy can be calculated from (1) by replacing g by h and h by x. It follows that xy can be calculated if any of the commutators $[g, x], [g, y], [h, x], [h, y]$ differ from e, thus (3) follows from (4).

Since $gx = xg$ and $hy = yh$, xy and yh are known. Then

$$(gx)(hy) \in S_5 = \{(gx)(hy), (hy)(gx)\} = \{ghxy, yxhg\}.$$

Now $[gh, xy] = [gh, yx] = e$. Let $S_6 = gh\{xy, yx\} = \{ghxy, ghyx\}$. If $hygx = ghyx$ then $gh = hg$, and thus $S_5 \cap S_6 = \{ghxy\}$ is determined. On inserting h^{-1}, g^{-1}, and $ghxy$ for g, h, m in (1), it follows that xy is determined. The lemma is proved. $\qquad\square$

For another proof of this result see [37]. There are related results on "half homomorphisms", see [256].

Proof (of Theorem 1.9, (ii), (iii)). The theorem follows directly from Lemma 1.5, since by Theorem 1.10 the regular 3-character determines the triples xyz such that at least one of xyz and xzy equals e, and thus z^{-1} is either xy or yx. For if as in the proof above a fixed pair $g, h \in G$ is taken, an arbitrary choice may be made of one member of the pair $\{gh, hg\}$ as gh. If in fact gh is chosen the above indicates a constructive method to produce G. If hg is chosen then the group $(G^{op}, *)$ is produced, with $x * y = yx$ for all $x, y \in G$. $\qquad\square$

In [169] it is proved that if G is a finite group with a faithful representation ρ of degree 2 with character χ, an explicit set of representing matrices can be constructed from $\chi^{(2)}$. There a similar choice needs to be made so that either a set $\{\rho(g) : g \in$

G} of representing matrices for G or a set $\{\rho(g)^t : g \in G\}$ which represent G^{op} is obtained.

Remark 1.3 The proofs that the group determinant determines a group assume implicitly that if a polynomial is given which is a group determinant the variable x_e is indicated. This is not an important restriction. For if a polynomial $\varphi(x_1, \ldots, x_n)$ is a group determinant of a group G, and the variable x_i is chosen arbitrarily as that corresponding to the element e and a group multiplication is determined as above, then the group obtained is isomorphic to G. This follows from the next Lemma.

Lemma 1.6 *Suppose that G is a group. For any element s in G define the multiplication $(*)$ on G by*

$$g * h = gs^{-1}h.$$

Then $(G, *)$ is a group isomorphic to G with identity element s.

Proof The map $\alpha : g \to gs$ from G to $(G, *)$, satisfies

$$\alpha(g) * \alpha(h) = (gs)s^{-1}(hs) = ghs = \alpha(gh).$$

It is clear that α is an isomorphism, and since $g * s = g$, s is the identity element in $(G, *)$.   □

Let $\mathcal{L}_G$ be the group matrix with respect to the operation $(*)$, and $\mathcal{M}_G$ be the group matrix of G. It follows that $\det(\mathcal{L}_G)$ is E-equivalent to $\det(\mathcal{M}_G)$.

1.11   Summary of Further Developments

1. k-characters
   There are several instances of the introduction of k-characters in recent work, and the authors for the most part seem to have been unaware of their connections to Frobenius' original work.

   (a) Helling's characterization of characters. While the famous theorem of Brauer [162, p. 127] is usually referred to as the characterization of characters, it gives a criterion for a class function to be a generalized character. Helling gave a criterion for a class function on a finite group to be a genuine character (with restrictions on the underlying field), and his criterion may be directly stated in terms of k-characters. (Recently Collins in [62] has given a version of Brauer's theorem for modular characters. It may be an interesting problem to see whether there is a modular version of Helling's result.)
   (b) The work of Wiles [296] and Taylor [275] leading up to the proof of Fermat's Theorem involved "pseudocharacters". The context is of finite dimensional representations of an infinite group over a field which is not

necessarily algebraically closed. They define a pseudocharacter to be a class function f such that $\delta^k f = 0$ for some positive integer k. A collection of such pseudocharacters is obtained, and from them by a limiting process a pseudocharacter is constructed with values in an algebraically closed field, which is a genuine character by an extension of Helling's result obtained in [275] using matrix invariants.

(c) In a series of papers [36, 37] and [38] Buchstaber and Rees define a Frobenius n-homomorphism (n a positive integer) to be a trace-like function f on an algebra such that that $\delta^n f = 0$. They showed that if A, B are commutative algebras and $f : A \rightarrow B$ is an n-homomorphism then f induces a homomorphism from the symmetric algebra $Symm^n(A)$ to B and use this to obtain an extension to symmetric spaces of the well-known result of Gelfand-Kolmogorov identifying a space with the function algebra on it. More recently Khudaverdian and Veronov in [192] reproved some of the results of Buchstaber and Rees. In particular they introduce a Berezinian (arising in the context of mathematical physics), which may also be thought of as a generating function.

(d) In probability theory the FKG inequality is well-known and has many applications. It can be interpreted in terms of the covariance of two probabilities on a space. The covariance occurs as the second of a set of "higher correlations" or **cumulants** corresponding to each positive integer k, but counterexamples exist to a direct generalization of the inequality for $k > 2$. In [241] Richards introduced a modified set of such higher correlations and gave a "proof", unfortunately with a gap, that the result can be generalized to the cases $k = 3, 4, 5$. His "higher correlations" (for $k = 3, 4, 5$) essentially use the operations δ^3, δ^4 and δ^5, suitably modified. Sahi in [249] explained that a more general conjecture can be expressed in terms of the operations δ^n. So far the conjecture has been proved by Sahi only in special cases. In Sahi's work he introduces an approach which considers all the inequalities together by means of a generating function.

2. 2-characters

The question of whether the information in the irreducible 1- and 2-characters determines a group was answered in the negative in [173]. Properties determined by this information were also investigated in a preprint by Mckay and Sibley [213]. Further studies were carried out in [174] and [184] where the Weak Cayley table of a group was introduced: groups G and H have the same irreducible 1- and 2-characters if and only if they have the same weak Cayley tables. Humphries in [150] examined the group of Weak Cayley Table isomorphisms, and the category of Weak Cayley morphisms was examined in [180].

3. Extended k-characters and k-classes

In [169] there are defined 2- and 3-character tables for a group. These include extended k-characters given for $k = 2, 3$. Vazirani in [284] gave a general definition of an "immanent" k-character and explained how a k-character table can be constructed for arbitrary k, as well as making connections with representations of wreath products and Schur functions. More recently, Humphries and Rode [154] have examined the S-ring which arises from the k-classes on a group. These form the natural domain for the k-characters, but whereas the information in the center of the group ring, generated by the conjugacy classes (i.e. the 1-classes) of a group is identical to that contained in the character table, if $k > 1$ the k-characters and the algebra generated by the k-classes contain different information. For example, whereas neither the information in the 2-characters nor that in the 2-class algebra determines the derived length of a group, the combined information does.

4. Group matrices and group determinants

   (a) Group matrices appeared in papers on control theory in the 1970s, for example in [280].
   (b) The **symmetric group matrix** of a group G is the matrix obtained from X_G by setting x_g equal to $x_{g^{-1}}$ for all g in G. It is a symmetric matrix. Its determinant is the **symmetric group determinant**. Sjogren in [258] examined the symmetric group determinant for the case for a group of odd order, and related this to graph theory.
   (c) Humphries in [150] used results on group matrices to examine presentations of infinite groups.
   (d) Waldron in papers on wavelets and finite frames used the fact that a "Gramian" corresponding to a symmetric set of vectors is a group matrix.
   (e) Cooper and Walsh in [66] introduced group matrices and group determinants in the context of the geometry of three-manifolds.

5. Probability

   The book [82] discussed many objects from group representation theory. In certain cases the transition matrix of a Markov chain is a group matrix, and the Fourier transform may be regarded as a diagonalization of a group matrix. There are many more connections which will be discussed in Chap. 7.

6. Group matrices and group rings

   Taussky-Todd in [273] gives an account of how group matrices can be used in the theory of group rings. These ideas have been taken up recently by Hurley and applied to coding theory in [159, 161], and [160].

   Subsequent chapters describe more fully these developments. Throughout the book it will be understood that unless explicit mention is made of the underlying field this will be taken as $\mathbb{C}$.

Chapter 2
Further Group Matrices and Group Determinants

Abstract The construction and properties of group matrices are analyzed in more detail. The group matrix of a cyclic group with a certain ordering is a circulant matrix. The book by Davis (Circulant Matrices, Chelsea, New York, 1994) gives a comprehensive account of circulants and the chapter is designed to provide a far reaching extension and generalization of the results there. If an arbitrary subgroup H of a group G is taken, it is shown that with an appropriate ordering the group matrix X_G is a block matrix in which each block is of the form $X_H(\underline{u})$ where the elements in $\underline{u}$ are all distinct and of the form x_{g_i}. This leads to methods to partially diagonalize the group matrix which facilitate the calculation of the group determinant. Group matrix versions of the ring of representations and the Burnside ring of a group are described using supermatrices. A description of projective group matrices, corresponding to projective representations, is given. Work of L. E. Dickson on group matrices and group determinants over a finite field is also described.

2.1 Introduction

The group matrix of a finite group G has appeared in Chap. 1 and some of its properties are given there. In particularly, the construction of the group matrix via the Ward quasigroup associated to a group is described and its representation in the form of block circulants related to the cosets of any cyclic subgroup is given. In this chapter the construction and properties of group matrices are analyzed in more detail. A comprehensive account of circulant matrices (equivalent to special forms of group matrices for cyclic groups) has appeared in the book [72] and the work here may be considered as extending the discussion there to group matrices for arbitrary finite groups. An attempt has been made to use elementary tools so far as it is possible.

Originally, the term "group matrix" was used only for the matrices described in Frobenius' original paper [107], i.e. the matrix $\{x_{gh^{-1}}\}$ which is the full group matrix. As indicated in Chap. 1, it became standard to use the term group matrix for

© Springer Nature Switzerland AG 2019

K. W. Johnson, *Group Matrices, Group Determinants*
and Representation Theory, Lecture Notes in Mathematics 2233,
https://doi.org/10.1007/978-3-030-28300-1_2

any matrix of the form

$$X_G^\rho = \sum_{g \in G} \rho(g) x_g$$

where ρ is a representation of G and there is selected a specific set of matrices $\{\rho(g)\}_{g \in G}$, where for some r, $\rho(g) \in GL(r, \mathbb{C})$. In particular the full group matrix corresponds to the right regular representation. For arbitrary ρ with associated character χ it easily follows that $Tr(X_G^\rho) = \sum \chi(g) x_g$. In this chapter the term group matrix is also extended to any matrix which is obtained from a group matrix by replacing the variables $\{x_g\}$ by any elements of a ring, for example the x_g could themselves be matrices. If a distinction needs to be made, the group matrix $X_G^\rho(\underline{x}) = \sum_{g \in G} \rho(g) x_g$ will be referred to as a **strict** group matrix and a group matrix $X_G^\rho(\underline{u})$ where the elements in $\underline{u}$ are elements of a ring will be referred to as a **generalized** group matrix.

Section 2.2 discusses the structure of strict group matrices X_G^ρ for a fixed ρ and a fixed choice of representing matrices $\{\rho(g)\}_{g \in G}$. It is shown that if ρ is faithful then they form a copy of the group algebra. If ρ_1 and ρ_2 are representations of G, then a construction of particular forms of $X_G^{\rho_1 + \rho_2}$ and $X_G^{\rho_1 \otimes \rho_2}$ from $X_G^{\rho_1}$ and $X_G^{\rho_2}$ is described, and in order to do this for the latter a "dual tensor product" of group matrices is defined. The material here also leads into that in Chap. 7, in that certain specializations of the variables give group matrices which are diagonalizable, leading to techniques of harmonic analysis.

In Sect. 2.3, the construction of full group matrices is examined further. Since $X_G(\underline{x})$ is obtained from the multiplication table for $Wa(G)$ by replacing g by x_g for all g in G, any construction of $Wa(G)$ automatically leads to one for $X_G(\underline{x})$. If an arbitrary subgroup H of G is taken, it is shown that with an appropriate ordering X_G is a block matrix in which each block is of the form $X_H(\underline{u})$ where the elements in $\underline{u}$ are all distinct and of the form x_{g_i}. The ordering behind this result goes back to a paper of Fite [99], but there only a weaker result is given. This result extends that given in Chap. 1 for the case where H is cyclic. The group matrix of a direct product $H \times K$ is then discussed, and not surprisingly it can be described in a relatively simple way. In the case where H is a normal subgroup of G it is shown that, in the above construction, for each block of the form $X_H(\underline{u})$ the elements of $\underline{u}$ are of the form x_{g_i} where the g_i lie in the same coset of H, but in general it is difficult to see very much symmetry if no further assumptions are made. In the case of a split extension of H by K it can be shown that the pattern of X_G can be described in terms of X_H and X_K. A less restrictive assumption is that a group G is an exact product of subgroups H and K which need not be normal. It is shown in this situation that X_G can also be described in terms of X_H and X_K.

If G has an ascending chain of subgroups which do not necessarily have any properties of normality, it is indicated how the group matrix of G can be constructed by using this chain, essentially by iterating the above process.

The group matrix of an abelian group can be described easily, since the group is a direct product of cyclic groups but even in this case the factorization of the group determinant plays a significant role in number theory (see [200] where a full page discussion is given to it and [250]). This factorization is quickly obtained from the discussion here (and is also directly related to the character table).

A full group matrix for G is a symmetric matrix only in the case where G is abelian of exponent 2. However if x_g and $x_{g^{-1}}$ are identified for all g in G the resulting matrix which will be denoted by X_G^{Symm} is necessarily symmetric. This appears in the work of Sjogren [258] and of Cooper and Walsh [66] which will be discussed in Chap. 10. In this case the irreducible factors of its determinant are polynomials with real coefficients.

The form of X_G (or equivalently) $Wa(G)$ is dependent on the chosen ordering of G. If an ordering is taken which lists the elements by conjugacy classes it is shown that the set of commutators can be read off from the diagonal blocks of $Wa(G)$. There is a further way of ordering G to obtain another form of X_G which was introduced by Dickson in order to discuss factorization of the group determinant over a field with finite characteristic. Given a subgroup H of G of index r a set of $r \times r$ matrices $\{M(h)\}_{h \in H}$ is constructed such that X_G becomes $X_H(M(h_i h_j^{-1}))$.

A digression is made to give a short discussion of an analogous "loop matrix" for the Octonion loop O_{16}. In this case there is a cyclic subgroup H of order 4 and the construction of the loop matrix by ordering the elements via the left cosets of H produces a block matrix but the condition that the blocks are circulants is not satisfied. However in this case those blocks which are not circulants are reverse circulants.

In Sect. 2.4 methods which either fully or partially diagonalize X_G are discussed. In the case where no assumption is made about the subgroup H, it is shown that, when X_G is constructed via the cosets of H in the form indicated above, there is a matrix P such that $P^{-1} X_G P$ is a block diagonal matrix with blocks which are group matrices each of which corresponds to a representation of the form $\text{Ind}_H^G \rho_i$ where ρ_i is an irreducible representation of H, and all such induced representations appear (this could be used as a naive definition of induced representations). This work has relevance to harmonic analysis on finite groups and will be discussed further in Chap. 7. Applications to factoring group determinants are then given. It is also explained how the determinant of O_{16} is as given in Chap. 1.

The Grothendieck ring of virtual representations of a group does not appear to have a counterpart in a set of group matrices. However if the supermatrices arising in physics are considered, "super group matrices" may be used to construct a concrete representation of this ring.

In Sect. 2.5 an interpretation of the group matrix in terms of the Hopf algebra structure on the group algebra is outlined.

The Burnside ring of a finite group obtained from permutation actions predated group representation theory and appears in many places. A natural question arises as to whether there is an interpretation of this ring in terms of group matrices. This is addressed in Sect. 2.6. If $n_i \Omega_i$ is an element of the Burnside ring which corresponds to a G-set, i.e. the n_i are all positive, then a group matrix $X_{n_i \Omega_i}$ which corresponds to this element may be constructed from the group matrices X_{Ω_i} which correspond to the Ω_i. If n_i is negative then supermatrices are needed (see above), and a super group matrix version of the Burnside ring is described.

Another basic idea at the foundations of representation theory (and homological algebra) is that of projective representations, pioneered by Schur [251]. A projective representation corresponds to a 2-cocycle α and Schur defined a full projective group matrix corresponding to such an α. Projective group matrices corresponding to arbitrary projective representations are discussed in Sect. 2.7. They also appear in the theory of wavelets, see Chap. 10, Sect. 10.7.

In Sect. 2.8 a short account is given of the work of Dickson on modular group matrices and determinants, where he uses the decomposition of the group matrix described above.

2.2   Arbitrary Group Matrices

A fundamental result is the following. For a finite group the group algebra KG of G over a field K has been defined in Chap. 1. This definition extends naturally to the case where K is an integral domain R.

Suppose that an ordering of the group G is fixed. Let the set of (full) group matrices of G with respect to this ordering and with elements in the domain R be denoted by $\mathfrak{X}_G(R)$. This set forms an algebra over R, the operations being matrix addition, multiplication and scalar multiplication. This algebra is isomorphic to RG, as the following lemma shows.

Lemma 2.1 *The function* $\sigma : RG \to \mathfrak{X}_G(R)$ *given by*

$$\sigma\left(\sum_{i=1}^{n} a_i g_i\right) = X_G(a_1, a_2, \ldots, a_n).$$

is an isomorphism.

Proof In the product

$$\left(\sum_{i=1}^{n} a_i g_i\right)\left(\sum_{i=1}^{n} b_i g_i\right)$$

the coefficient of g_k is $\sum_{g_i g_j = g_k} a_i b_j$. Thus this product is the same as $(\sum_{i=1}^{n} c_i g_i)$ where $\underline{c}$ is the convolution $\underline{a} * \underline{b}$. Then by Theorem 1.4, Chap. 1

$$X_G(a_1, a_2, \ldots, a_n) X_G(b_1, b_2, \ldots, b_n) = X_G(c_1, c_2, \ldots, c_n).$$

The fact that σ preserves addition and scalar multiplication follows directly. The kernel of σ is the inverse image of $X_G(0, 0, \dots, 0)$ which is 0 and σ is clearly onto.

$\square$

This result has been used in the theory of group algebras by Taussky and more recently by Hurley (see Chap. 10).

This may be extended to the following. Again suppose that an ordering of the group G is fixed. Let ρ is any faithful representation of G. Let $\mathfrak{X}_G^\rho(R)$ be the set of matrices of the form $X_G^\rho(R)$.

Lemma 2.2 *The function $\sigma^\rho : RG \rightarrow \mathfrak{X}_G^\rho(R)$ given by*

$$\sigma^\rho(\sum_{i=1}^n a_i g_i) = X_G^\rho(a_1, a_2, \dots, a_n).$$

is an isomorphism.

The proof is almost identical to that above, although the fact that the kernel of σ^ρ is trivial needs the faithfulness of ρ.

Given a group G and a fixed ordering of its elements $\{e = g_1, \dots, g_n\}$, the corresponding group matrix X_G is determined by its first column. For any vector $\underline{u} = (u_1, \dots, u_n)$ let $X_G(\underline{u})$ denote the group matrix with first column $\underline{u}^t$. Explicitly, this is constructed by replacing the element x_{g_i} in $X_G(x_{g_1}, x_{g_2}, \dots, x_{g_n})$ by u_i for all i. The entries in $\underline{u}$ may be elements in another group, elements of a commutative ring or even from a non-commutative ring, an important example being a ring of matrices. Forms of the following theorem seem to have been stated in various contexts-for example [280].

If $X_G^\rho(\underline{x})$ is a strict group matrix, with $\rho(e) = r$, the coefficient of x_e^r in $\det(X_G^\rho(\underline{x}))$ is 1, which implies that $X_G^\rho(\underline{x})$ is nonsingular, i.e. $X_G^\rho(\underline{x})^{-1}$ exists. The transpose of a matrix M will be denoted by M^T.

Theorem 2.1 *If G is a finite group with a fixed ordering $\{g_i\}_{i=1}^n$ and $X = X_G^\rho(\underline{x})$ is an arbitrary strict group matrix corresponding to the representation*

$$(\rho, \{\rho(g)\}_{g \in G})$$

(abbreviated by ρ), then

(a) *any power $(X)^n$ is a group matrix corresponding to ρ where n is an integer.*
(b) *The sum and product of $X_G^\rho(x)$ and $X_G^\rho(\underline{y})$ are group matrices corresponding to ρ.*
(c) *The transpose of X is a group matrix for G^{op} corresponding to the representation $(\widetilde{\rho} = \{\rho^T(g)\}_{g \in G})$.*

Proof

(a) Let ρ be an arbitrary representation of G with a specified set of matrices $\{\rho(g)\}_{g\in G}$. As above the corresponding group matrix is

$$X_G^\rho(\underline{x}) = \sum_{g\in G} \rho(g)x_g.$$

Then

$$X_G^\rho(\underline{x})X_G^\rho(\underline{y}) = X_G^\rho(\underline{z}) \text{ where } \underline{z} = \underline{x} * \underline{y}$$

This follows by direct calculation:

$$\sum_{g\in G}\rho(g)x_g \sum_{h\in G}\rho(h)y_h = \sum_{k\in G}\sum_{gh=k}\rho(g)x_g\rho(h)y_h = \sum_{k\in G}\sum_{gh=k}\rho(gh)x_g y_h$$

$$= \sum_{k\in G}\rho(k)z_k = X_G^\rho(\underline{z}). \tag{2.2.1}$$

It is clear that any positive integral power of X is a group matrix $X_G^\rho(\underline{u})$ from (2.2.1). The Cayley-Hamilton theorem applied to an arbitrary non-singular matrix may be used to express the inverse as a linear combination of non-negative powers of the matrix, multiplied by the inverse of $\det(X)$ and it follows that $(X_G^\rho)^{-1}$ is a group matrix (whose entries are rational functions of the entries of X_G^ρ).

(b) This follows from

$$X_G^\rho(\underline{x}) + X_G^\rho(\underline{y}) = \sum_{g\in G}\rho(g)(x_g + y_g)$$

and (2.2.1).

(c) This is clear.

$\square$

If the elements of $\underline{u}$ are arbitrary, it may be that $X_G^\rho(\underline{u})$ is singular. The Moore-Penrose pseudoinverse $(X)^+$ of a matrix X is a polynomial in X and its Hermitian transpose X^* [72, p. 46].

The following examples show that $(X_G(\underline{u}))^+$ is "like" a group matrix but with a possible change in the ordering of G. If G is cyclic with the usual ordering then the matrix $(X_G(\underline{u}))^+$ remains a circulant (this is shown in [72]).

Example 2.1 Suppose G is S_3 and ρ is the unique irreducible representation of degree 2. Then with respect to the ordering in Example (1.13), Chap. 1,

$$X_G^\rho = \begin{bmatrix} x_1 + \omega x_2 + \omega^2 x_3 & x_4 + \omega x_5 + \omega^2 x_6 \\ x_4 + \omega^2 x_5 + \omega x_6 & x_1 + \omega^2 x_2 + \omega x_3 \end{bmatrix},$$

with $\omega = e^{2\pi i/3}$. If the vector u is taken as $[x_1, x_2, x_3, x_1, x_2, x_3]$ then

$$X_G^\rho(u) = \begin{bmatrix} x_1 + \omega x_2 + \omega^2 x_3 & x_1 + \omega x_2 + \omega^2 x_3 \\ x_1 + \omega^2 x_2 + \omega x_3 & x_1 + \omega^2 x_2 + \omega x_3 \end{bmatrix}$$

is singular. Then

$$(X_G^\rho(u))^+ = \frac{1}{4}\begin{bmatrix} v_1 & v_2 \\ v_1 & v_2 \end{bmatrix}$$

where $v_1 = \frac{1}{x_1+\omega^2 x_2+\omega x_3}$ and $v_2 = \frac{1}{x_1+\omega x_2+\omega^2 x_3}$.

Now the full group matrix is

$$X_G = \begin{bmatrix} C(x_1, x_2, x_3) & C(x_4, x_6, x_5) \\ C(x_4, x_5, x_6) & C(x_1, x_3, x_2) \end{bmatrix}$$

thus

$$X_G(u) = \begin{bmatrix} C(x_1, x_2, x_3) & C(x_1, x_3, x_2) \\ C(x_1, x_2, x_3) & C(x_1, x_3, x_2) \end{bmatrix}$$

and

$$(X_G(u))^+ = \frac{1}{4\det(C(x_1, x_2, x_3))}\begin{bmatrix} C(z_1, z_3, z_2) & C(z_1, z_3, z_2) \\ C(z_1, z_2, z_3) & C(z_1, z_2, z_3) \end{bmatrix}$$

where

$$z_1 = x_1^2 - x_2 x_3, \ z_2 = x_2^2 - x_1 x_3, \ z_3 = x_3^2 - x_1 x_2.$$

This is a group matrix of the form $X_G(u')$ where $u' = [z_1, z_3, z_2, z_1, z_3, z_2]$.

Remark 2.1 Since the exponential of a matrix X is defined as

$$e^X = 1 + X + \frac{X^2}{2!} + \ldots + \frac{X^n}{n!} + \ldots$$

it follows from Theorem 2.1 that $e^{X_G^\rho(x)}$ is a group matrix corresponding to $(\rho, \{\rho(g)\})$. (Here the entries are infinite series in the x_{g_i}.)

Lemma 2.3 *Let ρ_1 and ρ_2 be representations of G with specified sets of representing matrices $\{\rho_1(g)\}_{g\in G}$ and $\{\rho_2(g)\}_{g\in G}$. Then the set*

$$\left\{\begin{bmatrix} \rho_1(g) & 0 \\ 0 & \rho_2(g) \end{bmatrix}\right\}_{g\in G}$$

is a set of representing matrices for $\rho_1 + \rho_2$ and thus the corresponding group matrix is

$$X_G^{\rho_1 + \rho_2}(\underline{x}) = diag(X_G^{\rho_1}(\underline{x}), X_G^{\rho_2}(\underline{x})).$$

Proof This is obvious. $\square$

For a finite group G the **delta function** δ_{gh} is the function which takes on the value 1 if $g = h$, 0 otherwise.

Definition 2.1 The **dual tensor product** $M_1 \otimes^* M_2$ of matrices M_1 and M_2 with entries which are linear combinations of $\{x_g\}$ is defined to be the matrix obtained from the usual Kronecker product $M_1 \otimes M_2$ by imposing the relations

$$x_g x_h = \delta_{gh} x_g \text{ for all } g, h \in G. \tag{2.2.2}$$

It may be seen that (2.2.2) is satisfied if each x_g is identified with the function e_g : $G \to \mathbb{C}$ defined by $e_g(h) = \delta_{gh}$, i.e if $\{x_g\}$ is regarded as the standard basis of the dual of $\mathbb{C}G$.

This will be discussed below in the context of the Hopf algebra structure on $\mathbb{C}G$.

Lemma 2.4 *Let ρ_1 and ρ_2 be representations of G. Then there is a set of representing matrices for $\rho_1 \otimes \rho_2$ such that*

$$X_G^{\rho_1 \otimes \rho_2}(\underline{x}) = X_G^{\rho_1}(\underline{x}) \otimes^* X_G^{\rho_2}(\underline{x}).$$

Proof A set of representing matrices for $\rho_1 \otimes \rho_2$ is $(\rho_1 \otimes \rho_2)(g) = \rho_1(g) \otimes \rho_2(g)$. Thus $X_G^{\rho_1 \otimes \rho_2}(\underline{x}) = \sum_{g \in G}(\rho_1(g) \otimes \rho_2(g))x_g$. Now

$$X_G^{\rho_1}(\underline{x}) \otimes^* X_G^{\rho_2}(\underline{x}) = (\sum_{g \in G}(\rho_1(g)x_g) \otimes^* (\sum_{g \in G}(\rho_2(g)x_g). \tag{2.2.3}$$

If the relations (2.2.2) are imposed on the collected terms of the right hand side of (2.2.3),

$$\sum_{g \in G}(\rho_1(g) \otimes \rho_2(g))x_g$$

is obtained. $\square$

Example 2.2 Let G be the Klein 4-group $V_4 = \{e, a, b, ab : a^2 = b^2 = (ab)^2 = e\}$ (with ordering as indicated). There is a representation ρ with group matrix

$$\begin{bmatrix} x_1 + x_2 & x_3 + x_4 \\ x_3 + x_4 & x_1 + x_2 \end{bmatrix}.$$

Then $\rho \otimes \rho$ has group matrix

$$\begin{bmatrix} x_1 + x_2 & 0 & 0 & x_3 + x_4 \\ 0 & x_1 + x_2 & x_3 + x_4 & 0 \\ 0 & x_3 + x_4 & x_1 + x_2 & 0 \\ x_3 + x_4 & 0 & 0 & x_1 + x_2 \end{bmatrix}.$$

It has already been indicated in Chap. 1 that the full group matrix is a patterned matrix, with its entries determined by its first row or column. This is no longer true for the matrices X_G^ρ.

Example 2.3 $G = S_3$ listed as $\{e, (1, 2, 3), (1, 3, 2), (1, 2), (13), (2, 3)\}$ and abbreviated as $\{1, \ldots, 6\}$. If $\underline{u} = \{u_i\}_{i=1}^6$ then

$$X_G(\underline{u}) = \begin{bmatrix} C(u_1, u_2, u_3) & C(u_4, u_6, u_5) \\ C(u_4, u_5, u_6) & C(u_1, u_3, u_2) \end{bmatrix}.$$

Let ρ be the unique non-linear representation with

$$\rho(1, 2, 3) = \begin{bmatrix} \omega & 0 \\ 0 & \omega^2 \end{bmatrix}, \quad \rho(1, 2) = \begin{bmatrix} 0 & 1 \\ 1 & 0 \end{bmatrix}.$$

With the ordering above

$$X_G^\rho(\underline{u}) = \begin{bmatrix} u_1 + \omega u_2 + \omega^2 u_3 & u_4 + \omega u_5 + \omega^2 u_6 \\ u_4 + \omega^2 u_5 + \omega u_6 & x_1 + \omega^2 u_2 + \omega u_3 \end{bmatrix}.$$

Note that $\rho \otimes \rho$ has group matrix

$$X_G^{\rho \otimes \rho}(\underline{u}) = \begin{bmatrix} u_1 + \omega^2 u_2 + \omega u_3 & 0 & 0 & u_4 + \omega^2 u_5 + \omega u_6 \\ 0 & u_1 + u_2 + u_3 & u_4 + u_5 + u_6 & 0 \\ 0 & u_4 + u_5 + u_6 & u_1 + u_2 + u_3 & 0 \\ u_4 + \omega u_5 + \omega^2 u_6 & 0 & 0 & u_1 + \omega u_2 + \omega^2 u_3 \end{bmatrix}.$$

It is an easy exercise to show that the above matrix is similar to a matrix with diagonal entries

$$\sum_{i=1}^6 u_i, \quad \sum_{i=1}^3 u_i - \sum_{i=4}^6 u_i, \quad X_G^\rho(\underline{u}).$$

This demonstrates that $\rho \otimes \rho$ is equivalent to $1_G \oplus sgn \oplus \rho$ (this is also deduced easily from character theory).

Compare with the equivalent representation given by

$$\sigma(1,2,3) = \begin{bmatrix} -1 & -1 \\ 1 & 0 \end{bmatrix}, \ \sigma(1,2) = \begin{bmatrix} 0 & 1 \\ 1 & 0 \end{bmatrix}.$$

where

$$X_G^{\sigma}(\underline{u}) = \begin{bmatrix} u_1 - u_2 + u_5 - u_6 & -u_2 + u_3 + u_4 - u_6 \\ u_2 - u_3 + u_4 - u_5 & u_1 - u_3 - u_5 + u_6 \end{bmatrix}$$

and $X_G^{\sigma \otimes \sigma}(\underline{u})$ is

$$\begin{bmatrix} u_1 + u_2 + u_5 + u_6 & u_2 + u_6 & u_2 + u_6 & u_2 + u_3 + u_4 + u_6 \\ -u_2 - u_5 & u_1 - u_5 - u_6 & -u_2 - u_3 + u_4 & -u_3 - u_6 \\ -u_2 - u_5 & -u_2 - u_3 + u_4 & u_1 - u_5 - u_6 & -u_3 - u_6 \\ u_2 + u_3 + u_4 + u_5 & u_3 + u_5 & u_3 + u_5 & u_1 + u_3 + u_5 + u_6 \end{bmatrix}.$$

In Chap. 1 the notation $C(u_1, \ldots, u_n)$ has already been used for the circulant whose first column is $(u_1, \ldots, u_n)$. Thus if $G = C_n$, listed as in Chap. 1, Example 1 then $X_G(\underline{u}) = C(\underline{u})$. To conform with the extended notation given above, if $|G| = n$ and $A_1, A_2, \ldots, A_n$ are $m \times m$ matrices then $X_G(A_1, A_2, \ldots, A_n)$ is used to denote the $nm \times nm$ matrix with the block A_k in the (i, j)th block whenever $X_G(\underline{u})$ has u_k in the (i, j)th position. Thus

$$C(A_1, A_2, \ldots, A_n) = \begin{bmatrix} A_1 & A_n & A_{n-1} & \ldots & A_2 \\ A_2 & A_1 & A_n & \ldots & A_3 \\ A_3 & A_n & A_1 & \ldots & A_4 \\ & & \cdots & & \\ A_n & A_{n-1} & A_{n-2} & \ldots & A_1 \end{bmatrix}.$$

2.2.1   A Connection with Invariant Theory

If an action of a group G on a polynomial $\phi(x_1, x_2, \ldots, x_r)$ is defined, the action being written as $\phi \to g(\phi)$ for $g \in G$ then ϕ is an **invariant** of G if $g(\phi) = \phi$ for all g in G. Invariant theory had been a major subject of nineteenth century mathematics, but after the famous Hilbert finiteness result of 1890 it began a decline and it appears that Frobenius held work on invariants in low regard. There are links to group representation theory and in fact many of the results on the representation theory of the symmetric group appeared in the paper [80] on invariant theory which appeared in 1892. An invariant usually arises from a representation of the group.

There has been a revival and generalization of the theory in modern times. A **relative invariant** (or semi-invariant) of G is a polynomial ϕ such that

$$g(\phi) = \alpha(g)\phi$$

where $\alpha(g)$ is a scalar which is called a multiplier. See for example [269].

Theorem 2.2 *Suppose that ρ is an arbitrary representation of G with corresponding determinant factor Θ_G^ρ. Then Θ_G^ρ is a relative invariant of the regular representation of G with multiplier $\alpha(g) = \det(\rho(g^{-1}))$.*

Proof Take the representation ρ of G and consider the group matrix $X_G^\rho = \sum_{g \in G} \rho(g)x_g$ and the corresponding group determinant factor

$$\Theta_G^\rho = \det(X_G^\rho).$$

Let $h \in G$ act on $X_G^\rho : \sum_{g \in G} \rho(g)x_g \to \sum_{g \in G} \rho(g)x_{gh}$. This extends to an action on Θ_G^ρ. Now

$$\sum_{g \in G} \rho(g)x_{gh} = \sum_{k \in G} \rho(kh^{-1})x_k = \left(\sum_{k \in G} \rho(k)x_k\right)\rho(h^{-1}) = X_G^\rho \, \rho(h^{-1})$$

Therefore the action of h on Θ_G^ρ may be expressed as

$$(\Theta_G^\rho)^h = \det(\rho(h^{-1}))\Theta_G^\rho. \tag{2.2.4}$$

This is equivalent to the statement that Θ_G^ρ is a relative invariant of G. □

Corollary 2.1 *If the trace of X_G^ρ is $\sum m_i x_{g_i}$ then the trace of $X_G^\rho \rho(h^{-1}) = \sum_{k \in G} \rho(kh^{-1})x_k$ is $\sum m_i x_{g_i h^{-1}}$.*

2.3 Representation of Group Matrices in Block Form

2.3.1 Group Matrices Built from Arbitrary Subgroups

Let H be a subgroup of G, with $H = \{e, h_2, \ldots, h_m\}$ and suppose that

$$G = eH + t_2H + \ldots + t_rH$$

where $mr = n$. Let the elements of G be ordered as

$$\{e, h_2, \ldots, h_m, t_2, t_2h_2, \ldots, t_2h_m, \ldots, t_r, t_rh_2, \ldots, t_rh_m\}. \tag{2.3.1}$$

This ordering is used by Fite in [99] . He remarks that the row sums in the blocks in the corresponding group matrix are the same, but he does not seem to be aware that the blocks are group matrices for H as given in the lemma below.

Lemma 2.5 *The group matrix X_G with respect to the above ordering is a block matrix $\{B_{ij}\}_{i,j=1}^r$ where $B_{ij} = X_H(u_1, u_2, ..., u_m)$ with $u_k = x_{t_i h_k t_j^{-1}}$, $k = 1, \ldots, r$.*

Proof Consider the block of $Wa(G)$ which is the intersection of the rows corresponding to the elements in the coset $t_i H$ and the columns corresponding to the coset $t_j H$. The element in the (k, l)th position in this block is

$$t_i h_k (t_j h_l)^{-1} = t_i h_k h_l^{-1} t_j^{-1}.$$

Therefore the block is obtained from $Wa(H)$ by replacing each occurrence of the element h_k by the element $x_{t_i h_k t_j^{-1}}$. This corresponds precisely to $X_H(u_1, u_2, .., u_r)$.

□

Note that in general the elements $\{u_k\}$ in the block described above do not necessarily lie in the same coset of H.

Example 2.4 Let $G = S_4$ and $H = \langle a, b \rangle = \langle (1, 2, 3, 4), (1, 4)(2, 3) \rangle$ (H is isomorphic to D_4). With H ordered as $\{e, a, a^2, a^3, b, ba, ba^2, ba^3\}$ and if an ordering of G is constructed as in (2.3.1), using the transversal $\{e, (123), (132)\}$, X_G may be written as a block matrix $\{B_{ij}\}_{i,j=1}^3$ where each B_{ij} is of the form $X_H(\underline{u}_{i,j})$ with

$$
\begin{aligned}
&\underline{u}_{1,1} = (1, 2, 3, 4, 5, 6, 7, 8) &&\underline{u}_{1,2} = (17, 12, 22, 15, 13, 24, 10, 19) \\
&\underline{u}_{1,3} = (9, 23, 16, 18, 21, 11, 20, 14) \quad &&\underline{u}_{2,1} = (9, 10, 11, 12, 13, 14, 15, 16) \\
&\underline{u}_{2,2} = (1, 20, 6, 23, 21, 8, 18, 3) &&\underline{u}_{2,3} = (17, 7, 24, 2, 5, 19, 4, 22) \\
&\underline{u}_{3,1} = (17, 18, 19, 20, 21, 22, 23, 24) \quad &&\underline{u}_{3,2} = (9, 4, 14, 7, 5, 16, 2, 11) \\
&\underline{u}_{3,3} = (1, 15, 8, 10, 13, 3, 12, 6)
\end{aligned}
$$

where x_{g_i} has been replaced by i for all i.

Except for B_{11}, the blocks no longer have the property that if the first column is $(x_{g_{i_1}}, \ldots, x_{g_{i_r}})$ then the k, lth element is $x_{g_{i_k} g_{i_l}^{-1}}$. It will be seen later that since G is an exact product of H and another subgroup this group matrix can be written in a more organized form.

2.3.2 Group Matrices of Direct Products

Let H and K be groups and consider the direct product $H \times K$. Let the elements of H and K be ordered as $\{e = h_1, \ldots, h_m\}$ and $\{e = k_1, \ldots, k_r\}$ respectively. Assume that the elements of $G = H \times K$ are ordered on the cosets of H as above, i.e.

$$\{h_1, \ldots, h_m, k_1 h_1, \ldots, k_1 h_m, \ldots, k_r h_1, \ldots, k_r h_m\}.$$

Lemma 2.6 *Let H and K be groups, and write the element (h, k) of $H \times K$ as hk. If the variable corresponding to the element $h_i k_j = k_j h_i$ is denoted by x_{ij}, then the group matrix $X_{H \times K}$ is of the form*

$$X_K(X_H(x_{11}, x_{21}, \ldots, x_{m1}), X_H(x_{12}, x_{22}, \ldots, x_{m2}), \ldots, X_H(x_{1r}, x_{2r}, \ldots, x_{mr})). \tag{2.3.2}$$

Proof From Lemma 2.5, $Wa(G)$ with respect to the above ordering consists of block matrices $\{B_{ij}\}_{i,j=1}^r$, where $B_{ij} = \{k_i h_s h_l^{-1} k_j^{-1}\}_{s,l=1}^r$ and where the first column of B_{ij} is $\{k_i h_s k_j^{-1}\}_{s=1}^m$. Since $k_i h_s k_j^{-1} = h_s k_i k_j^{-1}$ the elements of block B_{ij} are all in the coset $k_i k_j^{-1} H = H k_i k_j^{-1}$. The first column of this block is

$$(k_i, k_i h_2, \ldots, k_i h_m)$$

which corresponds to $(x_{1i}, x_{2i}, \ldots, x_{mi})$.

Thus X_G is obtained from the group matrix X_K by replacing each element x_{k_i} by the block matrix $X_H(x_{1i}, x_{2i}, \ldots, x_{mi})$.   □

Example 2.5 The group matrix of $C_3 \times C_3$ (relative to the ordering indicated above) is

$$\begin{bmatrix} B_1 & B_3 & B_2 \\ B_2 & B_1 & B_3 \\ B_3 & B_2 & B_1 \end{bmatrix} = C(B_1, B_2, B_3)$$

where

$$B_i = C(x_{1i}, x_{2i}, x_{3i}).$$

The group matrix of $S_3 \times S_3$ (with suitable ordering) is

$$\begin{bmatrix} C(B_1, B_2, B_3) & C(B_4, B_6, B_5) \\ C(B_4, B_5, B_6) & C(B_1, B_3, B_2) \end{bmatrix},$$

where

$$B_i = \begin{bmatrix} C(x_{i1}, x_{i2}, x_{i3}) & C(x_{i4}, x_{i6}, x_{i5}) \\ C(x_{i4}, x_{i5}, x_{i6}) & C(x_{i1}, x_{i3}, x_{i2}) \end{bmatrix}.$$

2.3.3   Groups with a Normal Subgroup

Lemma 2.7 *If $N \lhd G$ then if $T = \{t_1 = e, \ldots, t_r\}$ is a transversal to N in G, X_G may be represented as $\{B_{ij}\}_{i,j=1}^r$ where $B_{ij} = X_N(\underline{u})$ such that the elements of $\underline{u}$ all lie in the coset $Nt_i t_j^{-1}$.*

Proof Consider the ordering (2.3.1) of the elements of G in terms of the cosets of $H = N$. Then the subsets $t_i N t_j^{-1}$ are actually cosets of N and may be written $N t_i t_j^{-1}$. Applying Lemma 2.5, X_G has blocks $\{B_{ij}\}_{i,j=1}^r$ where $B_{ij} = X_H(u_1, u_2, \ldots, u_m)$ with $u_k = t_i h_s t_j^{-1}$, $s = 1, \ldots, r$. But since H is normal $t_i h_s t_j^{-1} = h_s' t_i t_j^{-1}$. The Lemma follows.   □

Example 2.6 Let $G = Q_8 = \{\pm 1, \pm i, \pm j, \pm k\}$. Take $N = \{1, i, -1, -i\}$ and take the ordering to be $\{1, i, -1, -i, j, -k, -1, k\}$. Then with respect to this ordering

$$X_G = \begin{bmatrix} C(1, 2, 3, 4) & C(7, 6, 5, 8) \\ C(5, 6, 7, 8) & C(1, 4, 3, 2) \end{bmatrix}.$$

If nothing more than the normality of N is assumed, all that can be said in general seems to be that if K is the quotient group G/N then X can be constructed from X_K by replacing the element in the $(k_i k_j^{-1})$th position by a matrix of the form $X_N(\underline{u})$ where the entries of $\underline{u}$ lie in $N t_i t_j^{-1}$, with the exception that the $(i, 1)$th block of X_G is $X_N(\underline{u})$ where (replacing x_i by i) $\underline{u} = (i(r-1)+1, \ldots, i(r-1))m)$.

In order to further describe the structure of group matrices "coordinates" for the elements of G with respect to a subgroup H are introduced by choosing an ordering for G of the form (2.3.1) and representing the jth element of the ith coset by $\langle i, j \rangle$.

2.3.4   Split Extensions

Let G be $K \ltimes H$, i.e H is a normal subgroup of G and K is a subgroup of order $G : H$ such that $K \cap H = \{e\}$. There is an action of K on H, $h \to h^k$. G may be described by the following operation on $K \times H$:

$$(k_1, h_1)(k_2, h_2) = (k_1 k_2, h_1^{k_2} h_2).$$

and in general $h^{k_1 k_2} = (h^{k_1})^{k_2}$.

Proposition 2.1 *The group matrix of $G = K \ltimes H$ where $|H| = m$ and $|K| = r$ is given by $X_G = \{B_{ij}\}_{i,j=1}^r$ where $B_{ij} = X_H(\underline{v})$ where $v_s = \langle k_i k_j^{-1}, h_s^{k_j^{-1}} \rangle$. Then X_G may be abbreviated as*

$$X_G = \langle X_K(k_1, \ldots, k_r), X_H(h_1^{k_j^{-1}}, \ldots, h_m^{k_j^{-1}}) \rangle.$$

The following example illustrates.

Example 2.7 Let H be the Klein 4-group $\{e, a, b, c\} \simeq C_2 \times C_2$ and $K = C_3 = \{e, \alpha, \alpha^2\}$.. Let the action of K on H be defined by $e^\alpha = e, a^\alpha = b, b^\alpha = c, c^\alpha = a$. In this case $G \simeq A_4$. It is clear that (for example from Lemma 2.6).

$$X_H = \begin{bmatrix} e & a & b & c \\ a & e & c & b \\ b & c & e & a \\ c & a & a & b \end{bmatrix}.$$

Then $X_G = \{B_{ij}\}_{i,j=1}^3$ where

$$
\begin{aligned}
B_{11} &= X_H(\langle e, e \rangle, \langle e, a \rangle, \langle e, b \rangle, \langle e, c \rangle) \\
B_{12} &= X_H(\langle \alpha^2, e \rangle, \langle \alpha^2, c \rangle, \langle \alpha^2, a \rangle, \langle \alpha^2, b \rangle) \\
B_{13} &= X_H(\langle \alpha, e \rangle, \langle \alpha, b \rangle, \langle \alpha, c \rangle, \langle \alpha, a \rangle) \\
B_{21} &= X_H(\langle \alpha, e \rangle, \langle \alpha, a \rangle, \langle \alpha, b \rangle, \langle \alpha, c \rangle) \\
B_{22} &= X_H(\langle e, e \rangle, \langle e, c \rangle, \langle e, a \rangle, \langle e, b \rangle) \\
B_{23} &= X_H(\langle \alpha, e \rangle, \langle \alpha, b \rangle, \langle \alpha, c \rangle, \langle \alpha, a \rangle) \\
B_{31} &= X_H(\langle \alpha^2, e \rangle, \langle \alpha^2, a \rangle, \langle \alpha^2, b \rangle, \langle \alpha^2, c \rangle) \\
B_{32} &= X_H(\langle \alpha, e \rangle, \langle \alpha, c \rangle, \langle \alpha, a \rangle, \langle \alpha, b \rangle) \\
B_{33} &= X_H(\langle e, e \rangle, \langle e, b \rangle, \langle e, c \rangle, \langle e, a \rangle).
\end{aligned}
$$

This may be further abbreviated by writing

$$X_H(\langle \alpha^i, u_1 \rangle, \langle \alpha^i, u_2 \rangle, \langle \alpha^i, u_3 \rangle, \langle \alpha^i, u_4 \rangle)$$

as $\langle \alpha^i, X_H(\underline{u}) \rangle$ and then

$$X_G = \begin{bmatrix} \langle e, X_H(e, a, b, c) \rangle & \langle \alpha^2, X_H(e, c, a, b) \rangle & \langle \alpha, X_H(e, b, c, a) \rangle \\ \langle \alpha, X_H(e, a, b, c) \rangle & \langle e, X_H(e, c, a, b) \rangle & \langle \alpha^2, X_H(e, b, c, a) \rangle \\ \langle \alpha^2, X_H(e, a, b, c) \rangle & \langle \alpha^2, X_H(e, c, a, b) \rangle & \langle e, X_H(e, b, c, a) \rangle \end{bmatrix}. \quad (2.3.3)$$

It is seen that if the second coordinate is ignored X_G becomes $X_K(e, \alpha, \alpha^2)$ and the expression

$$\langle X_K(e, \alpha, \alpha^2) \otimes X_H(e, a, b, c)^{\alpha^i} \rangle$$

abbreviates (2.3.3) further. For the benefit of the reader X_G is also set out as

$$
\begin{bmatrix}
1 & 2 & 3 & 4 & 9 & 12 & 10 & 11 & 5 & 7 & 8 & 6 \\
2 & 1 & 4 & 3 & 12 & 9 & 11 & 10 & 7 & 5 & 6 & 8 \\
3 & 4 & 1 & 2 & 10 & 11 & 9 & 12 & 8 & 6 & 5 & 7 \\
4 & 3 & 2 & 1 & 11 & 10 & 12 & 9 & 6 & 8 & 7 & 5 \\
5 & 6 & 7 & 8 & 1 & 4 & 2 & 3 & 9 & 11 & 12 & 10 \\
6 & 5 & 8 & 7 & 4 & 1 & 3 & 2 & 11 & 9 & 10 & 12 \\
7 & 8 & 5 & 6 & 2 & 3 & 1 & 4 & 12 & 10 & 9 & 11 \\
8 & 7 & 6 & 5 & 3 & 2 & 4 & 1 & 10 & 12 & 11 & 9 \\
9 & 10 & 11 & 12 & 5 & 8 & 6 & 7 & 1 & 3 & 4 & 2 \\
10 & 9 & 12 & 11 & 8 & 5 & 7 & 6 & 3 & 1 & 2 & 4 \\
11 & 12 & 9 & 10 & 6 & 7 & 5 & 8 & 4 & 2 & 1 & 3 \\
12 & 11 & 10 & 9 & 7 & 6 & 8 & 5 & 2 & 4 & 3 & 1
\end{bmatrix},
\tag{2.3.4}
$$

where the ordering of G is with respect to the left cosets of H as in (2.3.1).

Proof (of Proposition 2.1) From Lemma 2.5 on replacing t_j by k_j for all s X_G becomes $\{B_{ij}\}_{i,j=1}^r$ where $B_{ij} = X_H(v)$ with $v_s = k_i h_s k_j^{-1} = k_i k_j^{-1} h_s^{k_j^{-1}}$ for all s and using the coordinate notation v_s may be written as $\langle k_i k_j^{-1}, h_s^{k_j^{-1}} \rangle$. Then all the blocks in the column of the (block) table corresponding to k_j have second coordinate

$$X_H(h_1^{k_j^{-1}}, h_2^{k_j^{-1}}, \ldots, h_m^{k_j^{-1}})$$

and this justifies the abbreviated indicated in the statement of the Proposition. □

Example 2.8 Let $G = Fr_{21}$. This has already been discussed in Example 1.9. Then $G \simeq K \ltimes H$ where $H = C_7 = \langle a \rangle$ and $K = C_3 = \langle b \rangle$. If the elements of H are listed as $\{e, a, a^2, \ldots, a^6\}$, the action of K on H is given by the permutation $(2, 3, 5)(4, 7, 6)$. Then using notation consistent with (2.3.3) X_G can be represented as

$$
\begin{array}{lll}
\langle e, C_{11} \rangle & \langle b^2, C_{12} \rangle & \langle b, C_{13} \rangle \\
\langle b, C_{21} \rangle & \langle e, C_{22} \rangle & \langle b^2, C_{23} \rangle \\
\langle b^2, C_{31} \rangle & \langle b, C_{32} \rangle & \langle e, C_{33} \rangle
\end{array}
$$

where

$$C_{11} = C_{21} = C_{31} = C(1, 2, 3, 4, 5, 6, 7)$$
$$C_{12} = C_{22} = C_{32} = C(1, 3, 5, 7, 2, 4, 6)$$
$$C_{13} = C_{23} = C_{33} = C(1, 5, 2, 6, 3, 7, 4).$$

The group matrix $X_{K \ltimes H}$ can also be constructed by modifying $X_{K \times H}$ as follows. Represent $X_{K \times H}$ by

$$X_K(\langle k_1, X_H(h_1, h_2, \dots, h_m) \rangle, (\langle k_2, X_H(h_1, h_2, \dots, h_m) \rangle, \qquad (2.3.5)$$
$$\dots, \langle k_r, X_H(h_1, h_2, \dots, h_m) \rangle)).$$

Then $X_{K \ltimes H}$ is obtained by replacing the second coordinate in the jth column of (2.3.5) by $X_H(h_1^{k_j^{-1}}, h_2^{k_j^{-1}}, \dots, h_m^{k_j^{-1}})$, for $j = 1, \dots, r$.

2.3.5 Groups Which Have an Exact Factorization as a Product of Two Subgroups

An **exact factorization** of a group G in terms of two subgroups H and K occurs when both of the following hold

1. $H \cap K = \{e\}$ and
2. each element $g \in G$ may be expressed in the form hk where $h \in H, k \in K$.

Neither H nor K need be normal, and in fact many simple groups are exact products (for example $\mathcal{A}_5$). In particular each $g \in G$ may be written in a unique way as $g = hk$ with $h \in H$ and $k \in K$. If either H or K is normal in G, G is a split extension which is covered above. In the situation where H and K are arbitrary the operation on G can be represented as that given on the set $H \times K$ by

$$(k_1, h_1)(k_2, h_2) = (k_1 k_2^{h_1}, h_1^{k_2} h_2),$$

where two actions appear: $h \mapsto h^k$ and $k \mapsto k^h$ for any $h \in H, k \in K$. Here $\varphi(k) : h \mapsto h^k$ is a homomorphism from K into the group of permutations on the elements of H and similarly $\psi(h) : k \mapsto k^h$ is a homomorphism from H into the group of permutations on K. This is discussed fully in [181].

Example 2.9 The group $G = \mathcal{S}_4$ has been discussed in Example 2.4. G is the exact product KH where $H = D_4$ and $K = \mathcal{C}_3 = \langle (1, 2, 3) \rangle$. Neither H nor K is normal in G. The actions described above can be given as follows. The action of $(1, 2, 3)$ on H with the ordering of H listed above is given by the permutation $(2, 4, 7)(3, 6, 8)$. This describes completely the action of K on H. The action of H on K has kernel

$$N = \{e, (1, 2)(3, 4), (1, 3)(2, 4), (1, 4)(2, 3)\}$$

and an element in the coset $N(13)$ acts as by interchanging $(1, 2, 3)$ and $(1, 3, 2)$. Then a different description of X_G is obtained in terms of the coordinate notation

$$\begin{bmatrix} X_H(\underline{v}_{11}) & X_H(\underline{v}_{12}) & X_H(\underline{v}_{13}) \\ X_H(\underline{v}_{21}) & X_H(\underline{v}_{22}) & X_H(\underline{v}_{23}) \\ X_H(\underline{v}_{31}) & X_H(\underline{v}_{32}) & X_H(\underline{v}_{33}) \end{bmatrix},$$

where

$$\underline{v}_{11} = (\langle 1, 1\rangle, \langle 1, 2\rangle, \langle 1, 3\rangle, \langle 1, 4\rangle,$$
$$\langle 1, 5\rangle, \langle 1, 6\rangle, \langle 1, 7\rangle, \langle 1, 8\rangle),$$
$$\underline{v}_{21} = (\langle 2, 1\rangle, \langle 2, 2\rangle, \langle 2, 3\rangle, \langle 2, 4\rangle,$$
$$\langle 2, 5\rangle, \langle 2, 6\rangle, \langle 2, 7\rangle, \langle 2, 8\rangle),$$
$$\underline{v}_{31} = (\langle 3, 1\rangle, \langle 3, 2\rangle, \langle 3, 3\rangle, \langle 3, 4\rangle,$$
$$\langle 3, 5\rangle, \langle 3, 6\rangle, \langle 3, 7\rangle, \langle 3, 8\rangle),$$
$$\underline{v}_{12} = (\langle 3, 1\rangle, \langle 2, 4\rangle, \langle 3, 6\rangle, \langle 2, 7\rangle,$$
$$\langle 2, 5\rangle, \langle 3, 8\rangle, \langle 2, 2\rangle, \langle 3, 3\rangle),$$
$$\underline{v}_{22} = (\langle 1, 1\rangle, \langle 3, 4\rangle, \langle 1, 6\rangle, \langle 3, 7\rangle,$$
$$\langle 3, 5\rangle, \langle 1, 8\rangle, \langle 3, 2\rangle, \langle 1, 3\rangle),$$
$$\underline{v}_{32} = (\langle 2, 1\rangle, \langle 1, 4\rangle, \langle 2, 6\rangle, \langle 1, 7\rangle,$$
$$\langle 1, 5\rangle, \langle 2, 8\rangle, \langle 1, 2\rangle, \langle 2, 3\rangle),$$
$$\underline{v}_{13} = (\langle 2, 1\rangle, \langle 3, 7\rangle, \langle 2, 8\rangle, \langle 3, 2\rangle,$$
$$\langle 3, 5\rangle, \langle 2, 3\rangle, \langle 3, 4\rangle, \langle 2, 6\rangle),$$
$$\underline{v}_{23} = (\langle 3, 1\rangle, \langle 1, 7\rangle, \langle 3, 8\rangle, \langle 1, 2\rangle,$$
$$\langle 1, 5\rangle, \langle 3, 3\rangle, \langle 1, 4\rangle, \langle 3, 6\rangle),$$
$$\underline{v}_{33} = (\langle 1, 1\rangle, \langle 2, 7\rangle, \langle 1, 8\rangle, \langle 2, 2\rangle,$$
$$\langle 2, 5\rangle, \langle 1, 3\rangle, \langle 2, 4\rangle, \langle 1, 6\rangle).$$

It may be seen that the columns of the block matrix have second coordinate which exhibit the action of K on H, and the first coordinate of each column reflects the action of H on K.

The general situation is as follows. Let G be the exact product KH and consider the elements of G ordered according to the left cosets of H with the elements of a transversal being the elements of K. Then the (s, t)th element of the (i, j)th block is

$$(k_i, h_s)(k_j, h_t)^{-1} = k_i h_s h_t^{-1} k_j^{-1} = (k_i, h_s h_t^{-1})(k_j^{-1}, e) \qquad (2.3.6)$$

$$= (k_i (k_j^{-1})^{h_s h_t^{-1}}, (h_s h_t^{-1})^{k_j^{-1}}). \qquad (2.3.7)$$

Theorem 2.3 *If the group G is the exact product of groups H and K of respective orders m and r, and an ordering of G is produced by taking an ordering of H and taking the elements of K as a transversal, then $X_G = \{B_{ij}\}_{i,j=1}^r$ where*

$$B_{ij} = \{\langle k_i (k_j^{-1})^{h_s h_t^{-1}}, (h_s h_t^{-1})^{k_j^{-1}} \rangle\}_{s,t=1}^m.$$

Thus B_{ij} may be written as

$$X_H(\langle k_i(k_j^{-1})^{h_1}, h_1^{k_j^{-1}} \rangle, \langle k_i(k_j^{-1})^{h_2}, h_2^{k_j^{-1}} \rangle, \ldots, \langle k_i(k_j^{-1})^{h_m}, h_m^{k_j^{-1}} \rangle). \qquad (2.3.8)$$

Proof This follows directly from (2.3.6) □

Example 2.10 S_4 can also be represented as an exact product HK where $H = S_3 = \{e, (1, 2, 3), (1, 3, 2), (1, 2), (1, 3), (2, 3)\}$ and $K = C_4 = \langle(1, 2, 3, 4)\rangle$.

Then the action of $(1, 2, 3, 4)$ on H is given by the permutation $h_i \mapsto h_{\sigma(i)}$ where $\sigma = (2, 4, 6, 3)$. This determines the action of K on H. The action of $(1, 2, 3)$ on K is given by the permutation $k_i \mapsto k_{\tau(i)}$ where $\tau = (2, 4, 3)$. The action of $(1, 2)$ on K is given by $(3, 4)$. This determines the action of H on K. Then $X_G = \{B_{ij}\}_{i,j=1}^4$ where $B_{ij} = X_H(\underline{v}_{ij})$ and

$$\underline{v}_{11} = (\langle 1, 1 \rangle, \langle 1, 2 \rangle, \langle 1, 3 \rangle, \langle 1, 4 \rangle, \langle 1, 5 \rangle, \langle 1, 6 \rangle),$$

$$\underline{v}_{21} = (\langle 2, 1 \rangle, \langle 2, 2 \rangle, \langle 2, 3 \rangle, \langle 2, 4 \rangle, \langle 2, 5 \rangle, \langle 2, 6 \rangle),$$

$$\underline{v}_{31} = (\langle 3, 1 \rangle, \langle 3, 2 \rangle, \langle 3, 3 \rangle, \langle 3, 4 \rangle, \langle 3, 5 \rangle, \langle 3, 6 \rangle),$$

$$\underline{v}_{41} = (\langle 4, 1 \rangle, \langle 4, 2 \rangle, \langle 4, 3 \rangle, \langle 4, 4 \rangle, \langle 4, 5 \rangle, \langle 4, 6 \rangle),$$

$$\underline{v}_{12} = (\langle 4, 1 \rangle, \langle 2, 3 \rangle, \langle 3, 6 \rangle, \langle 3, 2 \rangle, \langle 2, 5 \rangle, \langle 4, 4 \rangle),$$

$$\underline{v}_{22} = (\langle 1, 1 \rangle, \langle 3, 3 \rangle, \langle 4, 6 \rangle, \langle 4, 2 \rangle, \langle 3, 5 \rangle, \langle 1, 4 \rangle),$$

$$\underline{v}_{32} = (\langle 2, 1 \rangle, \langle 4, 3 \rangle, \langle 1, 6 \rangle, \langle 1, 2 \rangle, \langle 4, 5 \rangle, \langle 2, 4 \rangle),$$

$$\underline{v}_{42} = (\langle 3, 1 \rangle, \langle 1, 3 \rangle, \langle 2, 6 \rangle, \langle 2, 2 \rangle, \langle 1, 5 \rangle, \langle 3, 4 \rangle),$$

$$\underline{v}_{13} = (\langle 3, 1 \rangle, \langle 4, 6 \rangle, \langle 2, 4 \rangle, \langle 4, 3 \rangle, \langle 3, 5 \rangle, \langle 2, 2 \rangle),$$

$$\underline{v}_{23} = (\langle 4, 1 \rangle, \langle 1, 6 \rangle, \langle 3, 4 \rangle, \langle 1, 3 \rangle, \langle 4, 5 \rangle, \langle 3, 2 \rangle),$$

$$\underline{v}_{33} = (\langle 1, 1 \rangle, \langle 2, 6 \rangle, \langle 4, 4 \rangle, \langle 2, 3 \rangle, \langle 1, 5 \rangle, \langle 4, 2 \rangle),$$

$$\underline{v}_{43} = (\langle 2, 1 \rangle, \langle 3, 6 \rangle, \langle 1, 4 \rangle, \langle 3, 3 \rangle, \langle 2, 5 \rangle, \langle 1, 2 \rangle),$$

$$\underline{v}_{14} = (\langle 2, 1 \rangle, \langle 3, 4 \rangle, \langle 4, 2 \rangle, \langle 2, 5 \rangle, \langle 4, 5 \rangle, \langle 3, 3 \rangle),$$

$$\underline{v}_{24} = (\langle 3, 1 \rangle, \langle 4, 4 \rangle, \langle 1, 2 \rangle, \langle 3, 5 \rangle, \langle 1, 5 \rangle, \langle 4, 3 \rangle),$$

$$\underline{v}_{34} = (\langle 4, 1 \rangle, \langle 1, 4 \rangle, \langle 2, 2 \rangle, \langle 2, 5 \rangle, \langle 4, 5 \rangle, \langle 1, 3 \rangle),$$

$$\underline{v}_{44} = (\langle 1, 1 \rangle, \langle 2, 4 \rangle, \langle 3, 2 \rangle, \langle 3, 5 \rangle, \langle 1, 5 \rangle, \langle 2, 3 \rangle).$$

In this case, a modification of the operation on $K \times H$ produces the smallest simple non-associative Bol loop (see [181]).

2.3.5.1 The Construction of the Group Matrix of a Group from a Chain of Subgroups

Any chain of subgroups $H_1 \subset H_2 \subset H_3 \subset \ldots \subset H_r = G$ may be used to construct X_G. Starting with X_{H_1}, X_{H_2} may be constructed as above by listing the elements of H_2 on the cosets of H_1, then X_{H_3} can constructed by listing the elements of H_3 on the cosets of H_2 etc. The following illustrates how the group matrix of A_5 can be constructed in this manner via the chain $C_2 \subset V_4 \subset A_4 \subset A_5 = G$. Now G has an exact factorization $K \times H$ where $K = C_5$ and $H = A_4$. The group matrix of A_4 on the cosets of V_4 has been given in Example 2.7 and consider a listing of the elements in the order given there, specifically

$$\{e, a, b, c, \alpha, \alpha a, \alpha b, \alpha c, \alpha^2, \alpha^2 a, \alpha^2 b, \alpha^2 c\},$$

abbreviated as $\{1, \ldots, 12\}$. Let the elements of K be listed as powers of $\sigma = (1, 2, 3, 4, 5)$, abbreviated as $\{1, \ldots, 5\}$. The action of σ on H is given by the permutation

$$\sigma = (2, 7, 10, 11, 8)(4, 6, 9, 12, 5)$$

which determines the action of K on H. The action of $(1, 2)(3, 4)$ on K is given by the permutation $(2, 3)(4, 5)$ and the action of $(1, 2, 3)$ is given by the permutation $(3, 5, 4)$ which determine the action of K on H. Then $X_G = \{B_{ij}\}_{i,j=1}^5$ where each B_{ij} can be written as $X_H(\alpha_1, \ldots, \alpha_{12})$. Here each α_s can be written in terms of coordinates as $\langle k_s, \sigma^{1-j}(h_s) \rangle$. The vector $v_{ij} = (k_1, \ldots, k_{12})$ is described in the formula (2.3.8), and the first two columns are:

$$v_{i1} = (i, i, i \ldots i), i = 1, \ldots, 5$$
$$v_{12} = (5, 4, 2, 3, 3, 5, 2, 4, 4, 3, 2, 5) \quad v_{22} = (1, 5, 3, 4, 4, 1, 3, 5, 5, 4, 3, 1)$$
$$v_{32} = (2, 1, 4, 5, 5, 2, 4, 1, 1, 5, 4, 2) \quad v_{42} = (3, 2, 5, 1, 1, 3, 5, 2, 2, 1, 5, 3)$$
$$v_{52} = (4, 3, 1, 2, 2, 4, 1, 3, 3, 2, 1, 4)$$

Each of the B_{ij} may further be broken into blocks of V_4 group matrices, which again can be broken into 2×2 circulants.

2.3.6 The Group Matrix Using Conjugacy Class Ordering

Let $C_1, C_2, \ldots, C_r$ be the conjugacy classes of G. If C_i is any given class either $g^{-1} \in C_i$ for all $g \in C_i$ or there is a class $C_i^* = \{g^{-1} : g \in C_i\}$. Thus typically G will have conjugacy classes $C_1, C_2, \ldots, C_s$ such that for $1 \le i \le s$, $C_i = C_i^*$ and for $i > s$ pairs of distinct classes (C_i, C_i^*). Let each class be ordered arbitrarily, and let this induce an ordering of G by listing the classes $C_1, C_2, \ldots, C_s, C_{s+1}, C_{s+1}^*, C_{s+2}, C_{s+2}^*, \ldots$. If X_G^c is the group matrix with respect to this ordering, an element x_h is in the diagonal block $C_i C_i^*$ if and only if h is a commutator. This follows because every element of the product $C_i C_i^*$ is an element of the form $g^{-1}(k^{-1}gk) = [g, k]$. Moreover every commutator appears in a diagonal block.

Example 2.11 Let $G = A_4$, with the $C_1 = \{e\}$, $C_2 = \{(1, 2)(3, 4)\}^G$, $C_3 = \{(1, 2, 3)\}^G$ and $C_4 = \{(1, 3, 2)\}^G$. Then with respect to the ordering via classes X_G (identified with $Wa(G)$) becomes

$$
\begin{bmatrix}
1 & 2 & 3 & 4 & 9 & 10 & 11 & 12 & 5 & 6 & 7 & 8 \\
2 & 1 & 4 & 3 & 10 & 9 & 12 & 11 & 7 & 8 & 5 & 6 \\
3 & 4 & 1 & 2 & 11 & 12 & 9 & 10 & 8 & 7 & 6 & 5 \\
4 & 3 & 2 & 1 & 12 & 11 & 10 & 9 & 6 & 5 & 8 & 7 \\
5 & 6 & 7 & 8 & 1 & 4 & 2 & 3 & 9 & 11 & 12 & 10 \\
6 & 5 & 8 & 7 & 4 & 1 & 3 & 2 & 11 & 9 & 10 & 12 \\
7 & 8 & 5 & 6 & 2 & 3 & 1 & 4 & 10 & 12 & 11 & 9 \\
8 & 7 & 6 & 5 & 3 & 2 & 4 & 1 & 11 & 9 & 10 & 12 \\
9 & 11 & 12 & 10 & 5 & 8 & 6 & 7 & 1 & 2 & 3 & 4 \\
10 & 12 & 11 & 9 & 7 & 6 & 8 & 5 & 2 & 1 & 4 & 3 \\
11 & 9 & 10 & 12 & 8 & 5 & 7 & 6 & 3 & 4 & 1 & 2 \\
12 & 10 & 9 & 11 & 6 & 7 & 5 & 8 & 4 & 3 & 2 & 1
\end{bmatrix}. \tag{2.3.9}
$$

The set of commutators can be read off from the diagonal blocks in rows 2–4 and columns 2–4, in rows 5–8 and columns 5–8, and in rows 9–12 and columns 9–12. In this case the set of non-trivial commutators is the class C_2.

If $|G|$ is odd every non-trivial class C_i is distinct from C_i^* and each C_i may be ordered as $\{g_{i_1}, \ldots, g_{i_t}\}$ with a corresponding ordering of C_i^* as $\{g_{i_1}^{-1}, \ldots, g_{i_t}^{-1}\}$.

Example 2.12 Consider $G = Fr_{21}$ with the ordering as given in Example 1.9. Here the conjugacy classes are $C_1 = \{e\}, C_2 = \{2, 3, 5\}, C_3 = \{4, 6, 7\}, C_4 = \{8, \ldots 14\}$, and $C_4 = \{15, \ldots 21\}$. Then $Wa(G)$ ordered by classes has diagonal blocks of the form

$$C(1,2,3,4,5,6,7), \quad \begin{bmatrix} 1\,4\,2\,3\,5\,6\,7 \\ 7\,1\,4\,6\,3\,2\,5 \\ 5\,7\,1\,2\,6\,4\,3 \\ 6\,3\,5\,1\,4\,7\,2 \\ 2\,6\,3\,7\,1\,5\,4 \\ 3\,5\,7\,4\,2\,1\,6 \\ 4\,2\,6\,5\,7\,3\,1 \end{bmatrix}, \quad \begin{bmatrix} 1\,5\,6\,7\,3\,4\,2 \\ 2\,1\,5\,4\,7\,6\,3 \\ 3\,2\,1\,6\,4\,5\,7 \\ 4\,7\,3\,1\,5\,2\,6 \\ 6\,4\,7\,2\,1\,3\,5 \\ 7\,3\,2\,5\,6\,1\,4 \\ 5\,6\,4\,3\,2\,7\,1 \end{bmatrix}.$$

This confirms that G' is the subgroup of order 7. The matrix X_G^{symm} may be obtained from the group matrix given in Example 8 by replacing each block $B_{i,j}$ where $i > j$ with the transpose of the block $B_{j,i}$, together with suitable identification on the diagonal to obtain

$$\begin{bmatrix} C(1,2,3,4,4,3,2) & C(8,9,10,11,12,13,14)^t & C(8,10,12,14,9,11,13) \\ C(8,9,10,11,12,13,14) & C(1,3,4,2,2,4,3) & C(8,12,9,13,10,14,11)^t \\ C(8,10,12,14,9,11,13)^t & C(8,12,9,13,10,14,11) & C(1,4,2,3,3,2,4) \end{bmatrix}.$$

2.3.7 A Digression: The Octonion Loop

The Octonion loop O_{16} is the set $\{\pm 1, \pm e_1, \pm e_2, \ldots \pm e_7\}$ (with $e_i^2 = -1$), of basis octonions and their additive inverses in $\mathbb{K}$ (for the specifics see for example [8]). It is non-associative but satisfies the Moufang property $x(y.xz) = (x.yx)z$ which implies that the loop is diassociative (i.e. each subloop generated by two elements is a group) and in particular it has two-sided inverses. It contains the cyclic subgroup H of order 4 with elements listed as $\{1, e_1, -1, -e_1\}$ which is contained in the subgroup K isomorphic to Q_8 with elements listed as $\{1, e_1, -1, -e_1, e_2, -e_3, -e_2, e_3\}$. Let an ordering of the loop be completed by listing the coset $e_4 K$ in the order induced by that on K. The quasigroup with the operation of right division is not a Ward quasigroup (see [122]). With respect to the ordering indicated the multiplication table of this quasigroup is

$$\begin{bmatrix} C(1,2,3,4) & C(7,6,5,8) & C(11,10,9,12) & C(15,14,13,16) \\ C(5,6,7,8) & C(1,4,3,2) & R(15,14,13,16) & R(9,12,11,10) \\ C(9,10,11,12) & R(13,16,15,14) & C(1,4,3,2) & R(7,6,5,8) \\ C(13,14,15,16) & R(9,12,11,10) & R(7,6,5,8) & C(1,4,3,2) \end{bmatrix},$$

$$\tag{2.3.10}$$

where $R(a_1, a_2, a_3, a_4)$ denotes the **reverse circulant**

$$\begin{bmatrix} a_1 & a_2 & a_3 & a_4 \\ a_2 & a_3 & a_4 & a_1 \\ a_3 & a_4 & a_1 & a_2 \\ a_4 & a_1 & a_2 & a_3 \end{bmatrix}.$$

For each $i = 2, 3, 4$, the elements in the $(1, 1)$, $(1, i)$, $(i, 1)$, (i, i) blocks form a group, which is isomorphic to the quaternion group Q_8. This follows from diassociativity and the ordering forces the blocks in these positions to be circulants. It is interesting that all the blocks which are not forced by diassociativity to be circulants are reverse circulants.

2.4 Methods to Factorize the Group Determinant

2.4.1 Explicit Diagonalization of a Group Matrix

The following algorithm appears in modified form in [82, p. 48]. Suppose that G is a finite group of order n and that the set of irreducible representations $\{\rho_1, \ldots, \rho_r\}$ is known, and assume that for each ρ_i a set of representing matrices $\{\rho_i(g)\}_{g \in G}$ has been chosen. Let $\rho_i(g)_{jk}$ denote the element in the (j, k)th position of $\rho_i(g)$. Let d_i be the degree of ρ_i. Let the vector ψ_i of length d_i^2 be defined as

$$\psi_i(g) = \sqrt{\frac{d_i}{n}} (\rho_i(g)_{11}, \rho_i(g)_{21}, \ldots, \rho_i(g)_{d_i 1}, \rho_i(g)_{12}, \ldots, \rho_i(g)_{d_i d_i}).$$

It is seen that this is a listing of the elements in $\rho_i(g)$ by column 1, column 2, ..., column d_i. Then construct the column vector of length n

$$\phi(g) = (\psi_1(g), \psi_2(g), \ldots, \psi_r(g))^T.$$

Given an ordering of G, $\{g_1, g_2, \ldots, g_n\}$, construct the $n \times n$ matrix P whose columns are successively $\phi(g_1), \phi(g_2), \ldots, \phi(g_n)$. Let $X_G = X_G(u_1, u_2, \ldots, u_n)$.

Theorem 2.4 *The matrix $P X_G P^{-1}$ is a block diagonal matrix with successive blocks*

$$X_G^{\rho_1}, X_G^{\rho_2}, \ldots, X_G^{\rho_2}, \ldots, X_G^{\rho_r}, \ldots, X_G^{\rho_r},$$

where the matrix $X_G^{\rho_i}$ appears d_i times.

The proof follows directly from Fourier inversion and will appear in Chap. 7, see Theorem 7.1. If G is a cyclic group, then P is a multiple of the Fourier matrix F_n.

Define P as constructed above to be a **generalized Fourier matrix** for G. The result in Theorem 2.4 seems hard to find in the standard texts on group representation theory. Note that the calculation of PX_GP^{-1} does not need commutativity of the u_i and in particular the conclusion of the theorem holds if the u_i are square matrices of the same size.

Example 2.13 Let G be S_3 with the ordering

$$\{e, (1,2,3), (1,3,2), (1,2), (1,3), (2,3)\}.$$

The 1-dimensional matrices corresponding to the linear characters are forced, and a set of representing matrices for the 2-dimensional representation ρ are determined by

$$\rho_2(1,2,3) = \begin{bmatrix} \omega & 0 \\ 0 & \omega^2 \end{bmatrix}, \quad \rho_2(1,2) = \begin{bmatrix} 0 & 1 \\ 1 & 0 \end{bmatrix},$$

where $\omega = e^{2\pi i/3}$. The matrix P is calculated to be

$$\frac{1}{\sqrt{6}} \begin{bmatrix}
1 & 1 & 1 & 1 & 1 & 1 \\
1 & 1 & 1 & -1 & -1 & -1 \\
\sqrt{2} & \sqrt{2}\omega & \sqrt{2}\omega^2 & 0 & 0 & 0 \\
0 & 0 & 0 & \sqrt{2} & \sqrt{2}\omega & \sqrt{2}\omega^2 \\
0 & 0 & 0 & \sqrt{2} & \sqrt{2}\omega^2 & \sqrt{2}\omega \\
\sqrt{2} & \sqrt{2}\omega^2 & \sqrt{2}\omega & 0 & 0 & 0
\end{bmatrix}.$$

In the case of C_n, the above process is almost identical to the Finite Fourier Transform, but is not used in practice since the Cooley-Tukey algorithm is much faster (see Chap. 7). As might be expected, for an arbitrary finite group, the direct calculation of PX_GP^{-1} is also not very efficient, and it is useful to investigate other ways of diagonalization or partial diagonalization.

2.4.2 The Partial Diagonalization with Respect to a Given Subgroup

Let H be an arbitrary subgroup of the group G of index r and let $|H| = m$, $|G| = n$. Suppose that X_G is constructed as in Lemma 2.5 using the set $\{t_k\}_{k=1}^r$ of representatives for the left cosets of H and the listing $\{h_1 = e, h_2, \ldots, h_m\}$ of the elements of H. Then X_G is a block matrix where each block is of size $r \times r$ and the (i, j)th block is of the form

$$X_H(t_i h_1 t_j^{-1}, t_i h_2 t_j^{-1}, \ldots, t_i h_m t_j^{-1})$$

Let $Irr(H) = \{\rho_i\}_{i=1}^s$. Suppose that the generalized Fourier matrix P for H has been constructed as above, so that

$$PX_H P^{-1} = diag(X_H^{\rho_1}, X_H^{\rho_2}, \ldots, X_H^{\rho_2}, \ldots, X_H^{\rho_s}, \ldots, X_H^{\rho_s}),$$

where the block $X_H^{\rho_i}$ is repeated $deg(\rho_i)$ times. Consider the $n \times n$ block matrix $P_r = diag(P, P, \ldots, P)$ with r repeats of P, then the product $P_r X_G P_r^{-1}$ is also a block matrix whose (i, j)th block B_{ij} is of the form

$$X_{pd} = diag(X_H^{\rho_1}(\underline{u}), X_H^{\rho_2}(\underline{u}), \ldots, X_H^{\rho_2}(\underline{u}), \ldots, X_H^{\rho_s}(\underline{u}), \ldots, X_H^{\rho_s}(\underline{u})) \qquad (2.4.1)$$

where

$$\underline{u} = (t_i h_1 t_j^{-1}, t_i h_2 t_j^{-1}, \ldots, t_i h_m t_j^{-1}).$$

Example 2.14 Let $G = S_4$ and $H = S_3$. The ordering of G and the notation will be consistent with that in Example 2.10. Then X_{pd} has block B_{11} equal to

$$diag(x_{\langle 1,1 \rangle} + x_{\langle 1,2 \rangle} + x_{\langle 1,3 \rangle} + x_{\langle 1,4 \rangle} + x_{\langle 1,5 \rangle} + x_{\langle 1,6 \rangle},$$

$$x_{\langle 1,1 \rangle} + x_{\langle 1,2 \rangle} + x_{\langle 1,3 \rangle} + x_{\langle 1,4 \rangle} + x_{\langle 1,5 \rangle} + x_{\langle 1,6 \rangle}, Y, Y),$$

where

$$Y = \begin{bmatrix} x_{\langle 1,1 \rangle} + \omega x_{\langle 1,2 \rangle} + \omega^2 x_{\langle 1,3 \rangle} & x_{\langle 1,4 \rangle} + \omega x_{\langle 1,5 \rangle} + \omega^2 x_{\langle 1,6 \rangle} \\ x_{\langle 1,4 \rangle} + \omega^2 x_{\langle 1,5 \rangle} + \omega x_{\langle 1,6 \rangle} & x_{\langle 1,1 \rangle} + \omega^2 x_{\langle 1,2 \rangle} + \omega x_{\langle 1,3 \rangle} \end{bmatrix},$$

$$\omega = e^{2\pi i/3},$$

and the other blocks are produced similarly from the vectors v_{ij} in Example 2.10.

If μ is a permutation on $\{1, \ldots, n\}$ let $T(\mu)$ be the corresponding permutation matrix (see Definition 1.10).

Definition 2.2   Given a matrix X, the matrix X^μ is defined to be $T(\mu)^{-1} X T(\mu)$.

Informally, X^μ is obtained by permuting the rows and columns of X by μ. It is clear that X^μ is similar to X.

The matrix X_{pd} in the above example can be further diagonalized by constructing X_{pd}^μ, where the permutation $\mu(i)$ is given as follows for $i \in \{1, 2, \ldots, 24\}$:

$$\{1, 5, 9, 13, 17, 21, 2, 6, 10, 14, 18, 22, 3, 7, 11, 15, 19, 23, 4, 8, 12, 16, 20, 24\}.$$

This produces the block diagonal matrix with diagonal blocks $B_{11}, B_{22}, B_{33}, B_{44}$ where

$$B_{11} = \begin{bmatrix} y(v_{11}) & y(v_{12}) & y(v_{13}) & y(v_{14}) \\ y(v_{21}) & y(v_{22}) & y(v_{23}) & y(v_{24}) \\ y(v_{31}) & y(v_{32}) & y(v_{33}) & y(v_{34}) \\ y(v_{41}) & y(v_{42}) & y(v_{43}) & y(v_{44}) \end{bmatrix}$$

where if $v = (x_{v_1}, \ldots, x_{v_6})$, $y(v) = x_{v_1} + \ldots + x_{v_6}$,

$$B_{22} = \begin{bmatrix} z(v_{11}) & z(v_{12}) & z(v_{13}) & z(v_{14}) \\ z(v_{21}) & z(v_{22}) & z(v_{23}) & z(v_{24}) \\ z(v_{31}) & z(v_{32}) & z(v_{33}) & z(v_{34}) \\ z(v_{41}) & z(v_{42}) & z(v_{43}) & z(v_{44}) \end{bmatrix}$$

where if $v = (x_{v_1}, \ldots, x_{v_6})$, $z(v) = x_{v_1} + x_{v_2} + x_{v_3} - x_{v_4} - x_{v_5} - x_{v_6}$, and

$$B_{33} = B_{44} = \begin{bmatrix} M(v_{11}) & M(v_{12}) & M(v_{13}) & M(v_{14}) \\ M(v_{21}) & M(v_{22}) & M(v_{23}) & M(v_{24}) \\ M(v_{31}) & M(v_{32}) & M(v_{33}) & M(v_{34}) \\ M(v_{41}) & M(v_{42}) & M(v_{43}) & M(v_{44}) \end{bmatrix}$$

where if $\underline{v} = (x_{v_1}, \ldots, x_{v_6})$,

$$\underline{M(v)} = \begin{bmatrix} x_{v_1} + \omega x_{v_2} + \omega^2 x_{v_3} & x_{v_4} + \omega x_{v_5} + \omega^2 x_{v_6} \\ x_{v_4} + \omega^2 x_{v_5} + \omega x_{v_6} & x_{v_1} + \omega^2 x_{v_2} + \omega x_{v_6} \end{bmatrix}, \quad \omega = e^{2\pi i/3}.$$

In fact, the blocks B_{11}, B_{22}, B_{33} are group matrices for the induced representations (see the definitions below) 1_H^G, sgn_H^G and ρ_H^G where ρ is the 2-dimensional representation of S_3. That the blocks in this decomposition correspond to induced representations will be shown generally in the following proposition. It is easily verified in the example by taking traces.

2.4.3 Induction and Restriction of Representations

Given a finite group G with a subgroup H, and a representation $\rho : G \to GL(n, K)$ for some field K, there is naturally a homomorphism $\rho : H \to GL(n, K)$, called the **restriction** of ρ to H. On the other hand, if a representation of $\sigma : H \to GL(m, F)$ is given a representation of G may be constructed as follows.

Suppose $[G : H] = r$. Then let $\mathcal{T} = \{t_1, t_2, \ldots, t_r\}$ be a left transversal to H in G.

The induced representation constructs for each g in G an $mr \times mr$ matrix $\rho(g)$ which is an $r \times r$ block matrix with blocks of size $m \times m$. Let $q = t_i^{-1} g t_j$. For $i, j = 1, \ldots, r$, the ijth block of $\rho(g)$ is defined as $\sigma(q)$ if $q \in H$ and 0 otherwise.

The restriction of ρ to H will be denoted by $\mathrm{Res}_H^G \rho$ and the induced representation constructed above will be denoted by $\mathrm{Ind}_H^G \sigma$.

There are corresponding formulae for induction and restriction of characters (see [162]). If ϕ is any class function on H then

$$\mathrm{Ind}_H^G \phi = \frac{1}{|H|} \sum_{k \in G} \phi^\circ (k^{-1} g k),$$

where ϕ° is defined by $\phi^\circ(k) = \phi(k)$ if $k \in H$ and 0 otherwise. If ϕ is a character then $\mathrm{Ind}_H^G \phi$ is the character of the corresponding representation. Again, if ϕ is a class function on G then $\mathrm{Res}_H^G \phi$ is the usual restriction of ϕ as a function.

Proposition 2.2 *Let G be a group and H be a subgroup and suppose that X_{pd} has been constructed as in (2.4.1). Then the diagonal block $X_H^{\rho_i}(\underline{u})$ is a group matrix of the induced representation $\mathrm{Ind}_H^G \rho_i$, $i = 1, \ldots, s$.*

Proof It is sufficient to show that the character χ corresponding to $X_H^{\rho_i}(\underline{u})$ is equal to the character ψ of $\mathrm{Ind}_H^G \rho_i$. Let η_i be the character of ρ_i. The value of χ at an element $g \in G$ is the sum of the coefficients of the terms involving x_g which appear on the diagonal of $X_H^{\rho_i}(\underline{u})$, where $\underline{u} = (t_k h_1 t_k^{-1}, t_k h_2 t_k^{-1}, \ldots, t_k h_m t_k^{-1})$ as k runs from 1 to r. Now x_g appears in $\underline{u}$ if and only if $t_k h_\alpha t_k^{-1} = g$ for some α, or equivalently $t_k^{-1} g t_k = h_\alpha$. The corresponding coefficient of x_g is $\eta_i(h_\alpha)$.

Consider the formula for $\mathrm{Ind}_H^G \eta_i(g)$ which appears above

$$\mathrm{Ind}_H^G \eta_i(g) = \frac{1}{|H|} \sum_{k \in G} \eta_i^\circ (kgk^{-1}).$$

This may also be given as

$$\mathrm{Ind}_H^G \eta_i(g) = \frac{1}{|H|} \sum_{k \in G} \eta_i^\circ (k^{-1} g k). \tag{2.4.2}$$

Thus if $g = t_k h_\alpha t_k^{-1}$ then the term $\eta_i(h_\alpha)$ appears in the sum on the right-hand side of (2.4.2). Moreover, it follows that since $(t_k h_\beta)^{-1} g (t_k h_\beta) = h_\beta^{-1} h_\alpha h_\beta$ lies in H, $\eta_i(h_\beta^{-1} h_\alpha h_\beta) = \eta_i(h_\alpha)$ also appears in this sum, i.e. the left coset $t_k H$ contributes $|H| \eta_i(h_\alpha)$ to this sum. Now since the entries in $\underline{u}$ are clearly distinct, it has been shown that the value of sum of the coefficients of x_g in the diagonal of $X_H^{\rho_i}(\underline{u})$ is the sum in (2.4.2). Since the characters are equal, the representations must be equivalent. $\square$

Proposition 2.2 may be thought of as an explicit decomposition of the regular representation of a group into representations induced from a subgroup. It may be applied to give an easy proof of the following result which appeared in [4]. Other similar theorems follow.

Theorem 2.5 *Let G be a finite group with an abelian subgroup H of index r. Then all the irreducible representations of G are of dimension at most r.*

Proof This follows because every irreducible representation of G must occur as a constituent of a diagonal block of some $X_H^{\rho_i}(\underline{u})$ and since for H abelian $\deg(\rho_i) = 1$ for all i, all such blocks are of size $r \times r$.   □

2.4.4   Methods from General Determinant Theory

2.4.4.1   Williamson's Theorem

According to Cartier [49] the following theorem is called Williamson's theorem.

Theorem 2.6 *Given the block matrix A of the form*

$$\begin{bmatrix} A_{11} & A_{12} & \dots & A_{1n} \\ A_{21} & A_{22} & \dots & A_{2n} \\ & \dots & & \\ A_{n1} & A_{n2} & & A_{nn} \end{bmatrix},$$

where the matrices A_{ij} are pairwise commuting of size $r \times r$ then

$$\det(A) = \det\left(\sum_{\sigma \in S_n} (-1)^{sgn(\sigma)} A_{1\sigma(1)} A_{2\sigma(2)} \dots A_{n\sigma(n)} \right). \tag{2.4.3}$$

The proof indicated in [49] is by (a) assuming the A_{ij} are diagonalizable and (b) using a continuity argument in the case of non-diagonalizable matrices. The corollary below follows easily.

Corollary 2.2 *Let M be a matrix of size $nm \times nm$ consisting of circulant blocks A_{ij} of size $m \times m$, $i, j = 1 \dots n$. Thus*

$$M = \begin{bmatrix} A_{11} & A_{12} & \dots & A_{1n} \\ A_{21} & A_{22} & \dots & A_{2n} \\ & \dots & & \\ A_{n1} & A_{n2} & \dots & A_{nn} \end{bmatrix}.$$

Then the determinant of M is the determinant of the m × m circulant matrix

$$\sum_{\sigma \in S_n} (-1)^{sgn(\sigma)} A_{1\sigma(1)} A_{2\sigma(2)} \dots A_{n\sigma(n)}.$$

Corollary 2.3 *If G is an arbitrary group of order mr with a cyclic group of order m then $\Theta_G = \det(B)$ where B is an m × m circulant whose entries are polynomials of degree r.*

Example 2.15 In the case of the alternating group A_4 whose Ward quasigroup is given in (2.3.9) the group matrix may be partitioned into a 3 × 3 block matrix

$$\begin{bmatrix} A_{11} & A_{12} & A_{13} \\ A_{21} & A_{22} & A_{23} \\ A_{31} & A_{32} & A_{33} \end{bmatrix},$$

where the A_{ij} are group matrices of V_4 and therefore commute pairwise. The matrix inside the determinant in (2.4.3) may be calculated to be of the form

$$\widetilde{X} = X_{V_4}(\phi_1, \phi_2, \phi_2, \phi_2)$$

where ϕ_1 and ϕ_2 are polynomials of degree 3. Since $\widetilde{X}$ is similar to the diagonal matrix with entries $\phi_1 + \phi_2 + \phi_2 + \phi_2, \phi_1 + \phi_2 - \phi_2 - \phi_2, \phi_1 - \phi_2 + \phi_2 - \phi_2,$ $\phi_1 - \phi_2 - \phi_2 + \phi_2$, it follows that Θ_{A_4} factors as

$$(\phi_1 + 3\phi_2)(\phi_1 - \phi_2)^3.$$

The factor $\phi_1 + 3\phi_2$ splits into linear factors (corresponding to the three linear characters of A_4) and the factor $(\phi_1 - \phi_2)^3$ is the repeated factor corresponding to the irreducible 3 × 3 representation.

It may be an interesting question to find conditions on a group which ensure that the matrix in (2.4.3) has repeating entries as in the example above.

2.4.5 The Schur Complement

Consider the $n \times n$ matrix

$$M = \begin{bmatrix} A & B \\ C & D \end{bmatrix} \tag{2.4.4}$$

where A is a $k \times k$ submatrix. Suppose that A is invertible. Then Gaussian elimination may be carried out by

$$
\begin{bmatrix} I_k & 0 \\ -CA^{-1} & I_{n-k} \end{bmatrix} \begin{bmatrix} A & B \\ C & D \end{bmatrix} = \begin{bmatrix} A & B \\ 0 & D' \end{bmatrix}, \tag{2.4.5}
$$

where in effect linear combinations of the first k rows are added to the last $n - k$ rows. Here $D' = D - CA^{-1}B$ is called the **Schur complement** (see [33]). There the notation A/D is used for the Schur complement.

Lemma 2.8 *If M is defined as in (2.4.4) with A invertible then*

$$
\det(M) = \det(A) \det(A/D) \tag{2.4.6}
$$

Proof This follows directly from (2.4.5) by taking determinants.   □

A corollary is that if A, B, C, D are of size $n/2 \times n/2$ and commute pairwise then

$$
\det M = \det(AD - BC). \tag{2.4.7}
$$

The restriction on A can be avoided by a continuity argument. There is an extensive literature on Schur complements (see for example [229]). Ouellette has a quote that Frobenius in [109] used a weaker result of the same nature. In fact Frobenius seems to assume that his result is obvious and it is possible that he already knew the result of Lemma 2.8.

Remark 2.2 The work of Gelfand et al. [121] on noncommutative determinants includes many more general identities of the type (2.4.6).

It is not clear whether the many results given in [33] or [121] have significant applications to the calculation of group determinants. It seems hard to relate the general results on arbitrary determinants to those with the symmetry of a group determinant.

Definition 2.3 The **block Fourier matrix** F_{mn} is defined by

$$
F_{mn} = \frac{1}{\sqrt{n}} \begin{bmatrix} I & I & I & \cdots & I \\ I & \sigma I & \sigma^2 I & \cdots & \sigma^{n-1}I \\ I & \sigma^2 I & \sigma^4 I & \cdots \sigma^{2(n-1)}I \\ & & \cdots & \\ I & \sigma^{n-1}I & \sigma^{2n-2}I & \cdots \sigma^{(n-1)^2}I \end{bmatrix},
$$

where I is the $m \times m$ identity matrix, and $\sigma = e^{2\pi i/n}$.

Lemma 2.9 *The product* $F_{mn}C(A_1, A_2, \ldots, A_n)F_{mn}^{-1}$ *where the* A_i *are* $m \times m$ *matrices, is the block diagonal matrix with entries* $\{B_i\}_{i=1}^m$ *where*

$$B_i = A_1 + \sigma^{i-1}A_2 + \ldots + \sigma^{(i-1)(n-1)}A_n).$$

This is proved by direct multiplication.

2.4.5.1   The Factorization of Group Matrices of Direct Products

Theorem 2.7 *Let* $G = K \times H$ *where* K *and* H *are arbitrary and* $|K| = r$ *and* $|H| = m$. *Further let* $Irr(K)$ *be* $\{\rho_i\}_{i=1}^m$ *with* $\deg(\rho_i) = d_i$ *and* $Irr(H)$ *be* $\{\sigma_i\}_{i=1}^r$ *with* $\deg(\sigma_i) = e_i$. *Then the irreducible group matrices for* G *are all of the form* $X_K^{\rho_i}(X_H^{\sigma_j}(x_{11}, x_{21}, \ldots, x_{m1}), X_H^{\sigma_j}(x_{12}, x_{22}, \ldots, x_{m2}), \ldots, X_H^{\sigma_j}(x_{1r}, x_{2r}, \ldots, x_{mr}))$.

Proof Formula (2.3.2) expresses $X_{K \times H}$ as

$$X = X_K(X_H(x_{11}, x_{21}, \ldots, x_{m1}), X_H(x_{12}, x_{22}, \ldots, x_{m2}), \ldots,$$
$$X_H(x_{1r}, x_{2r}, \ldots, x_{mr})).$$

If P is a generalized Fourier matrix for H and if $\widehat{P}$ is the Kronecker product $I_r \otimes P = diag(P, P, \ldots, P)$ (k times) then from Theorem 2.4, $Y = \widehat{P}X\widehat{P}^{-1}$ is a block diagonal matrix with blocks

$$X_K(X_H^{\sigma_j}(x_{11}, x_{21}, \ldots, x_{m1}), X_H^{\sigma_j}(x_{12}, x_{22}, \ldots, x_{m2}), \ldots, X_H^{\sigma_j}(x_{1r}, x_{2r}, \ldots, x_{mr})).$$

Now if Q is a generalized Fourier matrix for K and $\widetilde{Q}$ is the Kronecker product $Q \otimes I_{e_i}$ i.e. the matrix obtained from Q by replacing each element a_{ij} by the diagonal matrix $diag(a_{ij}, \ldots, a_{ij})$ of size e_i then $\widetilde{Q}Y\widetilde{Q}^{-1}$ becomes a block diagonal matrix with blocks

$$X_K^{\rho_i}(X_H^{\sigma_j}(x_{11}, x_{21} \ldots, x_{m1}), X_H^{\sigma_j}(x_{12}, x_{22} \ldots, x_{m2}), \ldots, X_H^{\sigma_j}(x_{1r}, x_{2r} \ldots, x_{mr})). \tag{2.4.8}$$

The fact that the block matrix (2.4.8) is an irreducible matrix can be shown by calculating its character, and from the fact that the character table of G is the Kronecker product of the character tables of K and H.   □

Thus Theorem 2.7 gives an explicit factorization of the group determinant of $K \times H$ as the product of the determinants of the irreducible blocks with appropriate multiplicity. There is then the immediate corollary.

Corollary 2.4 *The group determinant of an abelian group factors into linear factors.*

This follows since an abelian group is a direct product of cyclic subgroups.

The factorization in the case of an abelian group may be expressed alternatively using the results of Frobenius which appeared in Chap. 1. In this case, the group determinant is the same as the determinant of the reduced group matrix, and the factors are all linear and in 1 : 1 correspondence with the irreducible characters of the abelian group A. If χ is an irreducible character of A, the corresponding factor is $\sum_{g \in A} \chi(g) x_g$. However, just as in the case where A is cyclic, the practical problem of calculating the factors which in this case is equivalent to the Finite Fourier Transform is an important problem in information science, and the methods to partially diagonalize described above become relevant.

2.4.6 More Examples

Example 2.16 Let G be the quaternion group $Q_8 = \{\pm 1, \pm i, \pm j, \pm k\}$, ordered as in Example 1.4, Chap. 1. Then (replacing x_r by r)

$$X_G = \begin{bmatrix} C(1,2,3,4) & C(7,6,5,8) \\ C(5,6,7,8) & C(1,4,3,2) \end{bmatrix}.$$

If F_4 is the Fourier matrix for C_4 and Q is the 8×8 matrix $diag(F_4, F_4)$ then

$$R = Q^{-1} X_G Q = \begin{bmatrix} \Lambda_{11} & \Lambda_{12} \\ \Lambda_{21} & \Lambda_{22} \end{bmatrix}.$$

If

$$u(n_1, n_2, n_3, n_4) = x_{n_1} + x_{n_2} + x_{n_3} + x_{n_4}$$
$$u'(n_1, n_2, n_3, n_4) = x_{n_1} - x_{n_2} + x_{n_3} - x_{n_4}$$
$$v(n_1, n_2, n_3, n_4) = x_{n_1} + ix_{n_2} - x_{n_3} - ix_{n_4}$$
$$v'(n_1, n_2, n_3, n_4) = x_{n_1}x - ix_{n_2} - x_{n_3} + ix_{n_4},$$

then

$$\Lambda_{11} = diag(u(1,2,3,4), v(1,2,3,4), u'(1,2,3,4), v'(1,2,3,4)),$$
$$\Lambda_{12} = diag(u(7,6,5,8), v(7,6,5,8), u'(7,6,5,8), v'(7,6,5,8)),$$
$$\Lambda_{21} = diag(u(5,6,7,8), v(5,6,7,8), u'(5,6,7,8), v'(5,6,7,8)),$$
$$\Lambda_{22} = diag(u(1,4,3,2), v(1,4,3,2), u'(1,4,3,2), v'(1,4,3,2)).$$

It follows that R is similar to a matrix of the form

$$diag(B_1, B_2, B_3, B_4)$$

where

$$B_1 = \begin{bmatrix} u(1,2,3,4) & u(7,6,5,8) \\ u(5,6,7,8) & u(1,4,3,2) \end{bmatrix}, \quad B_2 = \begin{bmatrix} v(1,2,3,4) & v(7,6,5,8) \\ v(5,6,7,8) & v(1,4,3,2) \end{bmatrix},$$

$$B_3 = \begin{bmatrix} u'(1,2,3,4) & u'(7,6,5,8) \\ u'(5,6,7,8) & u'(1,4,3,2) \end{bmatrix}, \quad B_4 = \begin{bmatrix} v'(1,2,3,4) & v'(7,6,5,8) \\ v'(5,6,7,8) & v'(1,4,3,2) \end{bmatrix}.$$

The factors of the group determinant are the four linear factors which arise from $\det(B_1)$ and $\det(B_3)$ and the quadratic factor which is $\det(B_2) = \det(B_4)$. Now

$$\det(B_2) = v(1,2,3,4)v(1,4,3,2) - v(7,6,5,8)v(5,6,7,8)$$
$$= (x_1 - x_3)^2 + (x_2 - x_4)^2 - (-((x_5 - x_7)^2 + (x_6 - x_8)^2))$$
$$= (x_1 - x_3)^2 + (x_2 - x_4)^2 + (x_5 - x_7)^2 + (x_6 - x_8)^2).$$

This confirms the factorization of the group determinant for Q_8 given in Chap. 1.

Example 2.17 Let $G = Fr_{20}$. Let X_G be constructed on the left cosets of the normal subgroup H of order 5 as follows. Consider G as a permutation group on $\{1,2,3,4,5\}$ generated by $\sigma = (1,2,3,4,5)$ and $(2,4,5,3)$. Choose as a system of representatives

$$\{y_1, y_2, y_3, y_4\} = \{e, (2,4,5,3), (2,5)(3,4), (2,3,5,4)\}$$

for the left cosets of H and order H as $\{e, \sigma, \sigma^2, \sigma^3, \sigma^4\}$. The coset $y_i H$ is referred to as the ith coset of H. Then

$$X_G = \begin{bmatrix} C(1,5,4,3,2)^1 & C(1,4,2,5,3)^4 & C(1,2,3,4,5)^3 & C(1,3,5,2,4)^2 \\ C(1,5,4,3,2)^2 & C(1,4,2,5,3)^1 & C(1,2,3,4,5)^4 & C(1,3,5,2,4)^3 \\ C(1,5,4,3,2)^3 & C(1,4,2,5,3)^2 & C(1,2,3,4,5)^1 & C(1,3,5,2,4)^4 \\ C(1,5,4,3,2)^4 & C(1,4,2,5,3)^3 & C(1,2,3,4,5)^2 & C(1,3,5,2,4)^1 \end{bmatrix},$$

where $C(1,5,4,3,2)^i$ is an abbreviation for the circulant $C(x_{g_1}, x_{g_5}, x_{g_4}, x_{g_3}, x_{g_2})$ with g_j the jth element of the ith coset of H. If F_5 is the Fourier matrix of order 5 and $P = diag(F_5, F_5, F_5, F_5)$, then

$$P^{-1}X_G P = \begin{bmatrix} \Lambda(1,5,4,3,2)^1 & \Lambda(1,4,2,5,3)^4 & \Lambda(1,2,3,4,5)^3 & \Lambda(1,3,5,2,4)^2 \\ \Lambda(1,5,4,3,2)^2 & \Lambda(1,4,2,5,3)^1 & \Lambda(1,2,3,4,5)^4 & \Lambda(1,3,5,2,4)^3 \\ \Lambda(1,5,4,3,2)^3 & \Lambda(1,4,2,5,3)^2 & \Lambda(1,2,3,4,5)^1 & \Lambda(1,3,5,2,4)^4 \\ \Lambda(1,5,4,3,2)^4 & \Lambda(1,4,2,5,3)^3 & \Lambda(1,2,3,4,5)^2 & \Lambda(1,3,5,2,4)^1 \end{bmatrix}$$

where $\Lambda(\alpha_1, \alpha_2, \alpha_3, \alpha_4, \alpha_5)^i$ is a diagonal matrix with jth entry

$$u_j(\alpha_1, \alpha_2, \alpha_3, \alpha_4, \alpha_5)^i = (x_{g_1} + \eta^j x_{g_5} + (\eta^j)^2 x_{g_4} + (\eta^j)^3 x_{g_3} + (\eta^j)^4 x_{g_2})$$

where $\eta = e^{2\pi i/5}$, $j = 0, 1, 2, 3, 4$, and $g_t = y_i(1, 2, 3, 4, 5)^{t-1}$. Note that $u_0^i = \sum_{g \in y_i H} x_g$.

It is possible by means of a similarity transformation permuting rows and columns to obtain from Proposition 2.2 a block diagonal matrix whose first block is

$$\begin{bmatrix} u_0^1 & u_0^4 & u_0^3 & u_0^2 \\ u_0^2 & u_0^1 & u_0^4 & u_0^3 \\ u_0^3 & u_0^2 & u_0^1 & u_0^4 \\ u_0^4 & u_0^3 & u_0^2 & u_0^1 \end{bmatrix} = C(u_0^1, u_0^4, u_0^3, u_0^2)$$

whose determinant, since it is a circulant, factors into linear factors. These correspond to the linear characters of G. The second block is

$$B_2 = \begin{bmatrix} u_1^1(1, 5, 4, 3, 2) & u_1^4(1, 4, 2, 5, 3) & u_1^3(1, 2, 3, 4, 5) & u_1^2(1, 3, 5, 2, 4) \\ u_1^2(1, 5, 4, 3, 2) & u_1^1(1, 4, 2, 5, 3) & u_1^4(1, 2, 3, 4, 5) & u_1^3(1, 3, 5, 2, 4) \\ u_1^3(1, 5, 4, 3, 2) & u_1^2(1, 4, 2, 5, 3) & u_1^1(1, 2, 3, 4, 5) & u_1^4(1, 3, 5, 2, 4) \\ u_1^4(1, 5, 4, 3, 2) & u_1^3(1, 4, 2, 5, 3) & u_1^2(1, 2, 3, 4, 5) & u_1^1(1, 3, 5, 2, 4) \end{bmatrix}.$$

The u_i^j are Fourier transforms of sets of variables. It may be shown that they are all independent (this essentially follows because the inverse Fourier transform can be applied) and thus the determinant of B_2 is an irreducible polynomial in the $\{u_i^j\}$, and also irreducible in the $\{x_i^j\}$. The other three blocks are similar to B_2. An explicit matrix representing g_i in the irreducible representation of degree 4 are obtained by setting $x_{g_i} = 1$ and $x_{g_j} = 0$ for $j \neq i$. The matrices corresponding to $(1, 2, 3, 4, 5)$ and $(2, 4, 5, 3)$ are respectively

$$\begin{bmatrix} \eta & 0 & 0 & 0 \\ 0 & \eta^2 & 0 & 0 \\ 0 & 0 & \eta^3 & 0 \\ 0 & 0 & 0 & \eta^4 \end{bmatrix}, \quad \begin{bmatrix} 0 & 0 & 0 & 1 \\ 1 & 0 & 0 & 0 \\ 0 & 1 & 0 & 0 \\ 0 & 0 & 1 & 0 \end{bmatrix}.$$

2.4.6.1 The Factorization of the Determinant of the "Loop Matrix" of the Octonion Loop

This matrix $X_{O_{16}}$ is a direct translation of the Ward quasigroup given in (2.3.10)

$$\begin{bmatrix} C(1, 2, 3, 4) & C(7, 6, 5, 8) & C(11, 10, 9, 12) & C(15, 14, 13, 16) \\ C(5, 6, 7, 8) & C(1, 4, 3, 2) & R(15, 14, 13, 16) & R(9, 12, 11, 10) \\ C(9, 10, 11, 12) & R(13, 16, 15, 14) & C(1, 4, 3, 2) & R(7, 6, 5, 8) \\ C(13, 14, 15, 16) & R(9, 12, 11, 10) & R(7, 6, 5, 8) & C(1, 4, 3, 2) \end{bmatrix},$$

by inserting corresponding variables. If P is the matrix

$$\begin{bmatrix} 1 & 1 & 1 & 1 \\ 1 & i & -1 & -i \\ 1 & -1 & 1 & 1 \\ 1 & -i & -1 & i \end{bmatrix}$$

it is a scalar multiple of the Fourier matrix F_4 and then by standard results from Chap. 1 if $C = C(x_1, x_2, x_3, x_4)$,

$$P^{-1}CP = \begin{bmatrix} v_1 & 0 & 0 & 0 \\ 0 & v_2 & 0 & 0 \\ 0 & 0 & v_3 & 0 \\ 0 & 0 & 0 & v_4 \end{bmatrix} \qquad (2.4.9)$$

where $v_1 = x_1 + x_2 + x_3 + x_4$, $v_2 = x_1 + ix_2 - x_3 + ix_4$, $v_3 = x_1 - x_2 + x_3 - x_4$ and $v_4 = x_1 - ix_2 - x_3 + ix_4$. If R is the reverse circulant $R(x_1, x_2, x_3, x_4)$, by direct calculation

$$P^{-1}RP = \begin{bmatrix} v_1 & 0 & 0 & 0 \\ 0 & 0 & 0 & v_2 \\ 0 & 0 & v_3 & 0 \\ 0 & v_4 & 0 & 0 \end{bmatrix} \qquad (2.4.10)$$

Then if the matrix Q is the 16×16 block diagonal matrix constructed from 4 copies of P the matrix $\tilde{X} = Q^{-1}X_{O_{16}}Q$ is obtained from (2.3.10) by replacing circulants and reverse circulants by the blocks indicated in (2.4.9) and (2.4.10). Now $\tilde{X}$ can further be transformed by row and column switches to a matrix of the form,

$$\begin{bmatrix} X_1 & 0 & 0 \\ 0 & X_2 & 0 \\ 0 & 0 & X_3 \end{bmatrix}$$

where X_1 consists of the non-zero entries in the rows and columns 1, 5, 9, 13, X_2 consists of the non-zero entries in the rows and columns 3, 7, 11, 15 and X_3 is the submatrix with entries in the remaining rows and columns. It is seen that X_1 and X_2 are group matrices for the Klein 4-group and hence factor into linear factors. The matrix X_3 may be written in the form

$$
\begin{bmatrix}
a_1 & 0 & b_1 & 0 & c_1 & 0 & d_1 & 0 \\
0 & a_2 & 0 & b_2 & 0 & c_2 & 0 & d_2 \\
-b_2 & 0 & a_2 & 0 & 0 & d_1 & 0 & -c_1 \\
0 & -b_1 & 0 & a_1 & d_2 & 0 & -c_2 & 0 \\
-c_2 & 0 & 0 & -d_1 & a_2 & 0 & 0 & b_1 \\
0 & -c_1 & -d_2 & 0 & 0 & a_1 & b_2 & 0 \\
-d_2 & 0 & 0 & c_1 & 0 & -b_1 & a_2 & 0 \\
0 & -d_1 & c_2 & 0 & -b_2 & 0 & 0 & a_1
\end{bmatrix},
$$

where $a_1 = (x_1 - x_3) - i(x_2 - x_4)$, $a_2 = \overline{a_1} = (x_1 - x_3) + i(x_2 - x_4)$, $b_1 = (x_5 - x_7) - i(x_6 - x_8)$ etc. Using a symbolic manipulation package $det(X_3) = (a_1 a_2 + b_1 b_2 + c_1 c_2 + d_1 d_2)^4$ and inserting the expressions for $a_1, \overline{a_1}$ etc. into the expression inside the parentheses the factor

$$
\left(\sum_{i=0}^{7}(x_{e_i} - x_{-e_i})^2\right)^4
$$

indicated in Chap. 1 is obtained. It does not seem to be easy to factor the determinant of X_3 without a symbolic manipulation package.

2.5   Virtual Representations via Supermatrices

For a group G with $Irr(G) = \{\chi_i\}_{i=1}^{r}$, the ring $K(G)$ of virtual characters (or generalized characters) may be defined to be the set of all linear combinations $\sum_{i=1}^{r} a_i \chi_i, a_i \in \mathbb{Z}$, with

$$
\sum_{i=1}^{r} a_i \chi_i + \sum_{i=1}^{r} b_i \chi_i = \sum_{i=1}^{r}(a_i + b_i)\chi_i
$$

and

$$
\left(\sum_{i=1}^{r} a_i \chi_i\right)\left(\sum_{i=1}^{r} b_i \chi_i\right) = \sum_{i,j=1}^{r} a_i b_j \chi_i \chi_j.
$$

In order to obtain an analogy of this ring via group matrices the following is introduced.

2.5.1 Superalgebras and Supermatrices

The discussion below includes material which is superfluous to the immediate discussion of group matrices. It is included because there seem to be interesting questions about group supermatrices over an arbitrary superalgebra. References are [14, 77] and [283].

Superalgebras arose in physics. A superalgebra is a $\mathbb{Z}_2$-graded algebra, i.e. it is an algebra over a commutative ring with a decomposition into "even" and "odd" pieces, with a multiplication operator which respects the grading.

The formal definition is as follows

Definition 2.4 Let K be a commutative ring. A superalgebra over K is a K-module A with a direct sum decomposition

$$A = A_0 \oplus A_1$$

together with a bilinear multiplication $A \times A \to A$ such that

$$A_i A_j \subset A_{i+j}$$

where the subscripts are read mod 2.

Usually, K is taken to be $\mathbb{R}$ or $\mathbb{C}$. The elements of A_i, $i = 1, 2$ are said to be **homogeneous**. The **parity** of a homogeneous element x, denoted by $|x|$, is 0 or 1 depending on whether it is in A_0 or A_1. Elements of parity 0 are said to be **even** and those of parity 1 are said to be **odd**. If x and y are both homogeneous, then so is the product and $|xy| = |x| + |y|$.

A superalgebra is **associative** if its multiplication is associative. It is **unital** if it has a multiplicative identity, which is necessarily even. It is usual to assume that superalgebras are both associative and unital.

A superalgebra A is **(super)commutative** if for all homogeneous $x, y \in A$,

$$yx = (-1)^{|x||y|}xy.$$

The standard example is an exterior algebra over K. Another example is the algebra A of symmetric and alternating polynomials, with A_0 the symmetric polynomials and A_1 being the alternating polynomials.

Definition 2.5 Let R be a superalgebra, which is unital and associative. Let p, q, r, s be nonnegative integers. A **supermatrix** of dimension $(r|s) \times (p|q)$ is an $(r + s) \times (p + q)$ matrix X with entries in R which is partitioned into a 2×2 block structure

$$X = \begin{bmatrix} X_{00} & X_{01} \\ X_{10} & X_{11} \end{bmatrix},$$

so that X_{00} has dimensions $r \times p$ and X_{11} has dimensions $s \times q$. An ordinary (ungraded) matrix may be interpreted as a supermatrix with $q = s = 0$.

Definition 2.6 A square supermatrix X has $(r|s) = (p|q)$.

This implies that X, X_{00} and X_{11} are all square in the usual sense.

An **even supermatrix** X has diagonal blocks X_{00} and X_{11} consisting of even elements of R, and X_{01} and X_{10} consisting of odd elements of R, i.e. it is of the form

$$\begin{bmatrix} even & odd \\ odd & even \end{bmatrix}.$$

An **odd supermatrix** X has diagonal blocks which are odd and the remaining blocks even, i.e. it is of the form

$$\begin{bmatrix} odd & even \\ even & odd \end{bmatrix}.$$

If the scalars R are purely even then there are no nonzero odd elements, so the even supermatrices are the block diagonal ones

$$X = \begin{bmatrix} X_{00} & 0 \\ 0 & X_{11} \end{bmatrix},$$

and an odd supermatrix is of the form

$$X = \begin{bmatrix} 0 & X_{01} \\ X_{10} & 0 \end{bmatrix}.$$

A supermatrix is homogeneous if it is either even or odd. The **parity**, $|X|$, of a non-zero homogeneous supermatrix X is 0 or 1 according to whether it is even or odd. Every supermatrix can be written uniquely as the sum of an even matrix and an odd one.

2.5.1.1 Operations

Let X, Y be supermatrices. $X + Y$ is defined entrywise, so that

$$X + Y = \begin{bmatrix} X_{00} + Y_{00} & X_{01} + Y_{01} \\ X_{10} + Y_{10} & X_{11} + Y_{11} \end{bmatrix}.$$

The sum of even matrices is even, and the sum of odd matrices is odd.

XY is defined by ordinary block matrix multiplication, i.e.

$$XY = \begin{bmatrix} X_{00}Y_{00} + X_{01}Y_{10} & X_{00}Y_{01} + X_{01}Y_{11} \\ X_{10}Y_{00} + X_{11}Y_{10} & X_{10}Y_{01} + X_{11}Y_{11} \end{bmatrix}.$$

If X and Y are both even or both odd, then XY is even, and if they differ in parity XY is odd.

The scalar multiplication differs from the ungraded case. It is necessary to define left and right scalar multiplication. **Left scalar multiplication** by $\alpha \in R$ is defined by

$$\alpha.X = \begin{bmatrix} \alpha X_{00} & \alpha X_{01} \\ \widehat{\alpha} X_{10} & \widehat{\alpha} X_{11} \end{bmatrix},$$

where $\widehat{\alpha} = (-1)^{|\alpha|}\alpha$. **Right scalar multiplication** is defined similarly

$$X.\alpha = \begin{bmatrix} X_{00}\alpha & X_{01}\widehat{\alpha} \\ X_{10}\alpha & X_{11}\widehat{\alpha} \end{bmatrix}.$$

If α is even then $\widehat{\alpha} = \alpha$ and both operations are the same as the ungraded versions. If α and X are homogeneous, then both $\alpha.X$ and $X.\alpha$ are homogeneous with parity $|\alpha| + |X|$. If R is supercommutative, then $\alpha.X = (-1)^{|\alpha||X|}X.\alpha$.

The supertranspose of the homogeneous supermatrix X is the $(p|q) \times (r|s)$ supermatrix

$$X^{st} = \begin{bmatrix} X_{00}^{t} & (-1)^{|X|}X_{10} \\ -(-1)^{|X|}X_{01} & X_{11}^{t} \end{bmatrix}$$

where M^{t} denotes the usual transpose of a matrix. This can be extended to arbitrary supermatrices by linearity. The supertranspose is not an involution: if X is an arbitrary supermatrix, then

$$(X^{st})^{st} = \begin{bmatrix} X_{00} & -X_{01} \\ -X_{10} & X_{11} \end{bmatrix}.$$

If R is supercommutative then for arbitrary supermatrices X, Y

$$(XY)^{st} = (-1)^{|X||Y|}Y^{st}X^{st}.$$

There is a new operation, the **parity transpose**. This is denoted by X^{π}. If X is a supermatrix, then

$$X^{\pi} = \begin{bmatrix} X_{11} & X_{10} \\ X_{01} & X_{00} \end{bmatrix},$$

and the following are satisfied

$$(X + Y)^\pi = X^\pi + Y^\pi,$$

$$(XY)^\pi = X^\pi Y^\pi,$$

$$(\alpha.X)^\pi = \widehat{\alpha}.X^\pi,$$

$$(X.\alpha)^\pi = X^\pi.\widehat{\alpha}$$

and in addition

$$\pi^2 = 1$$

$$\pi \circ st \circ \pi = (st)^3.$$

The supertrace of a square supermatrix is defined on homogeneous supermatrices by the formula

$$str(X) = tr(X_{00}) - (-1)^{|X|}tr(X_{11}).$$

If R is supercommutative then

$$str(XY) = (-1)^{|X||Y|}str(YX).$$

for homogeneous supermatrices X, Y.

The identity $(r/s) \times (r/s)$ supermatrix is

$$\begin{bmatrix} I_r & 0 \\ 0 & I_s \end{bmatrix}.$$

The inverse of the supermatrix X is

$$\begin{bmatrix} (X_{00} - X_{01}X_{11}^{-1}X_{10})^{-1} & -X_{00}^{-1}X_{01}(X_{11} - X_{10}X_{00}^{-1}X_{01})^{-1} \\ -X_{11}^{-1}X_{10}(X_{00} - X_{01}X_{11}^{-1}X_{10})^{-1} & (X_{11} - X_{10}X_{00}^{-1}X_{01})^{-1} \end{bmatrix}.$$

$$(2.5.1)$$

It may be seen that necessary conditions for the existence of an inverse are that the ordinary matrices in (2.5.1) which appear with exponent -1 are nonsingular.

The **Berezinian** or **superdeterminant** $Ber(X)$ of a square supermatrix X is only well-defined on even invertible supermatrices over a commutative superalgebra R. In this case

$$Ber(X) = \det(X_{00} - X_{01}X_{11}^{-1}X_{10})\det(X_{11})^{-1},$$

where det denotes the ordinary determinant of square matrices with entries in the commutative algebra R_0. The Berezinian satisfies similar properties to the ordinary determinant. In particular, it is multiplicative and invariant under the supertranspose. Moreover

$$Ber(e^X) = e^{str(X)}.$$

In particular, if R is purely even and X is even, then $Ber(X) = \det(X_{00}) \det(X_{11})^{-1}$.

2.5.2 Group Matrices Corresponding to Virtual Representations

For G a finite group let the set of irreducible representations of G be $Irr(G) = \{\rho_i\}_{i=1}^r$, and consider supermatrices of the form

$$\overline{X}_G = \begin{bmatrix} X_{1G} & 0 \\ 0 & X_{2G} \end{bmatrix},$$

where X_{1G} and X_{2G} are group matrices. $\overline{X}_G$ is to be regarded as an even supermatrix over a field. Call $\overline{X}_G$ a **group supermatrix**. Then multiplication of group supermatrices $\overline{X}_G$ and $\overline{Y}_G$ is the usual one:

$$\overline{X}_G \overline{Y}_G = \begin{bmatrix} X_{1G} Y_{1G} & 0 \\ 0 & X_{2G} Y_{2G} \end{bmatrix}.$$

The supertrace of $\overline{X}_G$ is $tr(X_{1G}) - tr(X_{2G})$ and $Ber(\overline{X}_G)$ is $\det(X_{1G}) \det(X_{2G})^{-1}$ where det is the ordinary determinant.

Consider a virtual representation of G with generalized character

$$\psi = \sum_{i=1}^r a_i \chi_i - \sum_{i=1}^r b_i \chi_i$$

where all the a_i and b_i are non-negative integers, and at most one member of $\{a_i, b_i\}$ is non-zero. The group matrix $X_G^{\sum a_i \rho_i}$ corresponding to $\sum_{i=1}^r a_i \chi_i$ is the block diagonal matrix of the form

$$diag(B_G^{\rho_1}, B_G^{\rho_2}, \dots, B_G^{\rho_r})$$

where the block $B_G^{\rho_i}$, $1 \le i \le r$ is

$$diag(X_G^{\rho_i}, \dots, X_G^{\rho_i})$$

with a_i copies of $X_G^{\rho_i}$. Now let $\overline{X}_G^{\psi}$ be the group supermatrix

$$\overline{X}_G^{\psi} = \begin{bmatrix} X_G^{\sum a_i \rho_i} & 0 \\ 0 & X_G^{\sum b_i \rho_i} \end{bmatrix}.$$

Then $str(\overline{X}_G^{\psi}) = \sum \psi(g) x_g$.

The ring of virtual group representations may be obtained by factoring out by the equivalence relation α on arbitrary group supermatrices defined by

$$\begin{bmatrix} X_{1G} & 0 \\ 0 & X_{2G} \end{bmatrix} \equiv_\alpha \begin{bmatrix} \widehat{X}_{1G} & 0 \\ 0 & \widehat{X}_{2G} \end{bmatrix}$$

if and only if X_{1G} is similar to $X_G^{\sum a_i \rho_i}$, X_{2G} is similar to $X_G^{\sum b_i \rho_i}$, $\widehat{X}_{1G}$ is similar to $X_G^{\sum \widehat{a}_i \rho_i}$, $\widehat{X}_{2G}$ is similar to $X_G^{\sum \widehat{b}_i \rho_i}$ such that $a_i - b_i = \widehat{a}_i - \widehat{b}_i$. It is necessary to embed the group matrix corresponding to a representation ρ as the α-equivalence class containing

$$\begin{bmatrix} X_G^{\rho+1} & 0 \\ 0 & X_G^1 \end{bmatrix},$$

where 1 is the trivial representation. Similarly, the supergroup matrix corresponding to a negative representation $-\rho$ is the α-equivalence class containing

$$\begin{bmatrix} X_G^1 & 0 \\ 0 & X_G^{\rho+1} \end{bmatrix}.$$

Then if $\overline{X}_{1G} \equiv_\alpha \overline{X}_{2G}$ then $Ber(\overline{X}_{1G}) = Ber(\overline{X}_{2G})$. The operations are defined to be

$$\begin{bmatrix} X_{1G} & 0 \\ 0 & X_{2G} \end{bmatrix} + \begin{bmatrix} Y_{1G} & 0 \\ 0 & Y_{2G} \end{bmatrix} = \begin{bmatrix} X_{1G} + Y_{1G} & 0 \\ 0 & X_{2G} + Y_{2G} \end{bmatrix}$$

and

$$\begin{bmatrix} X_{1G} & 0 \\ 0 & X_{2G} \end{bmatrix} \cdot \begin{bmatrix} Y_{1G} & 0 \\ 0 & Y_{2G} \end{bmatrix} = \begin{bmatrix} X_{1G} \otimes^* Y_{1G} & 0 \\ 0 & X_{2G} \otimes^* Y_{2G} \end{bmatrix}.$$

2.5.3   The Hopf Algebra Operations

For a finite group $G = \{g_1, \ldots, g_n\}$ the following defines a Hopf algebra. For a field K the group algebra KG consists of elements of the form $\underline{a} = \sum_{i=1}^{n} a_i g_i$ and the dual algebra is the space K^G of all maps from G to K. A multiplication operation $\mu : KG \otimes KG \to KG$ is defined by

$$\mu(\underline{a}_1 \otimes \underline{a}_2) = \underline{a}_1 \underline{a}_2$$

and a comultiplication $\Delta : KG \to KG \otimes KG$ is defined by

$$\Delta(\sum_{i=1}^{n} a_i g_i) = \sum_{i=1}^{n} a_i (g_i \otimes g_i).$$

It is explained in the preprint [51, p. 31] (this differs from the published version) that KG becomes a Hopf algebra with the addition of a unit and a counit and the obvious antipodism given by $g \to g^{-1}$. Moreover there is a dual Hopf algebra structure on K^G with multiplication pointwise multiplication of functions and coproduct

$$\Delta^*(u(g_1, g_2)) = u(g_1 g_2).$$

Cartier explains that this dual is the "ring of representative functions" $\mathcal{R}(G)$.

2.5.3.1   An Interpretation of the Group Matrix and Group Determinant in Terms of the Hopf Structure

The following is work with J.D.H. Smith.

Let G be a finite group. Consider the basis $\{e_g\}_{g \in G}$ for K^G where $e_g(h) = \delta_{gh}$. There is a dual pairing of KG with K^G by evaluation

$$(h, e_g) = e_g(h)$$

and $(K^G)^*$ can be identified with KG. Now let V be a G-module and let

$$\rho : G \to Aut_K(V)$$

be an explicit representation. Then consider the assignment

$$g \to \sum_{g \in G} e_g \otimes \rho(g).$$

Let $T(K^G)$ be the tensor algebra on K^G. Then consider the extension of ρ

$$\rho : KG \rightarrow End_{T(K^G)}(V).$$

Then $\sum_{g \in G} e_g \otimes \rho(g)$ can be regarded as lying in $End_{T(K^G)}(V)$. If d is the dimension of V, then $\sum_{g \in G} e_g \otimes \rho(g)$ may be regarded as a $d \times d$ matrix in $T(K^G)_d^d$.

Consider the right regular representation of $R(G)$ of G. This is an endomorphism of the (left) $T(k^G)$-module $T(K^G) \otimes KG$, explicitly $R(g) : KG \rightarrow KG, h \rightarrow hg$.

Then the (full) group matrix is

$$\sum_{g \in G} e_g \otimes R(g).$$

Now

$$KG = \oplus_{\rho \in Irr'(G)} \dim_K(V_\rho) \sum_{g \in G} e_g \otimes \rho(g).$$

Let $S(K^G)$ be the symmetric tensor algebra on K^G. Then over $S(K^G)$

$$\det(\sum_{g \in G} e_g \otimes R(g)) = \prod_{\rho \in Irr'(G)} \det[\sum_{g \in G} e_g \otimes \rho(g))]^{\dim_K(V_\rho)}.$$

In K^G the coproduct is defined by

$$\Delta_{K^G}(e_g) = \sum_{hk \in G} e_h \otimes e_k.$$

Proposition 2.3

$$(\sum_{g \in G} e_h \otimes \rho(h))(\sum_{g \in G} e_k \otimes \rho(k)) = \sum_{g \in G} \Delta_{K^G}(e_g) \otimes \rho(g).$$

Proof Now

$$(\sum_{h \in G} e_h \otimes \rho(h))(\sum_{h \in G} e_k \otimes \rho(k)) = \sum_{h,k \in G} (e_h \otimes \rho(h)) \otimes (e_k \otimes \rho(k))$$

$$= \sum_{h,k \in G} (e_h \otimes e_k) \otimes \rho(h)\rho(k)$$

$$= \sum_{g \in G} \Delta_{K^G}(e_g) \otimes \rho(g).$$

$\square$

Corollary 2.5

$$\det[\sum_{g\in G} e_h \otimes \rho(h)] \det[\sum_{g\in G} e_k \otimes \rho(k)] = \det[\sum_{g\in G} \Delta_{KG}(e_g) \otimes \rho(g)].$$

Note that the convolution property for a homogeneous polynomial $p(e_{g_1}, \ldots, e_{g_n})$ in $S(K^G[e_{g_1}, \ldots, e_{g_n}])$ is

$$p(e_{g_1}, \ldots, e_{g_n}) \otimes p(e_{g_1}, \ldots, e_{g_n}) = p(\Delta_{KG}(e_{g_1}), \ldots, \Delta_{KG}(e_{g_n})).$$

Then for group matrices $\sum_{g\in G} e_g \otimes \rho_1(g)$ and $\sum_{g\in G} e_g \otimes \rho_2(g)$ the group matrix corresponding to the representation to $\rho_1(g) \otimes \rho_2(g)$ is obtained by taking the Kronecker product of the matrices $\sum_{g\in G} e_g \otimes \rho_1(g)$ and $\sum_{g\in G} e_g \otimes \rho_2(g)$ and applying ∇_{KG} to the entries.

Example 2.18 Consider $G = C_2 = \{0, 1\}$. Then the group matrix of $R(G)$ is

$$\begin{bmatrix} e_0 & e_1 \\ e_1 & e_0 \end{bmatrix}.$$

The construction of the group matrix $X_{R\otimes R}$ for $R(G) \otimes R(G)$ first takes the Kronecker product

$$\begin{bmatrix} e_0 & e_1 \\ e_1 & e_0 \end{bmatrix} \otimes \begin{bmatrix} e_0 & e_1 \\ e_1 & e_0 \end{bmatrix} = \begin{bmatrix} e_0 \otimes e_0 & e_0 \otimes e_1 & e_1 \otimes e_0 & e_1 \otimes e_1 \\ e_0 \otimes e_1 & e_0 \otimes e_0 & e_1 \otimes e_1 & e_1 \otimes e_0 \\ e_1 \otimes e_0 & e_1 \otimes e_1 & e_0 \otimes e_0 & e_0 \otimes e_1 \\ e_1 \otimes e_1 & e_1 \otimes e_0 & e_0 \otimes e_1 & e_0 \otimes e_0 \end{bmatrix}$$

and then since $\nabla_{KG}(e_i \otimes e_j) = \delta_{ij} e_j$ the group matrix $X_{R\otimes R}$ is

$$\begin{bmatrix} e_0 & 0 & 0 & e_1 \\ 0 & e_0 & e_1 & 0 \\ 0 & e_1 & e_0 & 0 \\ e_1 & 0 & 0 & e_0 \end{bmatrix}.$$

Proposition 2.3 gives an explanation of the operation of $\otimes^*$ given above. The above example may be checked by calculating that the character of χ of $R(G) \otimes R(G)$ is the square of the regular character π where $\pi(0) = 2, \pi(1) = 0$, i.e. $\chi(0) = 4, \chi(1) = 0$. Since the character of the representation at an element i is the coefficient of e_i in the trace of the group matrix this confirms that the operation has constructed a group matrix for $R(G) \otimes R(G)$.

2.6   Group Matrices and the Burnside Ring

The Burnside ring of a finite group G is defined as follows. A **G-set** is a set Ω together with an action map $G \times \Omega \to \Omega$, $(g, x) \to g(x)$. The G-set Ω is **transitive** if for any pair (x_1, x_2) of elements of Ω there is an element $g \in G$ such that $g(x_1) = x_2$. The G-sets Ω_1 and Ω_2 are **equivalent** if there is a bijection $f : \Omega_1 \to \Omega_2$ such that $f(g(x)) = g(f(x))$ for all $x \in \Omega_1$, $g \in G$. Addition of G-sets is defined by disjoint union

$$\Omega_1 + \Omega_2 = \Omega_1 \cup \Omega_2 \tag{2.6.1}$$

with the obvious action of G, and multiplication by the direct product

$$\Omega_1.\Omega_2 = \Omega_1 \times \Omega_2 \tag{2.6.2}$$

with the action $g(x_1, x_2) = (g(x_1), g(x_2))$. The set $\mathcal{D}$ of equivalence classes $\{\Omega_i\}$ of transitive G-sets is taken, and the **Burnside Ring** $B(G)$ is the Grothendieck ring of $\mathcal{D}$. This may be thought of as the ring of formal linear combinations

$$\sum_{i=1}^{r} n_i \Omega_i$$

where $\{\Omega_i\}_{i=1}^{r}$ is a set of representatives of the equivalence classes of transitive G-sets with $n_i \in \mathbb{Z}$, with addition and multiplication extended from (2.6.1) and (2.6.2). The Burnside ring appeared before representation theory.

   In terms of group matrices, associated to the G-set Ω there is the representation $g \to T(g)$ where the permutation matrix T is described in Chap. 1, Definition 1.10.

   The group matrix corresponding to this is $\sum T(g) x_g$. Denote this group matrix by X_G^{Ω}. Let the (permutation) group matrices $X_G^{\Omega_1}$ and $X_G^{\Omega_2}$ be β-**equivalent** if and only if there is a permutation matrix V such that $V^{-1} X_G^{\Omega_1} V = V^{-1} X_G^{\Omega_2} V$. A group matrix corresponding to $\Omega_1 \cup \Omega_2$ is $diag(X_G^{\Omega_1}, X_G^{\Omega_2})$, and $X_G^{\Omega_1} \otimes^* X_G^{\Omega_2}$ is a group matrix corresponding to $\Omega_1 \times \Omega_2$. On going over to supermatrices, it is possible to produce the Burnside ring while remaining in the set of monomial matrices. Explicitly, the virtual G-set

$$\sum_{i=1}^{r} n_i \Omega_i$$

may be written as

$$\Lambda_1 - \Lambda_2,$$

where Λ_1 and Λ_2 are G-sets and at least one of which is non-empty. The corresponding supergroup matrix is

$$\overline{X}_G = \begin{bmatrix} X_G^{\Lambda_1} & 0 \\ 0 & X_G^{\Lambda_2} \end{bmatrix}.$$

It is necessary to embed an ordinary permutation representation Λ as the β-equivalence class of

$$\overline{X}_G = \begin{bmatrix} X_G^{\Lambda+1} & 0 \\ 0 & X_G^1 \end{bmatrix},$$

where 1 is the permutation representation on a singleton. Similarly a totally negative representation of the form $-\Lambda$ can be embedded as the β-equivalence class of

$$\overline{X}_G = \begin{bmatrix} X_G^1 & 0 \\ 0 & X_G^{\Lambda+1} \end{bmatrix}.$$

2.7   Projective Representations and Group Matrices

The theory of projective representations of a group was set out in two papers of Schur [251, 253]. He presents a remarkably complete account of the basic theory, incidentally bringing cohomology into the theory of groups for the first time. The definitions follow.

Given a finite group G, a set of $m \times m$ matrices $\{\kappa(g)\}_{g \in G}$ with entries in a field K forms a **projective representation** (also denoted by κ) if

$$\kappa(g_1)\kappa(g_2) = \alpha(g_1, g_2)\kappa(g_1 g_2) \text{ for all } g_1, g_2 \in G,$$

where $\alpha : G \times G \to K$ is a function which is usually called a **factor set**. In the following K will be taken to be $\mathbb{C}$. The representations κ_1 and κ_2 are **equivalent** if there exist elements $c(g_i)$ in K and an invertible matrix P in $M_m(K)$ such that

$$\kappa_1(g) = c(g)P^{-1}\kappa_2(g)P$$

for all $g \in G$. The associativity condition implies that each factor set α must satisfy

$$\alpha(g_1, g_2)\alpha(g_1 g_2, g_3) = \alpha(g_1, g_2 g_3)\alpha(g_2, g_3), \tag{2.7.1}$$

and if κ_1 and κ_2 are equivalent as above their factor sets α_1, α_2 are related by

$$\alpha_1(g_1, g_2) = c(g_1, g_2)\alpha_2(g_1, g_2)$$

where

$$c(g_1, g_2) = \frac{c(g_1)c(g_2)}{c(g_1 g_2)}.$$

In the language of cohomology, α is a 2-cocycle and $c(g_1, g_2)$ is a coboundary. Schur's theory involves a finite abelian group, the Schur multiplier M, which is isomorphic to the cocycles modulo the coboundaries, and an exact sequence

$$\{0\} \to M \to D \to G \to \{e\}. \tag{2.7.2}$$

Here D is called the "**Darstellungsgruppe**", or **covering group**. In (2.7.2)

$$M \subseteq D' \cap Z(D). \tag{2.7.3}$$

Covering groups are not uniquely determined (for example both D_4 and Q_8 are covering groups for the Klein group V_4). Each irreducible projective representation of G corresponds to an irreducible ordinary representation of D.

Explicitly, if $\mu : D \to G$ is the map in (2.7.2) and for each g in G a choice is made of an element $\tau(g)$ in D such $\mu(\tau(g)) = g$ then for each irreducible representation ρ of D with corresponding matrices $\{\rho(h)\}_{h \in D}$, the set of matrices

$$\{\kappa(g)\} = \rho(\tau(g))\}_{g \in G}$$

form a set of representing matrices for the corresponding projective representation of G. Ordinary representations will also be produced and the character table of D may be used to distinguish them. As a consequence of (2.7.3) the elements of M form singleton conjugacy classes of D and since the representing matrix on such elements must be a root of unity ω times the identity matrix, for each irreducible character χ of D

$$\chi(m) = \omega\chi(e).$$

Those characters for which $\chi(m) = \chi(e)$ for all m in M give rise to ordinary representations of G and the representations corresponding to the remaining characters give rise to projective representations of G. It is often the case that the Schur multiplier is C_2, and in this case there is only one non-trivial class of cocycles, so that the projective representations all correspond to this cocycle class. If the Schur multiplier is of higher order the cocycle of each representation may be determined directly from the set of representing matrices.

In [251] Schur shows the following: assume that α is a non-trivial cocycle of G.

1. The sum of the squares of the degrees of the irreducible projective representations corresponding to α is equal to $|G|$.

2. For an element g in G the **centralizer** $C_G(g) = (h \in G : hg = gh)$. He defines a conjugacy class C of G to be an α-class if for each g in G and each element h in $C_G(g)$, $\alpha(g, h) = \alpha(h, g)$. Then the number of irreducible projective representations of G which correspond to α is the number of α-classes.

Projective representations are natural objects to consider. In particular the work of Clifford on the restriction of irreducible linear representations to normal subgroups shows how projective representations occur in an essential way in the theory of linear representations. They also appear in physics as ray representations.

There is a very comprehensive account of the theory of projective representations by Karpilovsky [188]. Many results of the theory are parallel to those in the theory of linear representations, but sometimes the proofs are much more complicated

In particular there is a character theory for projective representations and this is surveyed in [57].

2.7.1 The Connection with the Group Determinant

In [251, p. 21], Schur specifically states that, in the same way that Frobenius used the group matrix and the group determinant to determine the ordinary irreducible representations of a group, the aim of the paper is to examine the problem of finding all projective representations of a finite group with the help of the theory of group matrices and group characters.

On p. 24 he considers a factor set α satisfying (2.7.1) for a group G and he defines the (full) **projective group matrix**

$$X_G^{\alpha} = \{\alpha(gh^{-1}, h)x_{gh^{-1}}\}.$$

The **Schur-Hadamard product** $A \circ B$ of $n \times n$ matrices $A = \{a_{ij}\}$ and $B = (b_{ij})$ is the matrix whose (i, j)th entry is $a_{ij}b_{ij}$. Then $X_G^{\alpha} = \Gamma \circ X_G$ where Γ is the matrix whose (g, h)th entry is $\alpha(gh^{-1}, h)$. He then explains that X_G^{α} may be written in the form

$$X_G^{\alpha} = \sum_{g \in G} \pi^{\alpha}(g)x_g$$

where the set $\{\pi^{\alpha}(g)\}_{g \in G}$ form a set of representing matrices for a projective representation of G of dimension $|G|$ with corresponding factor set α. Thus π^{α} may be regarded as the analogue of the right regular representation of G.

Given an arbitrary projective representation κ of the group G, with associated factor set α, and sets of variables $\{x_g\}_{g \in G}$ and $\{y_g\}_{g \in G}$ the **projective group matrix** which corresponds to κ

$$X_{\kappa}^{\alpha}(\underline{x}) = \sum_{g \in G} \kappa(g)x_g,$$

can be formed, with $X_\kappa(\underline{y})$ defined similarly. Then

$$X_\kappa(\underline{x})X_\kappa(\underline{y}) = \sum_{g\in G}\kappa(g)x_g\sum_{h\in G}\kappa(h)y_h = \sum_{k\in G}\sum_{gh=k}\kappa(g)\kappa(h)x_gy_h$$

$$= \sum_{k\in G}\sum_{gh=k}\alpha(g,h)\kappa(gh)x_gy_h.$$

Therefore, if the variables $\{z_g\}$ satisfy

$$z_k = \sum_{gh=k}\alpha(g,h)x_gy_h \qquad\qquad (2.7.4)$$

and

$$Z_\kappa = \sum_{g\in G}\kappa(g)z_g,$$

there is the multiplicative property

$$X_\kappa Y_\kappa = Z_\kappa.$$

The formula in (2.7.4) may be regarded as giving a "projective convolution" of the variables $\bar{x}$ and $\bar{y}$. It translates directly into a "projective convolution" of a pair (f_1, f_2) of functions from G to K

$$(f_1 * f_2)^\alpha(k) = \sum_{gh=k}\alpha(g,h)f_1(g)f_2(h).$$

This convolution depends on the choice of α in the cocycle class.

Example 2.19 Let G be the Klein group $V_4 = \{e, u, v, uv\}$. As mentioned above, the dihedral group D_4 is a covering group for G, i.e. there is a projection $\mu : D_4 \to G$ with kernel M, the Schur multiplier. In this case if

$$D_4 = \langle a, b : a^4 = b^2 = e, bab = a^3\rangle$$

then $M = \{e, a^2\}$. The character table of D_4 is given in Chap. 1, (1.5.3) as

Class	$\{e\}$	$\{a^2\}$	$\{a, a^3\}$	$\{b, a^2b\}$	$\{ab, a^3b\}$
χ_1	1	1	1	1	1
χ_2	1	1	1	-1	-1
χ_3	1	1	-1	1	-1
χ_4	1	1	-1	-1	1
χ_5	2	-2	0	0	0

It is seen that, according to the above criterion, the characters $\chi_1, \ldots, \chi_4$ correspond to ordinary representations of G and χ_5 corresponds to a projective representation κ of G. An explicit representation ρ_5 of D_8 corresponding to χ_5 is given by

$$a \to \begin{bmatrix} i & 0 \\ 0 & -i \end{bmatrix}, \quad b \to \begin{bmatrix} 0 & 1 \\ 1 & 0 \end{bmatrix}.$$

Matrices corresponding to κ may be obtained by choosing the map $\tau : G \to D_4$ as $\tau(e) = I, \tau(u) = a, \tau(v) = b, \tau(uv) = ab$. This produces

$$e \to I, \quad u \to \begin{bmatrix} i & 0 \\ 0 & -i \end{bmatrix}, \quad v \to \begin{bmatrix} 0 & 1 \\ 1 & 0 \end{bmatrix}, \quad uv \to \begin{bmatrix} 0 & i \\ -i & 0 \end{bmatrix}.$$

A projective group matrix corresponding to this choice of matrices is

$$X_G^\kappa = \begin{bmatrix} x_e + i x_a & x_b + i x_c \\ x_b - i x_c & x_e - i x_a \end{bmatrix}. \tag{2.7.5}$$

In this case κ is the unique irreducible projective representation (up to equivalence).

More recently an account of projective group matrices has appeared in [58], in connection with the theory of wavelets. These matrices are Gram matrices for sets of vectors in space. An account of this work appears in Chap. 10, Sect. 10.7.

The Full Projective Group Matrices

The ordinary full group matrix X_G of a finite group G is determined completely by the ordering of the elements of G, or equivalently by its first column. A version of the full projective group matrix $X_G^\alpha = \Gamma \circ X_G^\alpha$ defined above appears as $X_G^\alpha(f)$ in [58], where $f : G \to \mathbb{C}$ is a function, using the left division instead of right division. Although the $X_G^\alpha(f)$ is a matrix of constants, any results for such a matrix translate into those for the full group matrix X_G^α. The entries Γ may be chosen to be roots of unity and in the case where the Schur multiplier is C_2 they can be chosen as 1 or -1. However X_G^α depends on the choice of α: equivalent cocycles give different projective group matrices even if they are constructed from the same ordering of G.

It follows directly from the convolution property that if the ordering of G and the cocycle α is fixed that X_G^α is determined by its first row and column, i.e. it is a patterned matrix. By similar arguments to those for the ordinary full group matrix, any power of X_G^α, including the inverse, is also a full group matrix with respect to α, or in other words it has the same pattern as X_G^α.

A projective group determinant can be defined as $\det(X_G^\alpha)$. Many questions such as "Does the projective group determinant determine the group" may be formulated. In this case the proof given in Chap. 1 that the usual group determinant determines the group does not translate directly into a corresponding proof. The existence of

analogous objects to the k-characters which could be used to construct irreducible
factors of the projective group determinant should also lead to interesting results.

2.8   Group Matrices and Group Determinants in the Modular Case

Nearly all the discussion of the diagonalization of group matrices and the factorization of group determinants of finite groups over a field K has been restricted to
the case where K has either characteristic 0 or a finite characteristic p which does
not divide the group order. It appears that the sole author to discuss the case of
p dividing the group order is Dickson. There is, of course, a large body of work
on modular representations initiated by Richard Brauer but this seems to bypass
questions on the modular group determinant.

In three papers [89, 91, 92] Dickson investigates group characters, group matrices
and group determinants for a finite group G where the underlying field K has finite
characteristic p. Most of the results of Frobenius carry over if p does not divide
$|G|$, and the interesting case is when p divides $|G|$. A summary of the results in
[92] is given below. There seem to be many interesting continuations suggested by
Dickson's work, but the summary here does not attempt to go further than he went.

A group matrix or group determinant reduced modulo a prime p will be called
respectively a **modular group matrix** or **modular group determinant**.

2.8.1   Dickson's Construction of the Group Matrix as a Block Matrix

Dickson defines the group matrix using left division, but for consistency his results
are given here for the group matrix under right division. Given a group G and a
subgroup H of order m and index r, the Dickson ordering of G is defined as follows.
Let $T = \{e = t_1, t_2, \ldots, t_r\}$ be a left transversal to H in G, so that

$$G = t_1 H + t_2 H + \ldots + t_r H.$$

If an ordering of H is taken as $\{e = h_1, h_2, , , h_m\}$ then the Dickson ordering is

$$\{t_1, t_2, \ldots, t_r, t_1 h_2, t_2 h_2, \ldots, t_r h_2, \ldots, t_1 h_m, t_2 h_m, \ldots, t_r h_m\}.$$

Note that this ordering differs from that given in (2.3.1) above.

Proposition 2.4 *Let H be a subgroup of a group G of order m and index r. Let $T = \{e = t_1, t_2, \ldots, t_r\}$ be a left transversal to H in G as above*

$$G = t_1 H + t_2 H + \ldots + t_r H.$$

Then if X_G is constructed from this ordering it is a block matrix with blocks of size $r \times r$, the (i, j)th block indexed by the column of elements $\{t_1 h_i, t_2 h_i, \ldots, t_r h_i\}$ and the row $\{t_1 h_j, t_2 h_j, \ldots, t_r h_j\}$, so that the element in the (u, v)th position of the block is

$$t_u h_i (t_v h_j)^{-1} = t_u h_i h_j^{-1} t_v^{-1}.$$

If this block matrix is denoted by $M(h_i h_j^{-1})$, the full group matrix of G is the generalized group matrix

$$X_H(M(h_1), M(h_2), \ldots, M(h_m)).$$

In the following lemma the binomial coefficient $\frac{n!}{r!(n-r)!}$ will be denoted by $\binom{n}{r}$.

Lemma 2.10 *Suppose $C(a_1, \ldots, a_r)$ is a circulant with $r = p^\pi$. Let $P = P_{p^\pi}$ be the $r \times r$ matrix whose entry in the (i, j)th position is $\binom{j-1}{i-1}$, $i, j = 1, \ldots, r$ where if $j < i$ then $\binom{j-1}{i-1}$ is taken to be 0. Then $PC(a_1, \ldots, a_r)P^{-1}$ modulo p is a lower triangular matrix with the entry $\sum_{i=1}^{r} a_i$ in every place in the diagonal.*

It may be seen that the matrix P is a version of Pascal's triangle. It will be called a **Pascal triangle matrix**. The proof of the Lemma follows from standard combinatorics.

The following theorem implies that modulo p all p-groups of the same order have the same determinant.

Theorem 2.8 *Let H be a p-group of order p^π. Then there is an ordering of H and a constant matrix P such that modulo p, $PX_H P^{-1}$ is a lower triangular matrix with identical diagonal entries of the form $\sum_{h \in H} x_h$.*

Proof A sketch of the proof is as follows. Let K be a maximal subgroup of H. Then K is of index p in H and is normal. Construct X_H as above in the form $X_K(M(k_1), \ldots, M(k_s))$ where $s = p^{\pi-1} = |K|$. By induction assume that there is a matrix P_s such that

$$P_s X_K P_s^{-1}$$

is modulo p a lower triangular matrix with identical entries in the diagonal equal to $\sum_{k \in K} x_k$. Here X_K is the usual group matrix for K. Then if $Q = P_s \otimes I_p$ (the Kronecker product) the product

$$Q X_K(M(k_1), \ldots, M(k_s)) Q^{-1}$$

produces mod p a block matrix Y whose identical diagonal blocks are of the form $\sum_{i=1}^{s} M(k_i)$ and whose blocks above the diagonal are zero. The form stated in the Theorem is produced by taking the Pascal triangle matrix P_p of size p, forming the Kronecker product $R = P_p \otimes I_s$ and finally performing the similarity transformation RYR^{-1} mod p.   □

The following example illustrates.

Example 2.20 Consider the group $G = Q_8$. Let H be the subgroup $\{e, i, -1, -i\}$. Choose the transversal $\{e, k\}$. Then the Dickson ordering of G is

$$\{e, k, i, j, -1, -k, -i, -j\}.$$

With respect to this ordering

$$X_G = X_H(B_1, B_2, B_3, B_4) = C(B_1, B_2, B_3, B_4),$$

where (abbreviating x_i as i)

$$B_1 = \begin{bmatrix} 1 & 6 \\ 2 & 1 \end{bmatrix}, \ B_2 = \begin{bmatrix} 7 & 8 \\ 8 & 3 \end{bmatrix}, \ B_3 = \begin{bmatrix} 5 & 2 \\ 6 & 5 \end{bmatrix}, \ B_4 = \begin{bmatrix} 3 & 4 \\ 4 & 7 \end{bmatrix}.$$

Then

$$X_G' = (P_4 \otimes I_2)C(B_1, B_2, B_3, B_4)(P_4 \otimes I_2)^{-1}$$

$$= \begin{bmatrix} \sum_{i=1}^{4} B_i & 0 & 0 & 0 \\ B_2 + B_4 & \sum_{i=1}^{4} B_i & 0 & 0 \\ B_2 + B_3 & B_2 + B_4 & \sum_{i=1}^{4} B_i & 0 \\ B_2 & B_2 + B_3 & B_2 + B_4 & \sum_{i=1}^{4} B_i \end{bmatrix} \quad (\text{mod } 2)$$

and the further reduction is calculated by

$$(P_2 \otimes I_4)X_G'(P_2 \otimes I_4)^{-1} = \begin{bmatrix} s & 0 & 0 & 0 & 0 & 0 & 0 & 0 \\ \alpha & s & 0 & 0 & 0 & 0 & 0 & 0 \\ \beta & 0 & s & 0 & 0 & 0 & 0 & 0 \\ \gamma & \beta & \alpha & s & 0 & 0 & 0 & 0 \\ \zeta & \delta & \beta & 0 & s & 0 & 0 & 0 \\ \eta & \kappa & \gamma & \beta & \alpha & s & 0 & 0 \\ \lambda & \mu & \zeta & \delta & \beta & 0 & s & 0 \\ x_4 & \nu & \eta & \kappa & \gamma & \beta & \alpha & s \end{bmatrix},$$

where

$$s = \sum_{i=1}^{8} x_i, \ \alpha = x_2 + x_4 + x_6 + x_8, \ \beta = x_3 + x_4 + x_7 + x_8, \ \gamma = x_4 + x_8,$$

$$\delta = x_2 + x_3 + x_6 + x_7, \ \zeta = x_2 + x_3 + x_4 + x_5, \ \kappa = x_2 + x_4 + x_5 + x_7,$$

$$\eta = x_2 + x_4, \ \mu = x_3 + x_7, \ \lambda = x_3 + x_4, \ \nu = x_4 + x_7.$$

The group matrix X_G of an arbitrary group G with Sylow p-subgroup H may be put in block matrix form by ordering the elements of G via the left cosets of H as in Lemma 2.5, all the block matrices being of the form $X_H(\underline{u})$ for some ordered subset $\underline{u} = (g_{i_1}, \ldots, g_{i_{p^\pi}})$ of G. Then if $Q = diag(P, \ldots, P)$ (r copies of P) the product $QX_G Q^{-1}$ replaces each of the block matrices by the corresponding lower triangular matrix $PX_H(\underline{u})P^{-1}$. A series of row and column operations which sets

1. the r^2 $(1, 1)$ elements in each block in the first $r \times r$ matrix on the diagonal,
2. the r^2 $(2, 2)$ elements in each block in the second $r \times r$ matrix on the diagonal,

. . .

is a similarity operation which produces a block matrix $U^{-1} X_G U$ with $r \times r$ blocks whose blocks above the diagonal are zero and which has p^π identical matrices D on the diagonal. Hence the following holds.

Theorem 2.9 *The modular group determinant of G with respect to a suitable ordering is of the form Θ^{p^π} where $\Theta = det(D)$.*

Example 2.21 Let $G = Fr_{21}$. The group matrix obtained by listing the elements on the cosets of the normal subgroup H of order 7 is given in Example 9 of Chap. 1. The group matrix of H is a circulant. If P_7 is the matrix described above, then $PX_H(u_1, \ldots, u_n)P^{-1}$ (modulo 7) is of the form

$$\begin{bmatrix} \sum_{i=1}^{7} u_i & 0 & 0 & 0 & 0 & 0 & 0 \\ \alpha & \sum_{i=1}^{7} u_i & 0 & 0 & 0 & 0 & 0 \\ \beta & \alpha & \sum_{i=1}^{7} u_i & 0 & 0 & 0 & 0 \\ \gamma & \beta & \alpha & \sum_{i=1}^{7} u_i & 0 & 0 & 0 \\ \delta & \gamma & \beta & \alpha & \sum_{i=1}^{7} u_i & 0 & 0 \\ \zeta & \delta & \gamma & \beta & \alpha & \sum_{i=1}^{7} u_i & 0 \\ u_7 & \zeta & \delta & \gamma & \beta & \alpha & \sum_{i=1}^{7} u_i \end{bmatrix}$$

where $\alpha = u_2 + 2u_3 + 3u_4 + 4u_5 + 5u_6 + 6u_7$, $\beta = u_3 + u_7 + 3u_4 + 3u_6 + 6u_5$, $\gamma = u_4 + 4u_5 + 3u_6 + 6u_7$, $\delta = u_5 + 5u_6 + u_7$, $\zeta = u_6 + 6u_7$.

The block diagonal matrices D given by performing the transformation on X_G indicated above are circulants of the form $C(\lambda, \mu, \upsilon)$, where λ, μ, υ are respectively $\sum_{i=1}^{7} x_i$, $\sum_{i=8}^{14} x_i$, $\sum_{i=15}^{21} x_i$. Now $\det(C(\lambda, \mu, \upsilon))$ factors as

$$(\lambda + \mu + \upsilon)(\lambda + \omega\mu + \omega^2\upsilon)(\lambda + \omega^2\mu + \omega\upsilon)$$

where $\omega^3 = e^{2\pi i/3}$ and hence the modular group determinant is the above factor raised to the power 7.

If H is normal, the partial diagonalization factorization of D can be carried out by the transforming matrix which splits the regular representation of G/H into irreducibles. Dickson assumed that K is algebraically closed, and indicates a canonical form for the modular group matrix only for this case.

It may be noted that Dickson's methods do not need the variables x_i to commute.

Chapter 3
Norm Forms and Group Determinant Factors

Abstract The usual approach to group representations in modern texts is via the theory of algebras and modules. This chapter is based on a less well-known constructive approach to the theory of algebras which uses the generalization to noncommutative algebras of a (multiplicative) norm, which can be applied to obtain results on group determinants. This continues a line of research which goes back to Frobenius. Significant results which have not been translated into modern abstract accounts are to be found there. An essential result is that given a norm-type form on an algebra with suitable assumptions, the form can be constructed rationally from the first three coefficients of its "characteristic equation". This, applied to the group determinant, shows that the 1-, 2- and 3-characters determine a group.

3.1 Introduction

Modern accounts of group representation theory for finite groups usually begin with the theory of algebras and modules (mainly over $\mathbb{R}$ or $\mathbb{C}$) and the study of systems of hypercomplex numbers became merged into this theory. The approach usually follows the lines indicated by E. Noether and set out by van der Waerden and others, given in the text [282]. A less well-known constructive approach to the theory of algebras has appeared in works of Hoehnke and the school of A. Bergmann which uses the generalization to noncommutative algebras of a (multiplicative) norm. This continues a line of research which goes back to Frobenius and Brandt. Significant results which have not been translated into modern abstract accounts are to be found there. It was initially intended that this book would contain a more comprehensive account of this work, but the attempt to do this was abandoned when the author realized that such an account should probably occupy a book of its own and be written by an expert in the theory of norm forms on algebras. This chapter gives an abbreviated account of the work, mainly from the point of view of Hoehnke, which is slanted towards the applications to the group determinant. In addition Bergmann has provided valuable input.

© Springer Nature Switzerland AG 2019   111
K. W. Johnson, *Group Matrices, Group Determinants*
and Representation Theory, Lecture Notes in Mathematics 2233,
https://doi.org/10.1007/978-3-030-28300-1_3

The algebras which appear in number theory are commutative and in this case there is a well-known correspondence between ideals and norms. A first step in the direction of norm forms for noncommutative algebras occurs in Frobenius' paper "Über die Darstellung der endlichen Gruppen durch lineare Substitutionen I" [108]. There he gives the following theorems, which he characterizes as "merkwürdig" (remarkable).

Theorem 3.1 *Let X be an $n \times n$ matrix with entries which are independent variables. Suppose the $n \times n$ matrix Y has entries which are linear combinations of these variables, and that*

$$\det(X) = \alpha \det(Y)$$

where α is a constant non-zero scalar, then either

$$Y = AXB \text{ or } Y = AX^T B$$

where A and B are constant matrices. If $n > 1$, only one of these cases can occur, and the matrices A and B are determined up to a scalar factor.

Theorem 3.2 *Suppose that the conditions of Theorem 3.1 are satisfied and that X and Y have the same characteristic polynomial. Then either*

$$Y = AXA^{-1} \text{ or } Y = AX^T A,$$

where A is a constant matrix.

Theorem 3.3 *Consider a symmetric matrix X whose (i, j)th entry is x_{ij}, and assume that the entries $\{x_{ij}; j \geq i\}$ are independent variables. Suppose that the entries in the symmetric matrix Y are linear functions of these variables, and that*

$$\det(X) = \alpha \det(Y)$$

where α is a constant non-zero scalar, then

$$Y = AXA^T,$$

where A is a constant matrix which is determined up to sign. If in addition the characteristic polynomials of X and Y are the same, then A is an orthogonal matrix.

Hoehnke referred to these theorems when he gave the opinion that the group determinant should contain sufficient information to determine a group (see Chap. 1, p. 35). There is an extensive body of results generalizing the above theorems which are surveyed in [131] and [132].

A **reduced basis** of an n-dimensional algebra A is a basis $\{u_i\}_{i=0}^{n-1}$ with $u_0 = 1$.

Hoehnke also comments in [146] that these results of Frobenius indicated that a central simple algebra is determined by its generic norm (see below), although this was not touched upon by Frobenius. In addition he remarks that in [28] on page 10, article 22, Brandt made the following comment, in the context of quaternion algebras (central simple algebras of dimension 4 over a field K) "Im Fall der reduzierten Basis sind die Multiplikationszahlen einer Maximalordnung bezüglich einer Minimalbasis merkwürdigerweise durch die Koefficienten der Normenform und ihrer Reciproken schon vollständig bestimmt..." (In the case of the reduced basis the structure constants of a maximal order with respect to a minimal basis are in a remarkable way determined by the coefficients of the norm form and their reciprocals...). Hoehnke indicates that this discovery reinforced Brandt in conjecturing that in general the structure and arithmetic of an arbitrary separable algebra are essentially determined by its norm form. He further comments that this conjecture has two parts: an algebraic and an arithmetic part and that the algebraic part is correct, in the light of the above theorems of Frobenius and their subsequent interpretations, but that explicit expressions have only been obtained in the case of a reduced norm of degree 3, i.e. for the nonions, and it is only in this case that the arithmetic problem has been solved.

Hoehnke made further comments on this approach in [143] and compared it to the work of Klein on the reduction of the solution of an algebraic equation to the solution of a "Kleinformproblem", which appeared in the book [196] on the icosahedron.

The following is a theorem of Hoehnke, and independently Jacobson [165] and Bergmann. Hoehnke comments in [146] that the theorem is indicated by the results of Frobenius which appear above.

Theorem 3.4 *Let A and B be central simple associative algebras of the same degree. The generic norm forms of A and B are similar (i.e. one may be transformed into the other by a non-singular linear transformation of a reduced basis) if and only if A is isomorpic to B or anti-isomorphic to B.*

Hoehnke [144] gives a generalization of Theorem 3.4 to algebras A over a field K which are direct sums of complete matrix algebras, where the norm form may be regarded as the determinant of a matrix which is a block diagonal matrix with blocks which are matrices with entries which are independent variables. The statement is similar, in that given block diagonal matrices X and Y, where the entries in Y are linear combinations of the entries in X, and such that X and Y have the same characteristic polynomial, then with a suitable ordering each block Y_i of Y is either of the form $T_i X_i T_i^{-1}$ or $T_i X_i^T T_i^{-1}$ for some scalar matrix T_i.

There are three different sets of publications where the theory of algebras and norm forms have been developed from early work: those of Jacobson in the USA, Hoehnke in (former) East Germany and the school of A. Bergmann in (former) West Germany. These are apparently independent of each other, at least in the early stages. Some preliminary ideas appeared in the book of Jacobson [164] but the main ideas seem to have appeared first in the work of Hoehnke. Since publication of Hoehnke's paper [144], originally submitted in 1958, was delayed until 1967, it is not surprising that several of his results appeared in print for the first time in papers of Jacobson or Bergmann. The most comprehensive treatment of the theory of norm forms on algebras appears in the papers of Bergmann and his school; [12, 15, 16], and [298], but some of Hoehnke's results appear to be unique to his papers.

Section 3.2 contains a general discussion of the results on algebras arising from their norm forms. It includes a discussion of the various approaches to multiplicative properties. A generalization of the result of Frobenius on the irreducible factors of the group determinant (Proposition 1.4, Chap. 1) is given. In Sect. 3.3 the applications to the group determinant and associated results for arbitrary algebras are presented including the result of Hoehnke which implies that if A is an algebra the structure constants of the Jordan algebra A^+ are determined by $\delta^1 f, \delta^2 f, \delta^3 f$ where f is a suitable trace-like map (the shorter direct proof of Buchstaber and Rees is also given). Then in Sect. 3.4 an outline appears of the approach of Bergmann and his school to the same material, on the lines of remarks he transmitted to the author and also to Hoehnke. The last section gives a brief summary of the relationship to other sets of invariants which were given for groups by Roitman and indicates that they can be expressed in terms of the work here, and also how under less restrictive assumptions on the characteristic of K many of the results hold, sometimes with modifications.

It would appear to be an interesting project to investigate applications of this work to geometry, number theory and physics (some of which are suggested in Chaps. 8 and 10).

3.2 Norm-Type Forms on Algebras

In Chap. 1, p. 6 the definitions of semisimplicity and separability of an algebra are given. For finite dimensional algebras semisimplicity is equivalent to the algebra being isomorphic to a cartesian product of simple algebras. In the following it will be assumed that A is a separable semisimple associative algebra of finite dimension over a (commutative) field K. Bergmann has remarked that much of the theory remains valid when assumption of associativity is removed, for example for alternative algebras, but this aspect will not be pursued. Let A be such an algebra of dimension n over K, and assume that A has a unit element. An example is the group algebra KG of a finite group G where K is either of characteristic 0 or its characteristic does not divide $|G|$.

Let $\mathcal{U} = \{u_1, u_2, \ldots, u_n\}$ be a basis for A. Each element α of A may be expressed uniquely as

$$\alpha = \alpha_1 u_1 + \ldots + \alpha_n u_n \tag{3.2.1}$$

with $\alpha_i \in K$, and

$$u_i u_j = \sum_k a_{ij}^k u_k \tag{3.2.2}$$

for all $i, j \in \{1, \ldots n\}$. The a_{ij}^k are usually known as the **structure constants** and because of associativity satisfy

$$\sum_k a_{ij}^k a_{kl}^h = \sum_k a_{jl}^k a_{ik}^h.$$

The structure constants of A relative to $\mathcal{U}$ form a 3-tensor $[a_{ij}^k]$.

Example 3.1 Let A be the group algebra of $C_2 = \{e, b\}$ over $\mathbb{C}$, and take $\mathcal{U} = \{u_1 = e, \ u_2 = b\}$ as a basis. Then

$$u_1^2 = u_1, \ u_1 u_2 = u_2 = u_2 u_1, \ u_2^2 = u_1 \qquad (3.2.3)$$

The 3-tensor $[a_{ij}^k]$ has entries as follows

$$a_{11}^1 = 1, \ a_{11}^2 = 0 = a_{12}^1, \ a_{12}^2 = 1, \qquad (3.2.4)$$

$$a_{21}^1 = 0, \ a_{11}^2 = 1 = a_{12}^1, \ a_{12}^2 = 0. \qquad (3.2.5)$$

This may perhaps be more easily envisaged by giving the 2×2 matrices (slices) $M_A(i)$ which are obtained by keeping the left lower coordinate i constant:

$$M_A(1) = \begin{bmatrix} 1 & 0 \\ 0 & 1 \end{bmatrix}, \ M_A(2) = \begin{bmatrix} 0 & 1 \\ 1 & 0 \end{bmatrix}. \qquad (3.2.6)$$

A basis change produces quite different slices. For example if $\mathcal{U}' = \{u_1' = e + b, u_2' = e - b\}$ the matrix $[a_{ij}^k]'$ has the following slices

$$M_A'(1) = \begin{bmatrix} 2 & 0 \\ 0 & 2 \end{bmatrix}, \ M_A(2) = \begin{bmatrix} 2 & 0 \\ 0 & -2 \end{bmatrix}. \qquad (3.2.7)$$

The 3-dimensional matrices, or 3-tensors, which arise in this context have been studied in [145].

Associated to an algebra A there is a Lie algebra $(A^-, [\ ])$ defined by

$$[x, y] = xy - yx \qquad (3.2.8)$$

and (if the characteristic of K is not 2) a Jordan algebra $(A^+, \circ)$ defined by

$$x \circ y = \frac{1}{2}(xy + yx). \qquad (3.2.9)$$

It follows directly that if a_{ij}^k are the structure constants of A with respect to $\mathcal{U}$ the structure constants of A^- with respect to $\mathcal{U}$ are $a_{ij}^k - a_{ji}^k$ and those of A^+ with respect to $\mathcal{U}$ are $\frac{1}{2}(a_{ij}^k + a_{ji}^k)$.

The definition of a representation of a group and their equivalence extends naturally to algebras. A **representation** of an algebra A over a field K is defined to be a homomorphism $f : A \to Hom_K(V, V)$ for some vector space V over K. Let M_m be the algebra of $m \times m$ matrices over K. If V is of finite dimension m over K then a matrix representation $\rho : A \to M_m$ is given by choosing a basis $\{v_i\}_{i=1}^m$ for V and taking $\rho(u_i)$ to be the matrix whose (i, j)th entry is c_{ij} where $f(u_i) = \sum_{i=1}^m c_{ij} v_j$, so that, for α in the form of (3.2.1), $\rho(\alpha)$ has (i, j)th entry $\sum_{i=1}^n \alpha_i c_{ij}$.

The representations $\rho_1 : A \to M_n$ and $\rho_2 : A \to M_n$ are **equivalent** if there exists a (constant) non-singular matrix $P \in M_n$ such that

$$P^{-1}\rho_1(a)P = \rho_2(a) \quad \forall\, a \in A. \tag{3.2.10}$$

A finite dimensional algebra A is a **Frobenius** algebra if there exists a non-degenerate bilinear form $\beta : A \times A \to K$ which is associative in the sense that $\beta(ab, c) = \beta(a, bc)$ (see for example [69, p. 414]). In particular, the group algebra of a finite group over $\mathbb{C}$ is Frobenius.

3.2.1 Multiplicative Properties of Norm Forms on Algebras

Multiplicative properties of norm forms, generalizing those discussed in Chap. 1, are important in the theory. They have appeared in different ways and with different definitions. A summary follows.

Let A be an algebra of dimension n over K, with a unit element. As above, if $\mathcal{U} = \{u_1, u_2, \ldots, u_n\}$ is a basis for A, each element α of A has the unique representation

$$\alpha = \alpha_1 u_1 + \ldots + \alpha_n u_n.$$

An important idea is that of a **generic element** of A, i.e.

$$\underline{x} = x_1 u_1 + \ldots + x_n u_n,$$

where $\{x_1, \ldots, x_n\}$ are commuting variables. A **form** F on A is a function which takes $\underline{x}$ into the polynomial algebra $K[x_1, \ldots, x_n]$. F may also be thought of as a function from A to K where for $\alpha = \alpha_1 u_1 + \ldots + \alpha_n u_n \in A$, $F(\alpha) = p(\alpha_1, \ldots, \alpha_n)$ for a fixed polynomial p. Usually F is taken to be homogeneous. The group determinant of a finite group G provides an example, where A is the group algebra $\mathbb{C}G$ and F is the group determinant (and hence is homogeneous of degree $|G|$). In the following $\underline{x}$ will be denoted by x.

1. Hoehnke defines a form F on a finite dimensional algebra A to be **komponierbar** if

$$F(xy) = F(x)F(y), \qquad (3.2.11)$$

where x and y are generic elements, and the product xy is by convolution, defined as follows. If A has a basis $\{u_1, \ldots, u_n\}$ and $u_i u_j = \sum a_{ij}^k u_k$ define the matrix $R(u_j)$ by

$$R(u_j)_{p,q} = a_{pj}^q.$$

If $x = x_1 u_1 + \ldots + x_n u_n$, $y = y_1 u_1 + \ldots + y_n u_n$, then

$$xy = z_1 u_1 + \ldots + z_n u_n$$

where

$$z_k = \sum_{i,j=1}^{n} a_{ij}^k x_i y_j. \qquad (3.2.12)$$

If $a = \sum_{i=1}^n a_i u_i$ and $b = \sum_{i=1}^n b_i u_i$ are arbitrary elements of A it follows that $F(ab) = F(a)F(b)$, so that F gives rise to a homomorphism from A into K.

For a group algebra of a finite group, where the basis elements are chosen to be the group elements, the formula (3.2.12) coincides with Frobenius' formula (1.3.4), Chap. 1. In this chapter convolution in this sense will be written as juxtaposition, since it coincides with multiplication in the polynomial algebra $A[x_1, \ldots, x_n]$.

Hoehnke also gives another definition of a form to be of norm-type. This will be given below. For the purposes of this chapter this is essentially equivalent to that above.

2. In [15], Bergmann defines a form F of degree n to be **multiplicative** as follows. He defines the "trace function" ϕ of F as $\mathrm{red}\Phi_\theta^{(1,m-1)}(g, 1, 1, \ldots, 1)$ (this is explained in Sect. 3.3 below) and he extends this to act on $A^{\otimes n}$ by

$$\phi(a_1 \otimes a_2 \otimes \ldots \otimes a_n) = (\phi(a_1) \otimes \phi(a_2) \otimes \ldots \otimes \phi(a_n)). \qquad (3.2.13)$$

This defines a function Φ on the algebra $S^n(A)$ of symmetric tensors and then F is multiplicative if Φ is a homomorphism. He shows that a consequence of his multiplicative property is that F satisfies the komponierbar property in (3.2.11).

3. An n-homomorphism has been described in Chap. 1, p. 39: If f is a trace-like function from an algebra A to an algebra B then f is an n-homomorphism if $\delta^{n+1} f = 0$, $\delta^n f \neq 0$. In generalizing the result of Gelfand-Kolmogorov on the equivalence of a space and the algebra of functions on it, they show that an n-homomorphism gives rise to a homomorphism from $S^n(A)$ to B (with some obvious restrictions on B). This work will be discussed in Chap. 8.

4. Khudaverdian and Voronov in [192] give a different proof of the result of [36] described in (3) above. They introduce a characteristic function of a linear map of algebras, related to the **Berezinian**, which has appeared in mathematical physics, which attaches an infinite series to a function f from A to B. This reduces to a rational function if f is an n-homomorphism. They assume that A is commutative, but remark that this restriction may be removed. This is discussed more in Chap. 8.

5. There is a "characterization of characters" result which states that if B is an algebraically closed field of characteristic 0, a trace-like function $f : A \to B$ is the character of an n-dimensional representation if and only if it is a Frobenius n-homomorphism [141, 172, 275]. From this work, it follows that if f is a trace-like function a norm form may be constructed from $\delta^n f$ and this norm form is komponierbar in the sense of (1). Therefore, in the case where B is an algebraically closed field the form F is multiplicative in the sense of (2) if and only if it is komponierbar.

Since the forms considered here will usually be on algebras over $\mathbb{C}$ or an algebraically closed field of characteristic 0, a form F satisfying either (3.2.11) or Bergman's "homomorphism" property in (2) will be called **multiplicative**.

The results on multiplicative norm forms which are most relevant to the objects in this book may be summarized as follows.

1. If a multiplicative norm form on an algebra A of dimension n which is non-degenerate in a sense described below is given, the form can be constructed rationally from the knowledge of the first three coefficients of a "characteristic equation", which are polynomials in n variables of degree 1, 2 and 3 respectively.

2. The associated Jordan algebra A^+ is determined by the above three coefficients.

3. If f is a Frobenius m-homomorphism on A then a multiplicative norm-type form F of degree m can be constructed on A, from the information in $\delta^1 f, \delta^2 f, \delta^3 f$.

4. Under certain restrictions, the s_i, $i = 4, 5, \ldots$ defined below may be constructed rationally from s_1, s_2, s_3.

3.2.2 Multilinearization

The **multilinearization** (or **polarization**) of a homogeneous polynomial f in n variables $\mathbf{u} = (u_1, \ldots, u_n)$ of degree d is the polynomial

$$M_f(\mathbf{u}^{(1)}, \ldots, \mathbf{u}^{(d)}) = F(\mathbf{u}^{(1)}, \ldots, \mathbf{u}^{(d)})$$

$$:= \frac{1}{d!} \frac{\partial}{\partial v_1} \cdots \frac{\partial}{\partial v_d} f(v_1 \mathbf{u}^{(1)} + \ldots + v_d \mathbf{u}^{(d)}), \qquad (3.2.14)$$

where $\mathbf{u}^{(1)}, \ldots, \mathbf{u}^{(d)}$ is a collection of indeterminates and $\mathbf{u}^{(i)} = (u_1^{(i)}, u_2^{(i)}, \ldots, u_n^{(i)})$. $F(\mathbf{u}^{(1)}, \ldots, \mathbf{u}^{(d)})$ is linear separately in each $\mathbf{u}^{(i)}$ and is symmetric in the $\{\mathbf{u}^{(i)}\}$.

It has the property that $F(\mathbf{u}, \ldots, \mathbf{u}) = f(\mathbf{u})$. The algorithm in (3.2.14) may be described more informally as selecting the coefficient of $v_1 v_2 \ldots v_d$ in $f(v_1 \mathbf{u}^{(1)} + \ldots v_d \mathbf{u}^{(d)})$, and dividing by $d!$.

The multilinearization of a homogeneous polynomial can be carried out on each monomial separately and then summing the results. There is no assumption that the indeterminates $\{u_j^{(i)}\}$ commute.

Example 3.2 If $f(x)$ is the quadratic form $ax^2 + bxy + cy^2$ in $\mathbf{x} = (x, y)$ the bilinear form obtained from f is $F(\mathbf{x}^{(1)}, \mathbf{x}^{(2)}) = ax^{(1)}x^{(2)} + \frac{b}{2}(x^{(1)}y^{(2)} + x^{(2)}y^{(1)}) + cy^{(1)}y^{(2)}$.

It is sometimes convenient to use lower subscripts for the polarized polynomial.

Example 3.3 Let $p(x) = x^k$. Then $M_p(x_1, \ldots, x_k) = \frac{1}{k!} \sum_{\sigma \in S_k} x_{\sigma(1)} \cdots x_{\sigma(k)}$.

Example 3.4 Let $q(x, y) = x^2 y$. Then

$$M_q(x_1, y_1, x_2, y_2, x_3, y_3)$$
$$= \frac{1}{6}(x_1 x_2 y_3 + x_2 x_1 y_3 + x_1 x_3 y_2 + x_3 x_1 y_2 + x_2 x_3 y_1 + x_3 x_2 y_1).$$

The reader is reminded that $Tr(B)$ is used to denote the trace of a matrix B.

Example 3.5 The Cayley-Hamilton polynomial: For a 2×2 matrix B this is

$$f_{CH}(A) = x^2 - Tr(B)x + \frac{1}{2}(Tr(B)^2 - Tr(B^2))I.$$

Its multilinearization $F_{CH}(x, y)$ is given by

$$2F_{CH}(B, C) = (xy + yx) - (Tr(B)y + Tr(C)x) + (Tr(B)Tr(C) - Tr(BC))I.$$

For a 3×3 matrix the scaled multilinearization $6F_{CH}(x, y, z)$ is

$$(xyz + xzy + yxz + yzx + zxy + zyx)$$
$$-(Tr(B)(yz + zy) + (Tr(C)(xy + yx) + (Tr(C)(xz + zx))$$
$$+((Tr(B)Tr(C) - Tr(BC))z + (Tr(B)Tr(D) - Tr(BD))y$$
$$+Tr(C)Tr(D) - Tr(CD))x)$$
$$-((Tr(B)Tr(C)Tr(D) - Tr(BC)Tr(D) - Tr(BD)Tr(C) - Tr(CD)Tr(B)$$
$$+Tr(BCD) + Tr(BDC))I.$$

If A, B, C are arbitrary 3×3 matrices then they satisfy the identity

$$F_{CH}(A, B, C) = 0.$$

If χ is a character of a group G and A, B, C are corresponding representing matrices for elements $g, h, k \in G$ the coefficient of I in the above is $\chi^{(3)}(g, h, k)$.

Now consider a generic element $\underline{x}$ of A,

$$\underline{x} = x_1 u_1 + \ldots + x_n u_n. \tag{3.2.15}$$

This element lies in $A_{K(x_1,\ldots,x_n)}$, i.e. the coefficients are now regarded as lying in the field of rational functions $K(x_1, \ldots, x_n)$. The set $\{e, \underline{x}, \underline{x}^2, \ldots, \underline{x}^n\}$ must be linearly dependent since A is n-dimensional over $K(x_1, \ldots, x_n)$ and thus $\underline{x}$ must satisfy a polynomial of minimal degree m (the "reduced characteristic polynomial" $\varphi(\lambda; \underline{x})$):

$$\varphi(\lambda; \underline{x}) = \lambda^m - s_1(\underline{x})\lambda^{m-1} + \ldots + (-1)^m s_m(\underline{x}) \cdot 1$$

where the coefficient $s_i(\underline{x}) \in K(x_1, \ldots, x_n)$ is a homogeneous polynomial of degree i in $x_1, \ldots, x_n$ and where now 1 denotes the identity element of $K(x_1, \ldots, x_n)$. The abbreviations S, W, U, N will be used as follows

$$S = s_1, \quad W = s_2, \quad U = s_3, \quad N = s_m.$$

S is called the reduced trace and N is called the **reduced norm** (**generic norm** in the works of Jacobson, **Hauptnorm** in the works of A. Bergmann).

Define the matrix $\Gamma(\underline{x}) \in M_n(K(x_1, \ldots, x_n))$ by

$$\Gamma(\underline{x}) = \sum R(u_i) x_i.$$

If $\underline{y} = y_1 u_1 + \ldots + y_n u_n$ is another generic element associated to the independent variables $\{y_1, \ldots, y_n\}$ and

$$\Gamma(\underline{y}) = \sum R(u_i) y_i \tag{3.2.16}$$

is defined analogously, it follows by direct calculation that $\Gamma(\underline{x}.\underline{y}) = \Gamma(\underline{x})\Gamma(\underline{y})$, i.e. Γ encodes the regular representation of A.

Since it follows automatically that $\det(\Gamma(\underline{x})) \det(\Gamma(\underline{y})) = \det(\Gamma(\underline{x}.\underline{y}))$, the n-ary form

$$F_{reg}(\underline{x}) = \det(\Gamma(\underline{x}))$$

is a multiplicative norm form. Now $\Gamma(1) = I_n$ and the characteristic polynomial $\det(\lambda I - \Gamma(\underline{x}))$ of $\Gamma(\underline{x})$ may be expressed as

$$\mu(\lambda; \underline{x}) = \det(\lambda I - \Gamma(\underline{x})) = \det(\Gamma(\lambda - \underline{x})). \tag{3.2.17}$$

It follows by the Cayley-Hamilton theorem that substituting $\Gamma(\underline{x})$ for λ in $\mu(\lambda; \underline{x})$ produces the zero matrix. Since φ is minimal $\varphi(\lambda; \underline{x})$ must divide $\mu(\lambda; \underline{x})$.

Suppose that $F_{reg}(\underline{x}) = \prod_i N_i(\underline{x})$ is the decomposition into irreducible factors $N_i(\underline{x})$ over K, normed by assuming $N_i(1) = 1$. Then

$$F_{reg}(\underline{x}.\underline{y}) = \prod_i N_i(\underline{x}.\underline{y}).$$

On rewriting

$$F_{reg}(\underline{x}.\underline{y}) = F_{reg}(\underline{x}) \cdot F_{reg}(\underline{y})$$

the following holds

$$F_{reg}(\underline{x}) \cdot F_{reg}(\underline{y}) = \prod_i N_i(\underline{x}) \cdot N_i(\underline{y}) = \prod_i N_i(\underline{x}.\underline{y}).$$

It follows that

$$N_i(\underline{x}.\underline{y}) = c \prod^{'} N_j(\underline{x}) \prod^{''} N_k(\underline{y}) \ \forall x, y, \tag{3.2.18}$$

where $c \in K$. On inserting $\underline{x} = 1$, $\underline{y} = 1$ into (3.2.18) it follows that c must be equal to 1. Further if $x = 1$ is again inserted into (3.2.18) it follows that $N_i(\underline{y}) = \prod_k N_k(\underline{y})$ and similarly $N_i(\underline{x}) = \prod_j N_j(\underline{x})$. Using the irreducibility of the N_i it follows that

$$N_i(\underline{x}.\underline{y}) = N_i(\underline{x}) \cdot N_i(\underline{y}) \tag{3.2.19}$$

i.e N_i is also a multiplicative norm-type form.

In the case of the reduced norm N, if A is semisimple over K it is shown in [15] that each irreducible factor N_i of M is contained in N exactly once.

The original definition by Hoehnke of a norm-type form is as follows.

Definition 3.1 Let A be an algebra over K with $F_{reg}(\underline{x}) = \prod_i N_i(\underline{x})$ the decomposition into irreducible factors. Any product of factors

$$F(\underline{x}) = \prod N_i(\underline{x}) \tag{3.2.20}$$

is a norm-type form.

Now suppose that F is an arbitrary form on A, and $\mathcal{U} = \{u_1, u_2, \ldots, u_n\}$ is a basis for A over K. For

$$\underline{x} = x_1 u_1 + \ldots + x_n u_n$$

a generic element of A define the **characteristic polynomial** $\varphi^F(\lambda; \underline{x})$ with respect to $\mathcal{U}$ to be the polynomial in λ, $x_1, \ldots, x_n$ given by $F(\lambda - \underline{x})$. If when the substitution $\underline{x}$ for λ in $\phi^F(\lambda; \underline{x})$ is made 0 is obtained, it follows from the minimality of N that N divides F.

Lemma 3.1 *Every norm-type form is multiplicative. If the field K has infinitely many elements, then every multiplicative form is of norm-type.*

Proof The first statement follows from (3.2.19). Conversely suppose that

$$\mu(\lambda; \underline{x}) = x^n - s_1(\underline{x})x^{n-1} + \ldots + (-1)^n s_n(\underline{x}))$$

where $\mu(\lambda; \underline{x})$ is as defined in (3.2.17). Consider the adjoint element $\bar{x}$ which is defined as

$$\bar{x} = (-1)^{n+1}(\underline{x}^{n-1} - s_1(\underline{x})\underline{x}^{n-2} + \ldots + (-1)^{n-1}s_{n-1}(\underline{x})).$$

It has the property that $\underline{x}\bar{x} = s_n(x) = F_{reg}(x)$ and if $F(x)$ is any multiplicative form of degree k then

$$F(x)F(\bar{x}) = F(x\bar{x}) = F(F_{reg}(x).1) = (F_{reg}(x))^k F(1) = (F_{reg}(x))^k.$$

$\square$

It follows that $F(x)$ is a product of irreducible factors of $F_{reg}(x)$.

3.3 Results on Multiplicative Norm Forms on Algebras

3.3.1 The Cubic Form and the Jordan Algebra

Let A be a finite dimensional algebra over K with basis $\mathcal{U} = \{u_1, u_2, \ldots, u_n\}$ and structure constants $\{a_{ij}^k\}$. If there is a trace-like linear map $S : A \to \mathbb{C}$ such that the bilinear form q defined by $q(x, y) = S(xy)$ is non-degenerate then A is Frobenius (this is equivalent to the definition given above—see [69]).

As above let A^+ be the Jordan algebra associated with A with product

$$x * y = \frac{1}{2}(xy + yx).$$

Then the structure constants of A^+ with respect to $\mathcal{U}$ are

$$a_{(ij)}^k = \frac{1}{2}(a_{ij}^k + a_{ji}^k). \tag{3.3.1}$$

The statement and proof of following theorem is due to Buchstaber and Rees [37] (although it is implied by work of Hoehnke).

Theorem 3.5 *Let A be a Frobenius algebra with a non-degenerate bilinear form arising from the trace-like map S. Then the structure constants $a^k_{(ij)}$ for A^+ are determined by $\delta^1(S)$, $\delta^2(S)$ and $\delta^3(S)$.*

Proof From the definition,

$$\delta^2(S)(u_i, u_j) = S(u_i)S(u_j) - S(u_iu_j)$$

Set $d_{ij} = S(u_iu_j)$. The non-degeneracy of the bilinear form arising from S implies that matrix $D = \{d_{ij}\}^n_{i,j=1}$ is non-singular. Then by the linearity of S

$$S(u_iu_j) = \sum_r a^r_{ij}S(u_r). \tag{3.3.2}$$

and

$$S(u_iu_ju_k) = a^r_{ij}a^s_{rk}S(u_s) \tag{3.3.3}$$

Hence

$$d_{ij} = S(u_iu_j) = \sum_r a^r_{ij}S(u_r) = S(u_i)S(u_j) - \delta^2(S)(u_i, u_j)$$

and D is determined by $\delta^1(S)$ and $\delta^2(S)$. The formula for $\delta^3(S)$ is

$$\delta^3(S)(u_i, u_j, u_k) = S(u_i)S(u_j)S(u_k) - S(u_i)S(u_ju_k) - S(u_j)S(u_iu_k)$$
$$- S(u_k)S(u_iu_j) + S(u_iu_ju_k) + S(u_ju_iu_k).$$

on using (3.3.2) and (3.3.3) it follows that

$$\delta^3(S)(u_i, u_j, u_k) = S(u_i)S(u_j)S(u_k) - S(u_i)\sum_r a^r_{jk}S(u_r)$$
$$- S(u_j)\sum_r a^r_{ik}S(u_r) - S(u_k)\sum_r a^r_{ij}S(u_r)$$
$$+ \sum_{r,s} a^r_{ij}a^s_{rk}S(u_s) + \sum_{r,s} a^r_{ji}a^s_{rk}S(u_s)$$

and using (3.3.1) this becomes

$$\delta^3(S)(u_i, u_j, u_k) = S(u_i)S(u_j)S(u_k) - S(u_i)\sum_r a^r_{jk}S(u_r)$$

$$- S(u_j)\sum_r a^r_{ik}S(u_r) - S(u_k)\sum_r a^r_{ij}S(u_r)$$

$$+ \sum_{r,s} a^r_{(ij)}a^s_{rk}S(u_s).$$

On rearranging, the following set of linear equations is obtained

$$\sum_r d_{rk}a^r_{(ij)} = d_{jk}S(u_i) + d_{ik}S(u_j) + d_{ij}S(u_k) \tag{3.3.4}$$

$$- S(u_i)S(u_j)S(u_k) + \delta^3(S)(u_i, u_j, u_k).$$

Equation (3.3.4) may be regarded as a system of linear equations for fixed i and j in the $a^r_{(ij)}$, $r = 1, \ldots, n$, with coefficient matrix D. Since, by the assumption on S, D is non-singular it follows that the system has a unique solution. This proves the theorem. $\qquad\square$

Theorem 3.5 can be used to complete the proof of Theorem 1.9, Chap. 1. For if G is a group and S is the trace of any representation which contains at least one copy of each irreducible representation (i.e. $S = \sum_{i=1}^r n_i\chi_i$ and $n_i > 0$ for all i where $Irr(G) = \{\chi_i\}_{i=1}^r$) then S is non-degenerate. This is proved in the following lemma.

Lemma 3.2 Let $S = \sum_{i=1}^r n_i\chi_i$ with $n_i > 0$ for all i. Let $G = \{g_1, g_2, \ldots, g_n\}$. Then if B is the matrix $\{S(g_ig_j)\}_{i,j=1}^n$ then $\det(B) \neq 0$.

Proof If $B' = \{S(g_ig_j^{-1})\}_{i,j=1}^n$ then $\det(B') = \pm\det(B)$ and it is sufficient to show that $\det(B') \neq 0$. Now B' may be obtained from X_G by replacing $x_{g_ig_j^{-1}}$ by $S(g_ig_j^{-1})$. This may be done by first replacing $x_{g_ig_j^{-1}}$ by $x_{S(g_ig_j^{-1})}$ and then replacing $x_{S(g_ig_j^{-1})}$ by $S(g_ig_j^{-1})$. Since S is constant on conjugacy classes, for a conjugacy class C_i the notation $S(C_i)$ is well-defined as the value of $S(g)$ for any $g \in C_i$. The first replacement produces a matrix X'_{red} in which the element x_{C_i} is replaced by $x_{S(C_i)}$. Thus the determinant of X'_{red} is the product of factors

$$\phi_i = (\sum_{j=1}^n \chi_i(g_j)x_{g_j})^{\chi_i(e)}.$$

The second replacement takes ϕ_i into

$$(\sum_{j=1}^{n} \chi_i(g_j)S(g_j))^{\chi_i(e)} = (\langle \overline{\chi}_i, S \rangle)^{\chi_i(e)}$$

which is non-zero by assumption. Therefore $\det(B')$ and hence $\det(B)$ are non-zero.

□

Hence, using the group elements for the basis of $\mathbb{C}G$, since the $a_{(ij)}^k$ are determined by $\delta^1(S)$, $\delta^2(S)$ and $\delta^3(S)$, this is equivalent to the knowledge of $g_i g_j + g_j g_i$ for all i, j which then is equivalent to the knowledge of the set $\{gh, hg\}$ for all $g, h \in G$, which by Lemma 1.5, Chap. 1 determines G.

3.3.2 Arbitrary Multiplicative Forms on Algebras

Theorem 3.5 may also be deduced as a consequence of a more general result of Hoehnke, which is explained in this section. Let $F(\underline{x})$ be an arbitrary multiplicative form of degree m on an algebra A of dimension n, basis $\{u_i\}_{i=1}^{n}$ with characteristic polynomial as above:

$$\phi^F(\lambda; \underline{x}) = F(\lambda - \underline{x}) = \lambda^m - s_1(\underline{x})\lambda^{m-1} + \ldots + (-1)^m s_m(\underline{x}).$$

Here each $s_i(\underline{x})$ is a homogeneous polynomial in $\{x_1, \ldots, x_n\}$ and the s_i are uniquely determined by the coefficients of the form $F(\underline{x})$ and the components e_i of $1 = \sum_i e_i u_i$. This follows since

$$F(\lambda.1 - \underline{x}) = F(\sum_i (\lambda e_i - x_i)u_i).$$

Let $\lambda_1, \ldots, \lambda_m$ be the zeros of $\varphi^F(\lambda; \underline{x})$ regarded as a polynomial in λ in a suitable splitting field, i.e.

$$\varphi^F(\lambda; x) = (\lambda - \lambda_1) \ldots (\lambda - \lambda_m) \tag{3.3.5}$$

where the λ_i are functions of $\{x_1, x_2, \ldots, x_n\}$. Further, for any polynomial $f(\mu)$ of degree i with coefficients in $K_{(x_1, x_2, \ldots, x_n)}$ suppose that over a splitting field

$$f(\mu) = a_0 \mu^i + a_1 \mu^{i-1} + \ldots + a_i = a_0(\mu - \mu_1) \ldots (\mu - \mu_i).$$

Then

$$f(\underline{x}) = a_0 \underline{x}^i + a_1 \underline{x}^{i-1} + \ldots + a_i = a_0(\underline{x} - \mu_1) \ldots (\underline{x} - \mu_i)$$

and

$$
\begin{aligned}
F(f(\underline{x})) &= (-1)^{ir} a_0^r F(\mu_1 - \underline{x}) \ldots F(\mu_i - \underline{x}) \\
&= (-1)^{ir} a_0^r (\mu_1 - \lambda_1) \ldots (\mu_1 - \lambda_m) \\
&\quad \ldots (\mu_i - \lambda_1) \ldots (\mu_i - \lambda_m) \\
&= f(\lambda_1) \ldots f(\lambda_m).
\end{aligned}
\tag{3.3.6}
$$

On replacing $f(\mu)$ by $\lambda - f(\mu)$ the following is obtained

$$
F(\lambda - f(x)) = (\lambda - f(\lambda_1)) \cdots (\lambda - f(\lambda_m))
\tag{3.3.7}
$$

and if now $f(\mu)$ is taken to be μ^i

$$
\begin{aligned}
F(\lambda - \underline{x}^i) &= \lambda^m - s_1(\underline{x}^i)\lambda^{m-1} + \ldots + (-1)^m s_m(\underline{x}^i) \\
&= (\lambda - \lambda_1^i) \cdots (\lambda - \lambda_m^i).
\end{aligned}
$$

Note the analogy between this result and Frobenius' well-known result on the eigenvalues of a polynomial of a matrix.

In particular

$$
s_1(\underline{x}^i) = \lambda_1^i + \ldots + \lambda_m^i, \quad i = 1, 2, \ldots
$$

The abbreviations introduced above for the reduced norm form N will be extended to an arbitrary form F:

$$
s_1 = S, \quad s_2 = W, \quad s_3 = U
$$

so that

$$
S(\underline{x}^i) = \sum_{j=1}^{m} \lambda_j^i.
$$

In Chap. 1, Sect. 1.8 the k-characters are introduced and it is shown that a factor of the group determinant can be constructed from the corresponding k-characters (Theorem 1.7). The following generalizes this work to the setting of an arbitrary norm form.

Newton's formula (1.8.5, Chap. 1) translates to

$$
S(\underline{x}^i) - s_1(\underline{x})S(\underline{x}^{i-1}) + \ldots + (-1)^i i s_i(\underline{x}) = 0.
\tag{3.3.8}
$$

Here the $s_i(x)$ are the elementary symmetric functions in the λ_j: $s_1 = \sum_j \lambda_j$, $s_2 = \sum_{j \neq k} \lambda_j \lambda_k, \ldots$ and it is understood that $s_i = 0$ for $i > m$.

If Char$(K) \nmid m$ an explicit formula for s_i in terms of the $S(\underline{x}^j)$ is as follows

$$s_i(\underline{x}) = \sum_{i_1 + 2i_2 + \ldots + mi_m = i} (-1)^{i_1 + i_2 + i_3 + \ldots} \frac{S(\underline{x})^{i_1} S(\underline{x}^2)^{i_2} \ldots S(\underline{x}^m)^{i_m}}{1^{i_1} 2^{i_2} 3^{i_3} \ldots m^{i_m} i_1! \ldots i_m!}. \quad (3.3.9)$$

Conversely without restriction on Char K, Waring's formula gives explicitly

$$S(\underline{x}^i) = i \cdot \sum_{i_1 + 2i_2 + \ldots + mi_m = i} (-1)^{i_2 + i_4 + i_6 + \ldots} \frac{(i_1 + \ldots + i_m - 1)!}{i_1! \cdots i_m!} s_1^{i_1}(\underline{x}) \ldots s_m^{i_m}(\underline{x}). \quad (3.3.10)$$

Particular consequences are

$$\left. \begin{array}{l} S(\underline{x}^2) = s_1^2 - 2s_2, \\ S(\underline{x}^3) = s_1^3 - 3s_1 s_2 + 3s_3, \\ S(\underline{x}^4) = s_1^4 - 4s_1^2 s_2 + 2s_2^2 + 4s_1 s_3 - 4s_4, \\ S(\underline{x}^5) = s_1^5 - 5s_1^3 s_2 + 5s_1 s_2^2 + 5s_1^2 s_3 - 5s_2 s_3 - 5s_1 s_4 + 5s_5. \end{array} \right\} \quad (3.3.11)$$

Inserting $\underline{x} = 1$ it follows that $F(\lambda - 1) = (\lambda - 1)^m$ and thus

$$s_k(1) = \binom{m}{k} \quad (k = 1, \ldots, m). \quad (3.3.12)$$

Setting $S(\underline{x}) = \sum_i c_i x_i$ $(c_i \in K)$ it follows that

$$S(\underline{x}.\underline{y}) = S(\sum x_i u_i \sum y_j u_j) = S(\sum_{i,j} x_i y_j u_i u_j)$$

$$= S \sum_{i,j,k} x_i y_j a_{ij}^k u_k = \sum_{i,j,k} c_k x_i y_j a_{ij}^k.$$

Now let $d_{ij} = \sum_k c_k a_{ij}^k$. Then

$$S(xy) = \sum_{i,j} d_{ij} x_i y_j \quad (3.3.13)$$

The matrix $\{d_{jk}\} = \{S(u_i u_j)\}$ has appeared above in the special case of a group algebra. It is the **discriminant matrix** of the form $F(x)$. Its determinant D_F is the **discriminant** of F. F is said to be **nondegenerate** if $D_F \neq 0$. Note that if N is the reduced norm of A, then D_N is called the discriminant of the algebra A, and A is separable precisely when $D_N \neq 0$ (see [81]).

Let $\Gamma(\underline{x})$ be the matrix of the regular representation of A. If $\underline{y}$ is a second generic element, then

$$\Gamma(\underline{x}.\underline{y}) = \Gamma(\underline{y}^{-1})\Gamma(\underline{y}.\underline{x})\Gamma(\underline{y})$$

and hence

$$|\Gamma(\lambda - \underline{x}.\underline{y})| = |\lambda I - \Gamma(\underline{x}\underline{y})| = |\lambda I - \Gamma(\underline{y}^{-1})\Gamma(\underline{y}.\underline{x})\Gamma(\underline{y})| = |\lambda I - \Gamma(\underline{y}.\underline{x})|.$$

This implies that $s_i(\underline{x}.\underline{y}) = s_i(\underline{y}.\underline{x})$ for all i. In particular $S(\underline{x}.\underline{y})$ is a symmetric bilinear form in the variables $\{x_i\}$, $\{y_i\}$. Now $S(\underline{x}.1) = S(\underline{x})$ and $\overline{1}$ may be expressed as $\sum_k e_k u_k$ and thus

$$\sum_k d_{jk} e_k = c_j. \tag{3.3.14}$$

Writing the inverse matrix to $\{d_{jk}\}$ as $\{d^{il}\}$, from (3.3.14)

$$\sum_j d^{jk} c_k = e_k.$$

Let

$$s_k(\underline{x}) = \sum_{i_1,i_2,\ldots,i_k} c_{i_1,i_2,\ldots,i_k} x_{i_1} x_{i_2} \cdots x_{i_k} \quad (k = 1,\ldots,m). \tag{3.3.15}$$

Then, multilinearizing (3.3.15) it follows that

$$s_k(\underline{x}^{(1)},\ldots,\underline{x}^{(k)}) = \sum_{i_1,i_2,\ldots,i_k} c_{(i_1,i_2,\ldots,i_k)} x_{i_1}^{(1)} x_{i_2}^{(2)} \cdots x_{i_k}^{(k)}$$

where

$$c_{(i_1,i_2,\ldots,i_k)} = \sum_{\sigma \in S_k} c_{i_{\sigma(1)},i_{\sigma(2)},\ldots,i_{\sigma(k)}}.$$

Since

$$S(\underline{x}.\underline{y}) = S(\underline{x})S(\underline{y}) - s_2(\underline{x},\underline{y})$$

it follows that

$$W(x,y) = s_2(x,y) = \sum_{i,j} c_{i,j}\, x_i y_j. \tag{3.3.16}$$

Hence

$$d_{jk} = c_i c_k - c_{(jk)}$$

From (3.3.8) the expression

$$S(x^2) = (S(x))^2 - s_2(x) \qquad (3.3.17)$$

is obtained, together with its polarized version

$$S(xy) = S(x)S(y) - W(x, y). \qquad (3.3.18)$$

It follows that

$$d_{jk} = c_j c_k - c_{(jk)}. \qquad (3.3.19)$$

Theorem 3.6 *Suppose that A is an algebra over K with basis $\mathcal{U} = \{u_1, u_2, \ldots, u_n\}$.
Let F be a multiplicative form with characteristic polynomial $\phi^F(\lambda; \underline{x})$ defined for
a generic element $\underline{x} = x_1 u_1 + \ldots + x_n u_n$. Then the polynomials $s_r(\underline{x})$ may be
expressed as*

$$s_r(x_1, \ldots, x_n) = \psi_\chi = \sum \frac{1}{r!} \delta^{(r)} S(u_{i_1}, u_{i_2}, \ldots, u_{i_r}) x_{i_1} x_{i_2} \ldots x_{i_r} \qquad (3.3.20)$$

where the summation runs over all r-tuples of elements of $\mathcal{U}$.

Proof The formula (3.3.9) for $s_i(\underline{x})$ is

$$s_i(\underline{x}) = \sum_{i_1 + 2i_2 + \ldots + mi_m = i} (-1)^{i_1 + i_2 + i_3 + \ldots} \frac{S(\underline{x})^{i_1} S(\underline{x}^2)^{i_2} \ldots S(\underline{x}^m)^{i_m}}{1^{i_1} 2^{i_2} 3^{i_3} \ldots m^{i_m} i_1! \ldots i_m!}.$$

Consider the coefficient of an arbitrary product

$$x_{j_1} x_{j_2} \ldots x_{j_i}$$

in the above formula, where repetitions are allowed. There is a bijection between
these coefficients and the elements of the symmetric group $\mathcal{S}_i$, given specifically as
follows. If $\sigma \in \mathcal{S}_i$ is expressed as a product of disjoint cycles by

$$\sigma = \sigma_1 \sigma_2 \ldots \sigma_r$$

where $\sigma_j = (\alpha_1, \alpha_2, \ldots, \alpha_{t_j})$ so that $t_1 + t_1 + \ldots + t_r = i$, there corresponds the
term

$$S_\sigma(u_{j_1}, u_{j_2}, \ldots, u_{j_i}) x_{j_1} x_{j_2} \ldots x_{j_i} \qquad (3.3.21)$$

of (3.3.9). It remains to show that the coefficient of $x_{j_1} x_{j_2} \ldots x_{j_i}$ in (3.3.20)

$$\sum \frac{1}{i!} \delta^{(r)} S(u_{j_1}, u_{j_2}, \cdots, u_{j_r})$$

where the sum is over all orderings of the i-tuple $(j_1, j_2, \ldots, j_i)$ is exactly the same as in (3.3.21). This follows from the formula for the number of permutations of type $1^{i_1} 2^{i_2} \ldots m^{i_m}$ in S_i, where $i_1 + 2i_2 + \ldots + m i_m = i$ given as

$$\frac{i!}{1^{i_1} 2^{i_2} 3^{i_3} \ldots m^{i_m} i_1! \cdots i_m!}.$$

Since $S(\underline{x} \cdot 1) = S(\underline{x})$, and if $1 = \sum_k e_i u_i$ it follows that, using (3.3.13)

$$\sum_k d_{jk} e_k = c_j. \tag{3.3.22}$$

Now assume that $D_F \neq 0$. On writing $\{d^{il}\} = \{d_{jk}\}^{-1}$, (3.3.22) implies that

$$\sum_j c_j d^{jk} = e_k. \tag{3.3.23}$$

Suppose that

$$s_k(x) = \sum_{i_1, \ldots, i_k} c_{(i_1 \ldots i_k)} x_{i_1} \ldots x_{i_k} \quad (k = 1, \ldots, m). \tag{3.3.24}$$

Here the $c_{(i_1 \ldots i_k)}$ are symmetric, i.e. $c_{(i_1 \ldots i_k)} = c_{(\sigma(i_1) \ldots \sigma(i_k))}$ where σ is an arbitrary element of S_k.   □

Then consider the polarization obtained by introducing the generic elements $x^{(1)}, x^{(2)}, \ldots, x^{(k)}$ of A and then

$$s_k(x^{(1)}, \ldots, x^{(k)}) = \sum_{i_1, \ldots, i_k} c_{(i_1, \ldots, i_k)} x_{i_1}^{(1)} \ldots x_{i_k}^{(k)} \tag{3.3.25}$$

where now

$$c_{(i_1, \ldots, i_k)} = \sum_{\sigma \in \mathfrak{S}_k} c_{\sigma(i_1) \ldots \sigma(i_k)}$$

where $\mathcal{S}_k$ is the symmetric group on $\{1, \ldots, k\}$. Thus $s_k(x^{(1)}, \ldots, x^{(k)})$ is a symmetric k-fold form.

As above $a_{(ij)}^k$ will be used to denote the structure constants of A^+

$$a_{(ij)}^k = \frac{1}{2}(a_{ij}^k + a_{ji}^k).$$ (3.3.26)

Let $\underline{x}$, $\underline{y}$, $\underline{z}$ be generic elements and consider the product

$$\underline{xyz} = \underline{x} \sum a_{jk}^l y_j z_k u_l$$ (3.3.27)

$$= \sum a_{il}^m a_{jk}^l x_i y_j z_k u_m$$ (3.3.28)

Also

$$\underline{xzy} = \sum a_{il}^m a_{kj}^l x_i y_j z_k u_m$$ (3.3.29)

and

$$\underline{xyz} + \underline{xzy} = \sum a_{il}^m a_{(kj)}^l x_i y_j z_k u_m$$ (3.3.30)

and thus

$$S(\underline{xyz}) + S(\underline{xzy}) = \sum c_m a_{il}^m a_{(kj)}^l x_i y_j z_k.$$ (3.3.31)

It follows that

$$2S(x^3) = \sum c_m a_{il}^m a_{(kj)}^l x_i x_j x_k$$ (3.3.32)

and using $d_{ij} = \sum_i c_k a_{ij}^k$ this may be expressed as

$$2S(\underline{x}^3) = \sum_{l,i,j,k} d_{il} a_{(kj)}^l x_i x_j x_k$$ (3.3.33)

which is a cubic form with symmetric coefficients

$$\sum_l d_{il} a_{(kj)}^l$$ (3.3.34)

The identity below is again a consequence of (3.3.8) using polarization

$$U(x, y, z) = S(x)S(y)S(z) - S(x)S(yz) - S(y)S(xz)$$
$$- S(z)S(xy) + S(xyz) + S(xzy).$$ (3.3.35)

and hence

$$S(xyz) + S(xzy) = U(x, y, z) + S(x)S(yz) + S(y)S(xz)$$
$$+ S(z)S(xy) - S(x)S(y)S(z), \qquad (3.3.36)$$

which leads to (on writing $c_{(ijk)}$ as c_{ijk})

$$\sum_l d_{il} a^l_{(jk)} = c_{hjk} + c_h d_{jk} + c_j d_{hk} + c_k d_{hj} - c_h c_j c_k, \qquad (3.3.37)$$

and thus

$$a^l_{(jk)} = c_j \delta_{lk} + c_k \delta_{lj} - e_l c_{jk} + \sum d^{li} c_{ijk} \qquad (3.3.38)$$

which is equivalent to the Buchstaber-Rees formula above. After an obvious change of indices

$$a^i_{(jk)} = c_j \delta_{ik} + c_k \delta_{ij} - e_i c_{jk} + \sum d^{ih} c_{hjk}. \qquad (3.3.39)$$

There is the following expression for $\underline{x}^i$ in terms of the structure constants:

$$2^{i-1} \underline{x}^i = \sum a^{k_1}_{(j_1 k_2)} a^{k_2}_{(j_2 k_3)} \cdots a^{k_{i-1}}_{(j_{i-1}, j_i)} x_{j_1} \cdots x_{j_i}. \qquad (3.3.40)$$

and therefore

$$2^{i-1} S(\underline{x}^i) = 2 \sum d_{j_1 k_2} a^{k_2}_{(j_2 k_3)} \cdots a^{k_{i-1}}_{(j_{i-1}, j_i)} x_{j_1} \cdots x_{j_i} \qquad (3.3.41)$$

which means that the coefficients of $S(\underline{x}^i)$ can be expressed in terms of the coefficients c_i, c_{jk} and c_{hjk}. But from Waring's formula (3.3.10) it can be deduced that

$$S(\underline{x}^i) = \Phi_i(s_1, \ldots, s_{i-1}) + (-1)^{i-1} i s_i(\underline{x}) \qquad (3.3.42)$$

where the Φ_i are rational functions. Then the following recursion relation holds

$$(-1)^i i s_i(x) = \Phi_i(s_1(x), \ldots, s_{i-1}(x))$$
$$- 2^{2-i} \sum d_{j_1 k_2} a^{k_2}_{(j_2 k_3)} \cdots a^{k_{i-2}}_{(j_{i-2}, k_{i-1})} a^{k_{i-1}}_{(j_{i-1}, j_i)} x_{j_1} \cdots x_{j_i}. \qquad (3.3.43)$$

Further from equations (3.3.23) and (3.3.19),

$$\sum_k c_{jk} e_k = \sum_k c_j c_k e_k - \sum_k d_{jk} e_k = (m-1) c_j,$$

hence

$$W(1, x) = s_2(1, x) = (m - 1)S(x). \tag{3.3.44}$$

On multiplication by e_h and summation over h equations (3.3.12), (3.3.23), (3.3.37) yield

$$2d_{jk} = -mc_j c_k + md_{jk} + 2c_j c_k + \sum_h e_h c_{hjk},$$

i. e.

$$\sum_h e_h c_{hjk} = (m - 2)c_{(jk)}$$

and

$$U(1, x, y) = s_3(1, x, y) = (m - 2)T(x, y). \tag{3.3.45}$$

The theorem below has now been proved.

Theorem 3.7 *Suppose that A is a finite-dimensional algebra over the field K. Let F be a non-degenerate multiplicative norm-type form of degree m where $Char\,K \nmid m!$ Then each of the forms $s_k(x)$ arising as the coefficients of $F(\lambda - x)$ can be calculated rationally from the coefficients c_{ijk} of the trilinear form $U(x, y, z) = s_3(x, y, z) = \sum c_{ijk} x_i y_j z_k$ associated to the cubic form $U = s_3(x)$.*

In particular, the theorem holds for the reduced norm of A.
Explicit formulae for s_4 and s_5 are

$$4s_4(\underline{x}) = (m - 2)s_2^2 - 2s_1 s_3 - 9u_4 \quad (Char(K) \nmid 4!), \tag{3.3.46}$$

$$4 \cdot 5s_5(\underline{x}) = -(2m - 1)(m - 2)s_1 s_2^2 + 6s_1^2 s_3 + 4(3m - 7)s_2 s_3$$
$$+ 5 \cdot 9s_1 u_4 + 4 \cdot 27 u_5 \quad (Char(K) \nmid 5!). \tag{3.3.47}$$

where

$$\left.\begin{array}{l} u_4(\underline{x}) = \sum_{i,k,q,r,l,p} c_{ikl} d^{lp} c_{pqr} x_i x_k x_q x_r, \\ u_5(\underline{x}) = \sum c_{ikl} d^{lp} c_{pqr} d^{rs} c_{stv} x_i x_k x_q x_t x_v, \end{array}\right\} \tag{3.3.48}$$

and there are similar expressions for the higher forms $u_i(\underline{x})$ $i > 5$, although their calculation becomes increasingly difficult. Note that here because of the condition on the characteristic the coefficients $c_{i_1 \dots i_k}$ may be assumed to be symmetric.

3.4   The Relationship to the Work of the School of A. Bergmann

Bergmann considers a form F, homogeneous of degree n on a finite dimensional vector space V with values in a field K. Then for all $z_1, z_2, \ldots, z_m$ in V,

$$F\left(\sum_{i=1}^{n} \lambda_i z_i\right) = \sum_{\mu_1+\mu_2+\ldots+\mu_m=m} \lambda_1^{\mu_1} \lambda_2^{\mu_2} \ldots \lambda_m^{\mu_m} F_{\mu_1,\mu_2,\ldots,\mu_m}(z_1, z_2, \ldots, z_m),$$

where the $F_{\mu_1,\mu_2,\ldots,\mu_m}$ are multihomogeneous polynomials of multidegree

$$(\mu_1, \mu_2, \ldots, \mu_m)$$

where each μ_i is a non-negative integer. This implies that associated to any form F of degree n there is a system of "multiforms" $(F_{\mu_1,\mu_2,\ldots,\mu_m})$. The function $F_{1,1,\ldots,1}$ is the complete polarization of F. It is a symmetric multilinear form. If $\operatorname{Char} K \nmid m!$ then F, and hence all the $F_{\mu_1,\mu_2,\ldots,\mu_m}$, can be reconstructed from $F_{1,1,\ldots,1}$ using

$$m! F(z) = F_{1,1,\ldots,1}(z, z, \ldots, z)$$

Thus, omitting μ_i from $(\mu_1, \mu_2, \ldots, \mu_m)$ whenever it is 0, n may be written as

$$n = \mu_{i_1} + \mu_{i_2} + \ldots + \mu_{i_r}$$

with $\mu_{i_1}, \mu_{i_2}, \ldots, \mu_{i_r}$ non-zero and then $F_{\mu_1,\mu_2,\ldots,\mu_m}$ will written as $F_{\mu_{i_1},\mu_{i_2},\ldots,\mu_{i_r}}$.
Now if $(\mu_1', \mu_2', \ldots, \mu_r') = (\mu_{\sigma(1)}, \mu_{\sigma(2)}, \ldots, \mu_{\sigma(r)})$ for $\sigma \in S_r$ is a permutation of $(\mu_1, \mu_2, \ldots, \mu_r)$ it follows that

$$F_{\mu_{i_1}',\mu_{i_2}',\ldots,\mu_{i_r}'}(z_{\sigma(1)}, \ldots, z_{\sigma(r)}) = F_{\mu_{i_1},\mu_{i_2},\ldots,\mu_{i_r}}(z_1, z_2, \ldots z_r)$$

and hence the functions are independent of the ordering of $(\mu_{i_1}, \mu_{i_2}, \ldots, \mu_{i_r})$. In [15] and [16] these functions are denoted by $\operatorname{red} \Phi^{(\mu_1,\mu_2,\ldots,\mu_r)}$ and are called **reduced functions**. In particular $\operatorname{red} \Phi^{(1,1,\ldots,1)}$ is the complete polarization. There are, of course, other ways of constructing polarizations (see 3.2.2 above).

Satz 1a and Satz 1b in [15] show the equivalence of the multiplicative property for a form F on an algebra A over K

$$F(\underline{xy}) = F(\underline{x}) F(\underline{y})$$

for generic $\underline{x}$, $\underline{y}$ and the property (3.2.13) above, together with an equivalent definition in terms of the reduced functions. Further comments are given in the survey [17].

Now if F is a multiplicative form on an algebra A, with values in K, the forms s_k given above are obtained from

$$F(\lambda 1 - \underline{x}) = \lambda^m - s_1(\underline{x})\lambda^{m-1} \pm \ldots + (-1)^m s_m(\underline{x}).$$

This is the Cayley-Hamilton polynomial for F. Then in terms of the reduced functions,

$$s_k(\underline{x}) = \text{red}\Phi^{(k,m-k)}(\underline{x}, \ldots, \underline{x}, 1, \ldots, 1) \ (\underline{x} \text{ occurs } k \text{ times, 1 appears } m-k \text{ times})$$

for $k = 1, \ldots, m - 1$ and

$$s_m(\underline{x}) = F(\underline{x}).$$

In the case of a group algebra of a finite group G, taking the group elements as the basis, then if χ is a character of G of degree m with associated group determinant factor θ then

$$\chi(g) = \text{red}\Phi_\theta^{(1,m-1)}(g, 1, 1, \ldots, 1)$$

and

$$\chi^k(g_1, \ldots, g_k) = \text{red}\Phi_\theta^{(1,\ldots,1,m-k)}(g_1, \ldots, g_k, 1, \ldots, 1)$$

with

$$\chi^m(g_1, \ldots, g_m) = \text{red}\Phi_\theta^{(1,\ldots,1)}(g_1, \ldots, g_m)$$

the complete polarization of θ. Now

$$s_k(\underline{x}) = \text{red}\Phi_\theta^{(k,m-k)}(\underline{x}, \ldots, \underline{x}, 1, \ldots, 1)$$

for $k = 1, \ldots m$. Thus the complete polarization of $s_k(\underline{x})$ is

$$\text{red}\Phi_\theta^{(1,\ldots,1)}(x_1, \ldots, x_r, 1, \ldots, 1),$$

i.e. $\chi^k(g_1, \ldots, g_k)$ is the directly related to the complete polarization of s_k. Thus the k-characters are associated to special reduced functions.

It may be of interest to examine other reduced functions in the context of the group algebra. There is also the question of whether there exists an extension of the theory to superalgebras.

It is remarkable that so much information about a group is given by the characters, i.e. $s_1(\underline{x})$, and that this was detected by Frobenius in such a short time. Although this book addresses to a certain extent the extra information in the $s_k(\underline{x})$ for arbitrary k, it is likely that there remains much to be explored in this direction.

3.5   The Connection with the "Average" Invariants of Roitman

In [244] Roitman described a complete system of invariants for a finite group G. This description is as follows. Let G be of order n, let m be a positive integer and let $A_I \subseteq \{1, \dots, m\}$. Define the element $e_G(m, A_I)$ of the ring

$$\mathbb{Z}(G^m) = \mathbb{Z}(G \times \dots \times G) \simeq \mathbb{Z}(G) \otimes \dots \otimes \mathbb{Z}(G)$$

by the equation

$$e_G(m, A_I) = \sum_{g \in G} \varepsilon_1(g) \otimes \varepsilon_2(g) \otimes \dots \otimes \varepsilon_m(g)$$

where $\varepsilon_i(g) = g$ if $i \in A_I$, e otherwise. For any element $\alpha = \sum_{g \in G} \alpha_i g_i$ of $K(G)$ define $tr(\alpha) = \alpha_1$. Thus if π is the regular character of G extended linearly to $K(G)$ then $\pi(\alpha) = n \cdot tr(\alpha)$. The statement of the theorem of Roitman is the following.

Theorem 3.8 *The positive integers*

$$tr((e_G(m, A_I) + \dots + e_G(m, A_{I_r})))^k$$

for m, r, k positive integers, such that

$$A_{I_j} \subseteq \{1, \dots, m\},\ 1 \le |A_{I_j}| \le 3 \ for\ 1 \le j \le r$$

and

$$A_{I_{j_1}} \cap A_{I_{j_2}} \cap A_{I_{j_3}} \cap A_{I_{j_4}} = \varnothing \ for\ any\ 1 \le j_1 \le j_2 \le j_3 \le j_4 \le r,$$

determine G up to isomorphism. Conversely, these integers are determined by the isomorphism class of G.

There is no obvious connection of these invariants to k-characters, but it is shown in [149] that they can be expressed in terms of 1-, 2- and 3-characters of G. In the case $Char\ K = 0$, which is the only case considered by Roitman, they may be expressed in terms of the form s_3. Furthermore, if in addition to the invariants in Theorem 3.8 it is assumed that n is known, then m may be set equal to $7n^3$ and k may be set equal to $6n^3$.

Chapter 4
S-Rings, Gelfand Pairs and Association Schemes

Abstract The construction by Frobenius of group characters described in Chap. 1 may be generalized to the case where a permutation group G acting on a finite set has the Gelfand pair property explained below. A general setting which encompasses and extends this is that of an association scheme. For any association scheme there is available a character theory, which in the case where the scheme arises from a group coincides with that of group characters. The development of the theory is described in this chapter. Firstly Schur investigated centralizer rings of permutation groups, then Wielandt defined S-rings over a group. The theory of association schemes provides a character theory even when a group is not present, and this can be applied to obtain a character theory for a loop or quasigroup. In particular a Frobenius reciprocity result was obtained for quasigroup characters, and it was realized that this is available for arbitrary association schemes. A further idea, that of fusion of characters of association schemes, leads to interesting results including a "magic rectangle" condition.

4.1 Introduction

The objects described in this chapter may be said to have evolved from the work of Frobenius [106], which is discussed in Chap. 1. The group character theory for a group G is contained in X_G^R, which was introduced there as the matrix obtained from the full group matrix X_G by setting x_g, for each $g \in G$, equal to x_{C_i} where C_i is the conjugacy class containing g. If for a fixed i the $(0, 1)$ matrix A_i is formed from X_G^R by setting

$$x_{C_i} = 1 \text{ and } x_{C_j} = 0 \text{ for } j \neq i,$$

the set of matrices $\{A_i\}$ generate a commutative algebra isomorphic to the class algebra, and the character table can be calculated by finding their eigenvalues and renormalizing (although this is not how Frobenius originally proceeded in that Frobenius showed that $\det(X_G^R)$ splits into linear factors which are in $1 : 1$

© Springer Nature Switzerland AG 2019
K. W. Johnson, *Group Matrices, Group Determinants*
and Representation Theory, Lecture Notes in Mathematics 2233,
https://doi.org/10.1007/978-3-030-28300-1_4

correspondence with the irreducible characters of G). Such sets of $(0, 1)$ matrices are present in many other situations, and a framework into which they fit is that of association schemes.

The theory of association schemes came about from the recognition that several superficially separate areas have the same basic results. These include parts of harmonic analysis, probability theory, experimental designs, graph theory, symmetric spaces and number theory, in addition to relevant aspects of permutation groups and group representations. It had gradually emerged that a common structural foundation could be developed independent of the particular contexts, and the results were often enriched by intuition from each area. When infinite objects are considered the approach via finite $(0, 1)$ matrices is hard to use, but other formulations are available.

A stimulus towards a unified treatment was the thesis of Delsarte [76], which introduced new methods, and the set of lecture notes [10] made a major contribution in this direction. References include [9, 32, 45, 47, 123] and [297]. There are several variations on the definition of an association scheme (see for example [42] p. 60), and other similar objects such as coherent configurations and table algebras have been investigated.

The description here follows the path taken in finite algebra. In Chap. 1 an example has already been given of how Frobenius produced a modified character table of $\mathcal{A}_4$, essentially by fusing classes using the action of $\mathcal{S}_4$ by conjugation. The first step towards more general structures was that of Schur in 1933 [255] and a description of his approach is given. The situation which he considered was a permutation group with a regular subgroup, the motivation being a character-free proof of a theorem of Burnside ([39], Theorem VIII p. 343) on the existence of primitive permutation groups with a regular cyclic subgroup of prime power order. Schur in [254] had earlier reproved a result of Burnside, using the action of a permutation group on a bilinear form by permuting the variables, and he used similar ideas in his later paper [255] in which he introduced the centralizer ring of a permutation group. In number theory and harmonic analysis the term Hecke algebra is used for the corresponding structure, the context often being generalized to infinite situations. Then Wielandt [293] provided a more abstract approach in terms of S-rings which are defined below. It was subsequently discovered that the original Burnside "proof" contains a gap, indicated in [223], and thus Schur's proof of the above theorem is the first.

An important concept appearing is that of a **Gelfand pair**. These were first introduced for infinite groups. For a finite group, the pair (G, H) where H is a subgroup of the group G is Gelfand precisely when the centralizer ring of the action of G on the cosets of H is commutative. For infinite groups more subtle considerations are involved and a fuller discussion of Gelfand pairs is deferred to Appendix A, where their connection to the theory of spherical functions on groups is discussed.

An account of the theory of association schemes and related objects is given below. As an initial step an attempt is made to compare the various definitions of an association scheme and related objects. Then the basic results are set out. Given any association scheme, there is a corresponding character theory, with row and column orthogonality relations which are analogous to those of group character tables. The $\mathcal{P}$ and $\mathcal{Q}$ matrices associated to an association scheme are shown to embody this orthogonality. There have been several attempts to obtain a duality which generalizes that between a commutative group and the set of functions on it, and duality relations are discussed for arbitrary schemes.

The application which gave rise to this book follows. If Q is a finite quasigroup a procedure generalizing that of Frobenius described in Chap. 1 produces a matrix analogous to the reduced group matrix X_G^R, which is essentially determined by the orbits of Q acting on $Q \times Q$ by left and right multiplication, and the set of incidence matrices A_i associated to these orbits gives rise to a (commutative) association scheme. This may be used to produce a character theory for a quasigroup Q which coincides with that of Frobenius in the case where Q is a group, and the procedure to do this is set out. Most notably there appears the result that to each quasigroup there is associated a permutation group whose centralizer ring is commutative, and thus any finite quasigroup gives rise to a Gelfand pair. This result does not seem to be widely known. The examination of the properties of quasigroup characters and methods to calculate them led to several new results on the general theory of association schemes, although in the initial papers the discussion was limited to those arising from quasigroups. Here their extension to arbitrary association schemes is given more explicitly. This includes induction, restriction and Frobenius reciprocity for characters of association schemes. A version in the quasigroup setting of Artin's theorem on characters of groups induced from cyclic subgroups is given. Also the idea of a fusion of a group normalized character table of an association scheme is introduced. This was first discussed in the context of quasigroup characters, and has reappeared recently in the theory of "supercharacters" of finite groups. It is associated to a "magic rectangle condition" on the group normalized character table. The fusion of character tables of groups will be discussed more fully in Chap. 9.

In Sect. 4.2 Schur's theory of centralizer rings of permutation groups is given, leading into an exposition of the theory of S-rings. Section 4.3 gives an account of the theory of association schemes, introducing Gelfand pairs and Gelfand models. The results on fusion and the magic rectangle condition are given, as well as those on induction and Frobenius reciprocity. Section 4.4 applies the theory of association schemes to produce the (combinatorial) character theory for finite loops and quasigroups, and shows that each quasigroup gives rise to a Gelfand pair. The properties of quasigroup characters and the information which they contain is then discussed. The consequences for quasigroups of the results on fusion and induction of characters are set out.

A list of all the generalizations of association schemes and related objects which have appeared would be hard to produce and it would include cellular algebras, cells, coherent configurations, colored graphs, strongly regular graphs, distance regular graphs. . . .

4.2   Centralizer Rings of Permutation Groups

Let G be a transitive permutation group acting on a finite set $\Omega = \{1, \ldots, n\}$. Let G act on $\Omega \times \Omega$ by

$$g(i, j) = (g(i), g(j))$$

and let $H = G_1$ denote the point stabilizer of 1, $H = \{g \in G : g(1) = 1\}$. There is a 1:1 correspondence between the orbits

$$\Delta_1, \Delta_2, \ldots, \Delta_k \tag{4.2.1}$$

of G under the above action on $\Omega \times \Omega$ and the orbits

$$\Gamma_1, \Gamma_2, \ldots, \Gamma_k \tag{4.2.2}$$

of H acting on Ω, given as follows

$$\Delta_i \leftrightarrow \Gamma_i = \{m : (1, m) \in \Delta_i\}. \tag{4.2.3}$$

This correspondence is not unique since it depends on the selection of the element 1. The Γ_i are called **suborbits**. Let $k_i = |\Gamma_i|$. From the transitivity of the action it follows that $k_i = |\Delta_i|/n$. If Δ_i^{op} is defined to be $\{(s, t) : (t, s) \in \Delta_i\}$ then there is a j such that $\Delta_j = \Delta_i^{op}$. This index j will be denoted by i'. If Λ is a subset of Ω and $g \in G$ let $g(\Lambda) = \{g(j) : j \in \Lambda\}$.

Recall (Definition 1.10, Chap. 1), that if a complex vector space V has basis $\{u_i\}_{i=1}^n$ then the action of G on V by $g(u_i) = u_{g(i)}$ is a linear representation with explicit matrices $T(g)$ given by $[T(g)]_{i,j} = 1$ if and only if $g(i) = j$, 0 otherwise.

The **centralizer ring S** of (G, Ω) is the set of $n \times n$ matrices (with arbitrary complex entries) which commute with each element of the set $\{T(g)\}_{g \in G}$. The $n \times n$ **incidence matrix** A_r is assigned to the subset Δ_r of $\Omega \times \Omega$ by

$$A_r(i, j) = 1 \text{ if } (i, j) \in \Delta_r, 0 \text{ otherwise.} \tag{4.2.4}$$

There is a second way of generating **S** from the action (G, Ω). The orbits Δ_i are regarded as relations on Ω, and $\Delta_i \circ \Delta_j$ is defined as the multiset

$$\{n_{(x,y)}(x, y)\}_{(x,y) \in \Omega \times \Omega},$$

where $n_{(x,y)}$ is the number of elements $z \in \Omega$ such that $(x, z) \in \Delta_i$ and $(z, y) \in \Delta_j$.

Lemma 4.1 $\Delta_i \circ \Delta_j = \sum p_{ij}^r \Delta_r$ where the p_{ij}^r are non-negative integers.

Proof Suppose that pairs (x_1, y_1), (x_2, y_2) lie in Δ_r. Then there exists $g \in G$ such that

$$g(x_1, y_1) = (g(x_1), g(y_1)) = (x_2, y_2).$$

It is sufficient to show that the number n_1 of $z \in \Omega$ such that

$$(x_1, z) \in \Delta_i, (z, y_1) \in \Delta_j \qquad (4.2.5)$$

must be equal to the number n_2 of $t \in \Omega$ such that

$$(x_2, t) \in \Delta_i, (t, y_2) \in \Delta_j. \qquad (4.2.6)$$

For $(x_1, z) \in \Delta_i, (z, y_1) \in \Delta_j$,

$$(g(x_1), g(z)) = (x_2, g(z)) \in \Delta_i, (g(z), g(y_1)) = (g(z), y_2) \in \Delta_j.$$

It follows that for each z satisfying (4.2.5), there is a $t = g(z)$ such that t satisfies (4.2.6). This shows that $n_2 \geq n_1$ and by symmetry $n_2 \geq n_1$, thus $n_1 = n_2$.   □

Lemma 4.2 *The $\{A_i\}$ represent the algebra generated by the $\{\Delta_i\}$ in the sense that if $\Delta_i \circ \Delta_j = \sum_{r=1}^{k} p_{ij}^r \Delta_r$ then $A_i A_j = \sum_{r=1}^{k} p_{ij}^r A_r$.*

Proof Let (k, l) lie in Δ_r. Define n_1 as in (4.2.5). It is sufficient to show that the element which appears in $A_i A_j$ in the (l, r) position is n_1. This is the dot product of the vector $\underline{u}_k$ in the kth row of A_i with the vector $\underline{v}_l$ in the lth column of A_j. From the definition of the A_i, each non-zero entry in $\underline{u}_k$ corresponds to a pair $(k, z) \in \Delta_i$ and each non-zero entry in $\underline{v}_l$ corresponds to a pair $(z, l) \in \Delta_j$ and so this dot product is precisely n_1.   □

Lemma 4.3 *The matrices $\{A_i\}_{i=1}^{k}$ satisfy the following*

(i) $\sum_{i=1}^{k} A_i = J$.
(ii) *The set $\{A_i\}_{i=1}^{k}$ is linearly independent.*
(iii) *For each i there is an i' such that $A_i^t = A_{i'}$.*
(iv) *Each row and column of A_i contains exactly k_i non-zero entries.*
(v) *For each i and j, $A_i A_j = \sum_{r=1}^{k} p_{ij}^r A_r$ where the p_{ij}^r are non-negative integers.*

Proof All the statements are straightforward except for (v), which follows from Lemma 4.2.   □

Proposition 4.1 *The set of incidence matrices $A_1, A_2, \ldots, A_k$ form an integral basis for the centralizer ring.*

Proof Let $B = \{B_{i,j}\}$ be an arbitrary complex matrix. By direct calculation, for each $g \in G$,

$$[T(g)BT(g)^{-1}]_{i,j} = B_{g(i),g(j)}. \tag{4.2.7}$$

Thus $T(g)BT(g)^{-1} = B$ if and only if for all $g \in G$, $B_{i,j} = B_{g(i),g(j)}$. It follows that for all $g \in G$, $T(g)$ commutes with $A_r, r = 1, \ldots, k$.

Now consider an arbitrary matrix B which commutes with $T(g)$ for all $g \in G$. If B has a non-zero entry c in the $(1, j)$ position, where $j \in \Gamma_r$, then by (4.2.7) B has entry c in each (i, j) position where $(i, j) \in \Delta_r$. It follows directly that $B = \sum_{r=1}^{k} c_r A_r$ for $c_r \in \mathbb{C}$. $\qquad\square$

It also follows that the centralizer ring may be characterized as the set of matrices with constant coefficients which commute with the group matrix $X_G^T = \sum T(g)x_g$ which corresponds to T, regarded as a permutation representation.

A different interpretation of **S** may be given in terms of double cosets of H. The G-set Ω is equivalent to the action of G on the right cosets $\{Hu_i\}_{i=1}^{k}$ of a subgroup H of G under the action

$$g(Hu_i) = Hu_i g. \tag{4.2.8}$$

An orbit of the induced action of H consists of the right cosets of H in the double coset $\alpha = Hu_i H$. In the finite case the double coset algebra generated by the D_α, where

$$D_\alpha = \frac{1}{|H|} \sum_{h,k \in H} h u_i k,$$

is a subalgebra of $\mathbb{C}G$ isomorphic to **S**. A proof of this is given in [10] (p. 107).

In the infinite case it is more usual to consider the dual situation, where a set $L(G)$ of functions $G \rightarrow \mathbb{C}$ is used instead of the group algebra and then the functions constant on double cosets also give an interpretation of the centralizer ring. In this case questions involving the topology for $L(G)$ come into play. This will be described in Appendix A.

Example 4.1 Let G be a group and consider the group $Mlt(G)$ of permutations on the set G which is generated by the set $\{\lambda(x) : h \rightarrow xh, \rho(x) : h \rightarrow hx\}$. It is not difficult to show that $M(G) = Mlt(G)$ is isomorphic to the split extension of the inner automorphism group $Inn(G)$ by the image of the regular representation $x \rightarrow \rho(x)$. This gives rise to a centralizer ring **S** from the action of $M(G)$ as a permutation group on the set G. Here **S** is always commutative, and is isomorphic to the class algebra of G.

4.3 S-Rings

The centralizer ring was introduced by Schur in the following situation. Let the permutation group G act transitively on the set $\Omega = \{1, 2, \ldots, n\}$. A **regular subgroup** $P \subseteq G$ is a transitive subgroup all of whose non-identity elements act fixed point freely. If G contains a regular subgroup P, a bijection may be set up between the elements of Ω and those of P, given by $p \leftrightarrow p(1)$ for $p \in P$ and using this bijection P can be identified with Ω. Each orbit Δ_r of the action of G on $\Omega \times \Omega$ can then be regarded as a subset of $P \times P$ and each suborbit Γ_i can be regarded as a subset of P. The **stabilizer** of an element i is $\{g \in G | g(i) = i\}$ and is denoted by $Stab_G(i)$.

Lemma 4.4 *If the orbits of* $H = Stab_G(1)$ *are* $\{\Gamma_1, \ldots, \Gamma_k\}$, *then the centralizer ring* **S** *of* (G, Ω) *is isomorphic to the subring of the group algebra of P generated by*

$$\overline{\Gamma}_1, \overline{\Gamma}_2, \ldots, \overline{\Gamma}_k \tag{4.3.1}$$

where if $\Gamma_i = \{x_1, x_2, \ldots, x_s\}$, $\overline{\Gamma}_i = x_1 + x_2 + \ldots + x_s$.

Proof Let $\Gamma_i = \{x_1, x_2, \ldots, x_s\}$ and $\Gamma_j = \{y_1, y_2, \ldots, y_t\}$. Consider the product

$$\overline{\Gamma}_i \overline{\Gamma}_j = \sum_{r=1}^{s} \sum_{m=1}^{t} x_r y_m. \tag{4.3.2}$$

The elements x_r and y_m correspond to elements $(e, x_r) \in \Delta_i$ and $(e, y_m) \in \Delta_j$. The lemma follows if it can be shown that the coefficient of $x_r y_m$ in (4.3.2) is equal to the coefficient of $(e, x_r y_m)$ in the product $\Delta_i \circ \Delta_j$. This is equal to the number of $z \in P$ such that $(e, z) \in \Delta_i$ and $(z, x_r y_m) \in \Delta_j$. But $(z, x_r y_m) \in \Delta_j$ implies that $(e, z^{-1} x_r y_m) \in \Delta_j$. This produces a bijection between pairs of elements (z, u) with $z \in \Gamma_i, u \in \Gamma_j$ and such that $zu = x_r y_m$ and the pairs of the form $\{(e, z), (z, x_r y_m)\}$ with $(e, z) \in \Delta_i, (z, x_r y_m) \in \Delta_j$. This proves the lemma. $\square$

It follows that **S** may be regarded as a subalgebra of $\mathbb{C}P$. The following definition is due to Wielandt.

Definition 4.1 An **S-ring** corresponding to a finite group G is a subring of $\mathbb{C}G$ which is constructed from a partition **S** $=\{\Gamma_1 = \{e\}, \Gamma_2, \ldots, \Gamma_k\}$ of the elements of G satisfying:

(1) If $\Gamma_i = \{g_1, \ldots, g_s\}$ then $\Gamma_i^* = \{g_1^{-1}, \ldots, g_s^{-1}\} = \Gamma_j$ for some j.
(2) $\overline{\Gamma}_i \overline{\Gamma}_j = \sum a_{ijk} \overline{\Gamma}_k$ where a_{ijk} is a non-negative integer for all i, j, k.

It is clear that the $\overline{\Gamma}_i$ form an integral basis for a ring which will also be denoted by **S**.

An S-ring which arises from the action of a permutation group on a set is called **Schurian**. Not all S-rings are Schurian, and in fact almost all S-rings are not Schurian. In [98] examples are given of non-Schurian rings over cyclic groups.

S-rings were used by Wielandt and others to extend Schur's results.

Definition 4.2 A transitive permutation group G on the set Ω is **imprimitive** if there exists a non-trivial partition $\{\Lambda_1, \Lambda_2, \ldots, \Lambda_t\}$ of Ω such that $g(\Lambda_i) = \Lambda_j$ for all $g \in G$ and all $i = 1, \ldots, t$. If no such partition exists then G is **primitive**.

Equivalently G is primitive if there is no non-trivial equivalence relation on Ω which is preserved by the action of G. It is **simply primitive** if it is primitive but not doubly transitive.

Definition 4.3 A group P is a **B-group** if whenever P is contained as a regular subgroup of a permutation group G then G must be either doubly transitive or imprimitive.

The study of B-groups was initiated by Wielandt, and his book [293] contains an account of much of the work on them. Their presence in a permutation group ensures that the group cannot be simply primitive.

Definition 4.4 An S-ring $\mathbf{S} = \{\Gamma_i\}_{i=1}^k$ on G is **primitive** if no subgroup K except for $\{e\}$ or G is such that $\overline{K}$ lies in $\mathbf{S}$. If such a K exists then $\mathbf{S}$ is **imprimitive**.

Definition 4.5 The **trivial S-ring** on G is $\mathbf{T} = \{\{e\}, G - \{e\}\}$.

It follows that if every S-ring on G is either imprimitive or trivial then G is a B-group. Examples of B-groups are

1. $C_{p^a} \times C_{p^b}$ where $a \neq b$.
2. D_n.
3. The non-abelian group of order p^3 and exponent p^2 for $p > 3$.

The known B-groups are given in the article by P. M. Neumann in the collected works of Wielandt ([294] pp. 9, 13–15). In particular the result on primitive groups in [46] implies that for almost all integers n every group of order n is a B-group. The proofs of some of the results are quite difficult, for example the result of Wielandt proving that D_n is a B-group involves the theorem of Dirichlet on primes in an arithmetic progression. Neumann gives the following list of finite non-B-groups:

(a) $X_1 \times X_2 \times \ldots \times X_k$ where each X_i has order the same order $m > 2$ and $k > 1$.
(b) Non-abelian simple groups.
(c) The non-abelian group of order p^3 and exponent p for $p > 3$.
(d) The non-abelian group of order 21.
(e) The non-abelian group of order 27 and exponent 9.
(f) The Frobenius group of order 992.

The most extensive accounts of the work on S-rings and B-groups are in [257, 293], and [271]. The article by Neumann mentioned above gives a valuable summary of the results on B-groups. Schur motivated his methods in [254] by a comment to the effect that the techniques should have wider applications, which is borne out by the large body of work on association schemes and related objects described below. S-rings have also been used to obtain results on circulant graphs [221]. They will appear later in the discussion of fusion and fission of group character tables. The double coset algebra approach to centralizer rings is given by Frame in [103].

The "rational" methods of Schur using S-ring techniques seem to have been more useful than the character theoretic methods first used by Burnside in investigating questions involving B-groups, and the papers which use arguments involving character theory have been unusually prone to error. Knapp in [197] gives a completion of Burnside's "proof" mentioned above which he acknowledges is not very different in concept from the proof of Schur in that he is close to using a dual argument to that of Schur.

When infinite groups are considered the known results on B-groups are all negative. For an element g of a group let the square root set of g be $\{x : x^2 = g\}$. If $g = e$ then call the corresponding square root set **principal** and otherwise define the square root set of g to be **non-principal**. It is shown in [44] that if G is a group of countably infinite order which satisfies the condition that G cannot be represented as the union of a finite number of translates of a finite number of non-principal square root sets together with a finite set then G is not a B-group. In fact such a group G can be embedded as a regular subgroup of the automorphism group of the "universal" random graph, which is simply primitive (see [41, 44] and [43]). In particular no countably infinite abelian group is a B-group.

4.3.1 Permutation Groups with Loop Transversals

Suppose G is a transitive permutation group on the set $\Omega = \{1, 2, \ldots r\}$ and $H = G_1$. Suppose further that there exists a transversal
$$\mathcal{T} = \{x_1 = e, x_2, \ldots, x_r\} \text{ to } H \text{ in } G,$$

$$G = He \cup Hx_2 \cup \ldots \cup Hx_r,$$

such that $x_i x_j^{-1}$ is fixed-point-free for all $i \neq j$. In this case an operation $(*)$ on $\mathcal{T}$ can be defined by

$$x_i * x_j = x_k \text{ where } x_i x_j = hx_k , h \in H.$$

It will be shown that $(\mathcal{T}, *)$ is a loop, i.e there is an identity element and there is left and right cancellation (see Chap. 1, p. 10). It follows directly that e is an identity element in $(\mathcal{T}, *)$. Right cancellation follows since

$$x_{i_1} * x_j = x_{i_2} * x_j$$

implies that $x_{i_1}x_j = hx_{i_2}x_j$ for some $h \in H$ and hence $x_{i_1} = hx_{i_2}$, forcing $i_1 = i_2$. Now if

$$x_i * x_{j_1} = x_i * x_{j_2},$$

then there exists $h \in H$ such that $x_i x_{j_1} = hx_i x_{j_2}$ and $x_i x_{j_1} x_{j_2}^{-1} x_i^{-1} = h$, or equivalently that $x_{j_1} x_{j_2}^{-1}$ is conjugate to an element which fixes e. But by the assumption on $\mathcal{T}$, $j_1 = j_2$ and thus left cancellation holds, and $(\mathcal{T}, *)$ is a loop Q. In this situation $\mathcal{T}$ is a **loop transversal** to H in G. The **multiplication group** $Mlt(Q)$ of an arbitrary quasigroup Q is the permutation group on the set Q generated by the set of maps $\{L(q), R(q)\}_{q \in Q}$ where $L(q) : x \to qx$ and $R(q) : x \to xq$. This generalizes the situation for groups described in Example 4.1. As is the case for groups $Mlt(Q)$ may be regarded as a group of permutations on $\mathcal{T}$, identifying i with the unique element x_i of $\mathcal{T}$ which takes 1 to i. The original group G is in general a proper subgroup of $Mlt(Q)$. In general different loops can be constructed in this way from a permutation group, by varying $\mathcal{T}$. An expanded discussion of this appears in [166]. There has been considerable attention to the related situation where G contains "connected transversals" to a point stablizer ([224]), where the conditions ensure that G contains $Mlt(Q)$.

A basis of the centralizer ring of $(Mlt(Q), \Omega)$ can be identified with corresponding subsets of Q in a similar way to the case where G has a regular subgroup (see below). In particular, if the loop Q constructed from $\mathcal{T}$ is commutative then $Mlt(Q) \subseteq G$ and the centralizer ring of (G, Ω) is commutative. Note that the centralizer ring of $(Mlt(Q), Q)$ is always commutative (see below).

A **B-S loop** is defined to be a loop Q such that any permutation group G which contains a loop transversal $\mathcal{T}$ giving rise to Q is either imprimitive or doubly transitive. The following theorem appeared in [166].

Theorem 4.1 *If Q is a loop which is the direct product of a cyclic group of even order and a commutative loop P of odd order then Q is B-S.*

4.3.2   Commutativity of the Centralizer Ring

Consider the general situation of the centralizer ring of a permutation group G acting transitively on the set Ω. An important simplification takes place when the centralizer ring is commutative.

A permutation representation $g \to \pi(g)$ may be regarded as a linear representation over $\mathbb{C}$ and its character χ splits into irreducible constituents $\pi = \sum_{i=1}^{s} c_i \sigma_i$ with $\sigma_i \in Irr(G)$. It is **multiplicity-free** if each $c_i = 1$.

Theorem 4.2 *If G acts on the set Ω via $g \to \pi(g)$, then the centralizer ring is commutative if and only if π, regarded as a linear representation, is multiplicity-free.*

Proof Suppose that π is multiplicity-free. Then the group matrix $X_\pi = \sum \pi(g)x_g$ is similar to a block diagonal matrix, $D_\pi = \text{diag}(X_{\sigma_1}, \ldots, X_{\sigma_s})$ corresponding to the decomposition of π into irreducible constituents. Now by Schur's lemma, (Chap. 1, Lemma 1.3) any matrix which commutes with the group matrix corresponding to an irreducible representation is of the form λI, a scalar multiple of the appropriate identity matrix. Thus any matrix which commutes with D_π must be the matrix $\text{diag}(\lambda_1 I_{n_1}, \ldots, \lambda_s I_{n_s})$ which is an ordinary diagonal matrix, where the λ_i are arbitrary scalars. It clearly follows that the ring generated by these diagonal matrices is commutative, which implies that the centralizer ring is commutative. Conversely, using Schur's Lemma, the matrices which commute with a matrix of the form

$$\begin{bmatrix} X_\sigma & 0 \\ 0 & X_\sigma \end{bmatrix}, \tag{4.3.3}$$

are of the form

$$\begin{bmatrix} \lambda_1 I & \lambda_2 I \\ \lambda_3 I & \lambda_4 I \end{bmatrix}, \tag{4.3.4}$$

where $\lambda_1, \lambda_2, \lambda_3, \lambda_4$ are arbitrary, and the algebra of such matrices is isomorphic to the algebra of 2×2 matrices and so cannot be commutative. It follows that if $c_i > 1$ for any i then **S** cannot be commutative. $\qquad \square$

For a proof which is independent of group matrices the reader is referred to [293] or [10].

Definition 4.6 If G is a finite permutation group acting on a set Ω with centralizer ring **S** which is commutative and H is the stabilizer of a point, then the pair (G, H) is a **Gelfand pair**.

For infinite groups the above definition is modified as follows: The restriction of an irreducible representation of G to H contains the trivial representation with multiplicity at most 1. It is easily seen that this is equivalent to the definition above for finite groups (using Theorem 4.2). The theory of Gelfand pairs is closely related to the topic of spherical functions in the classical theory of special functions, and to the theory of Riemannian symmetric spaces in differential geometry.

In Appendix A other sets of conditions which ensure that a pair (G, H) is a Gelfand pair are described. A related definition is

Definition 4.7 A (complex) representation ρ of a group G is called a **Gelfand model** if it is equivalent to the direct sum of all irreducible representations of G, each occurring with multiplicity 1.

4.4 Hecke Algebras

Let G be an arbitrary group with a subgroup H. Let $L(G)$ be the functions from G to $\mathbb{C}$. Then f is **left H-invariant** if $f(hg) = f(g)$ for all $h \in H$, $g \in G$, and f is **right H-invariant** if $f(gh) = f(g)$ for all $h \in H$, $g \in G$. If f is both left H-invariant and right H-invariant then f is H-invariant. Then it can be shown that f is H-invariant if and only if f is constant on the double cosets of H in G. If G is finite the **convolution product** on the set of H-invariant functions is defined as follows. Let f_1 and f_2 be H-invariant. Then

$$f_1 * f_2(g) = \frac{1}{|H|} \sum_{k \in G} f_1(k) f_2(k^{-1}g).$$

It may be seen that this is, up to a constant factor, the same expression as the convolution involved in the product of group matrices X_G and Y_G given in Chap. 1, Theorem 1.4. In the infinite case convolution is given by an integral.

Definition 4.8 The algebra of H-invariant functions of under the convolution product is the **Hecke algebra** $\mathfrak{H}(G, H)$ of G with respect to H.

$\mathfrak{H}(G, H)$ is anti-isomorphic to the centralizer algebra of G acting on the right cosets of H. It appears as an important tool in number theory.

4.5 Association Schemes

A general framework will now be introduced which subsumes the above work. Sources are [9, 10, 76, 123, 190] and [297]. Let Ω be a non-empty finite set and $\mathcal{R} = \{\mathcal{R}_0, \mathcal{R}_1, \ldots, \mathcal{R}_d\}$ be a collection of subsets of $\Omega \times \Omega$ which form a partition of $\Omega \times \Omega$ into $d + 1$ classes. The $\mathcal{R}_i$ may be thought of as relations on the set. Then a configuration $\mathcal{X} = (\Omega, \{\mathcal{R}_i\}_{0 \le i \le d})$, or briefly $\mathcal{X}$, is called an **association scheme** of class d if it satisfies the following properties:

(AS 1) $\mathcal{R}_0 = \{(x, x) \mid x \in \Omega\}$.
(AS 2)   For any $\mathcal{R}_i$, the set $\mathcal{R}_i^t = \{(x, y) \mid (y, x) \in \mathcal{R}_i\}$ is in $\mathcal{R}$.
(AS 3)   For every pair $(x, y) \in \mathcal{R}_h$, the number of $z \in \Omega$ such that $(x, z) \in \mathcal{R}_i$, $(z, y) \in \mathcal{R}_j$ is a constant p_{ij}^h depending only on h, i, j.

   The constants p_{ij}^h are called the **intersection numbers** (or **parameters**) of $\mathcal{X}$.
   If $\mathcal{X}$ in addition satisfies
(AS 4) $p_{ij}^h = p_{ji}^h$ for all h, i, j,
   it is said to be a **commutative association scheme**. Often in the literature the term association scheme includes commutativity.

If the **composition** of $\mathcal{R}_i \mathcal{R}_j$ is defined by

$$\mathcal{R}_i \circ \mathcal{R}_j = \{(x, y); \exists z : (x, z) \in \mathcal{R}_i, (z, y) \in \mathcal{R}_j\}$$

condition (AS 3) above is equivalent to

$$\mathcal{R}_i \circ \mathcal{R}_j = \Sigma p_{ij}^h \mathcal{R}_h,$$

i.e. the set of formal linear combinations $\sum n_i \mathcal{R}_i$ of elements of $\mathcal{R}$ form an algebra $\mathfrak{A}$ over Z, with addition defined by

$$\sum n_i \mathcal{R}_i + \sum m_i \mathcal{R}_i = \sum (n_i \mathcal{R}_i + m_i \mathcal{R}_i)$$

and

$$\sum n_i \mathcal{R}_i \cdot \sum m_i \mathcal{R}_i = \sum n_i m_i (\mathcal{R}_i \circ \mathcal{R}_i).$$

Condition (AS 4) is equivalent to $\mathcal{R}_i \circ \mathcal{R}_j = \mathcal{R}_j \circ \mathcal{R}_i$.

The following condition is often satisfied by the association schemes arising in the theory of experimental designs:

(AS 5)   An association scheme is symmetric if $\mathcal{R}_i = \mathcal{R}_i^t$ for all i.

Note that (AS 5) implies (AS 4).

A Comparison of Definitions Various definitions of an association scheme and other related structures are given in the literature. A short summary of the common ones is as follows.

 (I) A **coherent configuration** is defined to satisfy (AS 2) and (AS 3) above but (AS 1) is weakened to: the set $\Delta = \{(x, x)\}_{x \in \Omega}$ is a disjoint union of some of the $\mathcal{R}_j$.

A coherent configuration is **homogeneous** if (AS 1) is satisfied.

 (II) (Bannai-Ito-Terwilliger) An association scheme satisfies conditions (AS 1)-(AS 4) above.

(III) Cameron-Bailey. An association scheme satisfies (AS 1)-(AS 5) above.

More general structures of a similar nature have also appeared. In [10] C-algebras are used to discuss duality and Blau and coauthors have extended the discussion to table algebras (see [21] which presents an overview of the various definitions).

Here the term association scheme will be used consistent with definition (II). In the language of Cameron-Bailey the association schemes here would be called commutative homogeneous coherent configurations.

Now let $\mathcal{X} = (\Omega, \{\mathcal{R}_i\}_{0 \leq i \leq d})$ be an association scheme as defined by (II) and assume that $\Omega = \{1, \ldots, n\}$. To each $\mathcal{R}_i$ there corresponds an adjacency matrix A_i exactly as in (4.2.4). Specifically the $(j, k)^{th}$ entry of A_i is 1 if $(j, k) \in \mathcal{R}_i$ and 0 otherwise. The A_i generate a $d + 1$-dimensional commutative algebra $\mathcal{U}$ known as the **Bose-Mesner algebra** of the scheme. This algebra is isomorphic to $\mathfrak{A}$. There is also a further algebra isomorphic to $\mathfrak{A}$ obtained by forming the matrices $\{P_i\}_{i=0,\ldots,d}$ where P_j the $(d+1) \times (d+1)$ matrix whose $(i, h)^{\text{th}}$ entry is p_{ij}^h. This is the regular representation of A_i acting on $\{A_i\}_{i=0,\ldots,d}$.

Example 4.2 Let $\Omega = \{1, 2, 3, 4\}$ and let

$$
A_0 = I, \ A_1 = \begin{bmatrix} 0 & 0 & 1 & 0 \\ 0 & 0 & 0 & 1 \\ 1 & 0 & 0 & 0 \\ 0 & 1 & 0 & 0 \end{bmatrix}, \ A_2 = \begin{bmatrix} 0 & 1 & 0 & 1 \\ 1 & 0 & 1 & 0 \\ 0 & 1 & 0 & 1 \\ 1 & 0 & 1 & 0 \end{bmatrix}.
$$

Then the structure constants are determined by

$$
A_0 A_i = A_i, \ i = 0, 1, 2, \ A_1^2 = A_0, \ A_1 A_2 = A_2, \ A_2^2 = 2A_0 + 2A_1,
$$

and

$$
P_0 = I, \ P_1 = \begin{bmatrix} 0 & 1 & 0 \\ 1 & 0 & 0 \\ 0 & 0 & 1 \end{bmatrix}, \ P_2 = \begin{bmatrix} 0 & 0 & 1 \\ 0 & 0 & 1 \\ 2 & 2 & 0 \end{bmatrix}.
$$

A matrix A is **normal** if $AA^* = A^*A$. The matrices $\{A_i\}_{i=0,\ldots d}$ form a commuting set of normal matrices and by standard theory ([116] p. 291 Theorem 11) they have a common set of eigenspaces, $\{V_i\}_{i=0}^d$ where V_0 can be choosen to be the one-dimensional subspace with basis $\{v_{01} = [1, 1, \ldots, 1]\}$. This algebra has a set $\{E_i\}_{0 \le i \le d}$ of primitive orthogonal idempotents where E_i is the matrix corresponding to the projection $\mathfrak{P}_i$ onto V_i. Explicitly, if $\{v_{i1}, v_{i2}, \ldots, v_{it_i}\}$ form a basis for V_i then a basis for the underlying space V can be written as

$$
\{v_{01}, v_{11}, v_{12} \ldots, v_{1t_1}, \ldots, v_{d1}, \ldots, v_{dt_d}\}, \tag{4.5.1}
$$

and with respect to this basis $v \in V$ may be written as

$$
v = \sum_{j=0}^d \sum_{k=1}^{t_j} \alpha_{jk} v_{jk}.
$$

Then $\mathfrak{P}_i$ sends v to $\sum_{k=1}^{t_i} \alpha_{ik} v_{ik}$. If U is the matrix whose columns are the common eigenvectors of the A_i ordered as in (4.5.1) then the matrix E_i may be calculated as $U D U^{-1}$ where

$$
D_i = diag(0, 0, \ldots, 1, 1, \ldots, 1, 0, \ldots, 0),
$$

where there are t_i entries 1 which start in the position $t_0 + t_1 + \ldots + t_{i-1} + 1$. In particular $E_0 = \frac{1}{n} J$ where J is the all 1 matrix.

Example Consider the association scheme in Example 4.2 above. Then $v_{01}, v_{11}, v_{21}, v_{22}$ may be chosen as

$$[1, 1, 1, 1], [-1, 1, -1, 1], [0, -1, 0, 1], [-1, 0, 1, 0]$$

and $t_0 = 1$, $t_1 = 1$ and $t_2 = 2$. Thus

$$U = \begin{bmatrix} 1 & -1 & 0 & -1 \\ 1 & 1 & -1 & 0 \\ 1 & -1 & 0 & 1 \\ 1 & 1 & 1 & 0 \end{bmatrix}.$$

By direct calculation

$$E_0 = \frac{1}{4} J, \quad E_1 = \frac{1}{4} \begin{bmatrix} 1 & -1 & 1 & -1 \\ -1 & 1 & -1 & 1 \\ 1 & -1 & 1 & -1 \\ -1 & 1 & -1 & 1 \end{bmatrix}, \quad E_2 = \frac{1}{2} \begin{bmatrix} 1 & 0 & -1 & 0 \\ 0 & 1 & 0 & -1 \\ -1 & 0 & 1 & 0 \\ 0 & -1 & 0 & 1 \end{bmatrix}.$$

The rank of the matrix E_i is denoted by m_i and the m_i are referred to as the **multiplicities** of $\mathcal{X}$. Further,

$$A_i = \sum_{j=0}^{d} p_i(j) E_j,$$

where the $p_i(j)$ are the eigenvalues of A_i. **The $\mathcal{P}$-matrix P** is the matrix whose $(i, j)^{th}$ entry is $p_j(i)$. It is also called the **first eigenmatrix** of $\mathcal{X}$. Note that since A_i is normal, $p_i(j) = \overline{p_{i'}(j)}$, where $\overline{p_{i'}(j)}$ denotes the complex conjugate of $p_{i'}(j)$.

The association scheme definition generalizes that of the centralizer ring of a G-set, in that the orbits Δ_i obtained from a G-set whose centralizer ring is commutative may be regarded as the $\mathcal{R}_i$ of an association scheme. Consistent with the terminology for S-rings, an association scheme which arises in this way from a G-set it is called **Schurian**. Not all G-sets give rise to association schemes, since in general the centralizer ring is not commutative (as is the case for example if $G = \mathcal{S}_n$ acting on unordered triples for n large enough). The following appears in [42]. If $\mathcal{X}$ is a homogeneous coherent configuration on Ω the symmetrization $\mathcal{X}^{sym}$ is the "scheme" obtained from the partition of $\Omega \times \Omega$ whose parts are the unions $\mathcal{R}_i \cup \mathcal{R}_i^t$. This may not be an association scheme, and if it is $\mathcal{X}$ is said to be **stratifiable**. Theorem 4.2 has the following generalization indicated in [42].

Theorem 4.3 *If $\mathcal{X}$ arises from the centralizer ring $\mathbf{S}$ of the permutation group G then $\mathcal{X}$ is stratifiable if and only if the action of G is real multiplicity free, i.e. when* *it is decomposed into linear representations irreducible over* $\mathbb{R}$ *they are pairwise non-isomorphic.*

4.5.1   Character Tables of Association Schemes

The $\mathcal{P}$-matrix P of an association scheme $\mathcal{X}$ has row and column orthogonality as follows. Let k_j denote $p_j(0)$ and as above let m_j be the jth multiplicity of $\mathcal{X}$. Then

$$(\text{i}) \sum_{u=0}^{d} \frac{1}{k_u} p_u(i)\overline{p_u(j)} = \frac{|\Omega|}{m_i}\delta_{ij}; \qquad (4.5.2)$$

$$(\text{ii}) \sum_{u=0}^{d} m_u p_i(u)\overline{p_j(u)} = |\Omega|k_i\delta_{ij}. \qquad (4.5.3)$$

A proof of these is given below.

The relations (i) and (ii) above are usually referred to as the **orthogonality relations** of the **first** and **second** kind respectively, and P is referred to as the (association scheme) **character table** of $\mathcal{X}$.

As was indicated previously in Chap. 1, Sect. 1.3, Frobenius first constructed group characters for a group G from the class algebra of G. This is isomorphic to the Bose-Mesner algebra of the association scheme which arises where Ω is the set of elements of G and the $\mathcal{R}_i$ are the orbits of the diagonal action on $G \times G$ of the permutation group $M(G)$ which are in 1:1 correspondence with the conjugacy classes of G as in Example 4.1. In this case the group character table is obtained by renormalizing the matrix P

$$\chi_i(j) = \frac{\sqrt{m_i}}{k_j} p_j(i), \qquad (4.5.4)$$

then the matrix Ψ defined by $\Psi_{i,j} = \{\chi_i(j)\}$ is the character table of the group. The matrix Ψ can be constructed for an arbitrary commutative association scheme by the same formula (4.5.4) and will be referred to as the **group normalization** of the character table of the scheme. The following lemma sets out the orthogonality relations between the group normalized characters of a commutative association scheme. If χ_s and χ_t are rows of the table, then define the inner product $\langle \chi_s, \chi_t \rangle$ as

$$\langle \chi_s, \chi_t \rangle = \frac{1}{|\Omega|} \sum_{j=1}^{d} k_j \chi_s(j) \chi_t(j).$$

Lemma 4.5 *If $\mathcal{X}$ is an association scheme and $\{\chi_1, \chi_2, \ldots, \chi_r\}$ are the set of irreducible characters of $\mathcal{X}$ as defined in (4.5.4)*

$$(i)\ \langle \chi_i, \chi_j \rangle = \delta_{ij}$$

$$(ii)\ \sum_{i=1}^{d} \chi_i(j)\chi_i(l)) = \delta_{jl}\frac{|\Omega|}{k_j}\ otherwise.$$

These follow directly from the relations (4.5.2) and (4.5.3).

Example 4.3 Let G be $\mathcal{S}_3$, and consider the association scheme corresponding to the conjugacy classes. The adjacency matrices may be described as follows. Let J_m be the $m \times m$ matrix all of whose entries are 1. The adjacency matrices of the scheme are

$$A_0 = [I_6],\ A_1 = \begin{bmatrix} J_3 - I_3 & 0 \\ 0 & J_3 - I_3 \end{bmatrix},$$

$$A_2 = \begin{bmatrix} 0 & J_3 \\ J_3 & 0 \end{bmatrix}.$$

The $\mathcal{P}$-matrix is

$$\begin{bmatrix} 1 & 2 & 3 \\ 1 & 2 & -3 \\ 1 & -1 & 0 \end{bmatrix} \tag{4.5.5}$$

and the character table (previously given in Chap. 1, (1.5.2)) is

	C_0	C_1	C_2
χ_1	1	1	1
χ_2	1	1	-1
χ_3	2	-1	0

(4.5.6)

where the rows represent irreducible characters and the columns are indexed by the conjugacy classes $C_0 = \{e\}$, $C_1 = \{(i, j, k)\}$ and $C_2 = \{(i, j)\}$. It will be shown below how quasigroup character tables can be defined analogously.

A further example of a character table of a Schurian association scheme is the following. Let G be $\mathcal{S}_5$ acting on the set Ω of unordered pairs $\{i, j\}_{1 \leq i, j \leq 5}$. Here $H = G_{\{1,2\}}$ is $\mathcal{S}_2 \times \mathcal{S}_3$ and Ω may be described in terms of the Petersen graph.

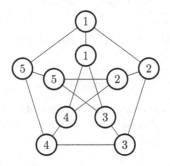

There are three classes, described by the relations $\mathcal{R}_0 = id$, $\mathcal{R}_1 \equiv$ disjointness, $\mathcal{R}_2 \equiv$ non-empty intersection, and there is an edge $(\{i, j\}, \{k, l\})$ of the graph if and only if $\{i, j\}\mathcal{R}_1\{k, l\}$. In this case the $\mathcal{P}$-matrix P is

$$\begin{bmatrix} 1 & 3 & 6 \\ 1 & -2 & 1 \\ 1 & 1 & -2 \end{bmatrix} \qquad (4.5.7)$$

with multiplicities $1, 4, 5$. This is an example where there is a loop transversal to H in G (see [167]). The group normalized table is

Class size →	1	3	6
χ_1	1	1	1
χ_2	2	$-\frac{4}{3}$	$\frac{1}{3}$
χ_3	$\sqrt{5}$	$\frac{\sqrt{5}}{3}$	$-\frac{\sqrt{5}}{3}$

4.5.2 Duality in Association Schemes

Suppose $\mathcal{X} = (\Omega, \{\mathcal{R}_i\}_{0 \leq i \leq d})$ is an association scheme, with incidence matrices $\{A_i\}$ and primitive idempotents $\{E_i\}$, which are the matrices of the projection operators as above. The following properties of the two sets of generators indicate a duality between the Bose-Messner algebra generated by the A_i and the algebra generated by the E_i under the Schur-Hadamard product: $(a_{ij}) \circ (b_{ij}) = (a_{ij}b_{ij})$.

For a matrix $A = (a_{ij})_{m \times m}$ let $\tau(A) = \sum_{i,j=1,\ldots,m} a_{ij}$. Suppose $|\Omega| = n$. A list of properties of the A_i follows:

P1: $A_0 = I_n$.

P2: $A_0 + A_1 + \ldots + A_d = J_n$.

P3: $A_i^T = A_j$ for some $j \in \{0, 1, \ldots, d\}$.

P4: $A_i A_j = \sum_{k=0}^{d} p_{ij}^k A_k$ for all i, j.

P5: $A_i A_j = A_j A_i$ for all i, j.

P6: $A_i \circ A_j = \delta_{ij} A_i$.

P7: $Tr(A_0) = n$, $Tr(A_i) = 0$, $i > 0$.

P8: $\tau(A_i) = nk_i$

P9: $A_i = \sum_{k=0}^{d} p_i(k) E_k$

The following are "dual" properties of the E_i:

D1: $E_0 = \frac{1}{n} J$.

D2: $E_0 + E_1 + \ldots + E_d = I$.

D3: $E_i^T = E_j$ for some $j \in \{0, 1, \ldots, d\}$.

D4: $E_i \circ E_j = \frac{1}{n} \sum_{k=0}^{d} q_{ij}^k E_k$ for all i, j.

D5: $E_i \circ E_j = E_j \circ E_i$

D6: $E_i E_j = \delta_{ij} E_i$.

D7: $Tr(E_i) = m_i$.

D8: $\tau(E_0) = n$, $\tau(E_i) = 0$, $i > 0$.

D9: $E_i = \frac{1}{|n|} \sum_{k=0}^{d} q_i(k) A_k$.

The q_{ij}^k are called the **Krein parameters** of the scheme, and are non-negative real numbers. Often they are rational (see [10], p. 70 for an example where they are not).

The matrix Q whose $(k, i)^{th}$ entry is $q_i(k)$ is called the **second eigenmatrix** or **Q-matrix** of $\mathcal{X}$. The orthogonality relations follow from the lemma below.

Lemma 4.6 *The orthogonality relations (4.5.2) and (4.5.3) are consequences of*

$$PQ = QP = nI, \qquad (4.5.8)$$

which in turn follows from the relationship

$$q_j(i)/m_j = \overline{p_i(j)}/k_i. \qquad (4.5.9)$$

Proof By direct calculation if A and B are any $n \times n$ matrices

$$\tau(A \circ B) = Tr(AB^T) = Tr(A^T B). \qquad (4.5.10)$$

Using properties D9 and P6 above

$$E_j \circ A_i = (\frac{1}{|n|} \sum_{k=0}^{d} q_j(k)A_k) \circ A_i$$

$$= \frac{1}{|n|} q_j(i) A_i,$$

Then using (4.5.10)

$$\tau(E_j \circ A_i) = Tr(E_j A_i^T) = Tr(E_j A_{i'}),$$

where P3 has been used. Now by P9

$$A_{i'} = \sum_{k=0}^{d} p_{i'}(k) E_k,$$

and therefore

$$E_j A_{i'} = p_{i'}(j) E_j,$$

and

$$Tr(E_j A_{i'}) = Tr(p_{i'}(j)E_j) = p_{i'}(j)m_j = m_j \overline{p_i(j)}, \qquad (4.5.11)$$

where the last step uses the fact that if v_j is an eigenvector of A_i, with eigenvalue $p_i(j)$, v_j is also an eigenvector of A_i^t with eigenvalue $\overline{p_i(j)}$. Now since $\tau(A_i) = nk_i$, it follows that

$$\tau(E_j \circ A_i) = \tau(\frac{1}{|n|} q_j(i) A_i) = q_j(i)k_i \qquad (4.5.12)$$

and (4.5.9) follows from (4.5.11) and (4.5.12).   □

If the association scheme arises from the S-ring of conjugacy classes of a group, then the matrix Q may be obtained from the transpose of the character table by multiplying each column by the degree of the corresponding character.

Example 4.4 Let $G = S_4$. The character table is

Class order	1	3	8	6	6
χ_1	1	1	1	1	1
χ_2	1	1	1	-1	-1
χ_3	3	-1	0	1	-1
χ_4	3	-1	0	-1	1
χ_5	2	2	-1	0	0

The $\mathcal{P}$-matrix is

$$\begin{bmatrix} 1 & 3 & 8 & 6 & 6 \\ 1 & 3 & 8 & -6 & -6 \\ 1 & -1 & 0 & 2 & -2 \\ 1 & -1 & 0 & -2 & 2 \\ 1 & 3 & -4 & 0 & 0 \end{bmatrix}$$

and the $\mathcal{Q}$-matrix is

$$\begin{bmatrix} 1 & 1 & 9 & 9 & 4 \\ 1 & 1 & -3 & -3 & 4 \\ 1 & 1 & 0 & 0 & -2 \\ 1 & -1 & 3 & -3 & 0 \\ 1 & -1 & -3 & 3 & 0 \end{bmatrix}.$$

Example 4.5 Let $G = S_3$. Continuing from Example 4.3, the matrices E_i are

$$E_0 = \frac{1}{6}J_6$$

$$E_1 = \frac{1}{6}\begin{bmatrix} J_3 & -J_3 \\ -J_3 & J_3 \end{bmatrix}$$

$$E_2 = \frac{1}{6}\begin{bmatrix} C(4,-2,-2) & 0 \\ 0 & C(4,-2,-2) \end{bmatrix}.$$

(where $C(4,-2,-2)$ is the circulant defined in Chap. 1). The Krein parameters are given by the products

$$E_0 \circ E_i = \frac{1}{6}E_i, \ E_1 \circ E_1 = \frac{1}{6}E_0, \ E_1 \circ E_2 = \frac{1}{6}E_2, \ E_2 \circ E_2 = \frac{2}{3}E_0 + \frac{2}{3}E_1 + \frac{1}{3}E_2.$$

There is a direct construction of A_i from the group matrix of S_3 by setting $x_g = 1$ for $g \in C_i$ and $x_g = 0$ otherwise. It may be considered that the dual algebra generated by the E_i under $\circ$ is related to the character ring (see Chap. 2, Sect. 2.5). In this example if the three irreducible characters $\chi_0 = 1$, $\chi_1 = sgn$, χ_2 given above there are the following product relations

$$\chi_0 \circ \chi_i = \chi_i, \chi_1 \circ \chi_1 = \chi_0, \chi_1 \circ \chi_2 = \chi_2, \chi_2 \circ \chi_2 = \chi_0 + \chi_1 + \chi_2.$$

where here the constants are necessarily integers. There have been several attempts to use the "duality" between the two structures extending the duality theory for abelian groups but the situation still seems somewhat murky (see [10] where Kawada-Delsarte duality is explained and where also Tannaka-Krein duality for association schemes, extending that for groups, is discussed).

4.6   Fusion of Character Tables

Suppose that $\mathcal{X} = (\Omega, \{\mathcal{R}_i\}_{0 \leq i \leq d})$ is an association scheme, with adjacency matrices $\{A_i\}_{0 \leq i \leq d}$. Let $P_{\mathcal{X}}$ be the $\mathcal{P}$-matrix of $\mathcal{X}$, and let the rows of $P_{\mathcal{X}}$, regarded as characters of $\mathcal{X}$, be denoted by $\lambda_0, \ldots, \lambda_d$, where $\lambda_i(j) = p_j(i)$ is an eigenvalue of A_j.

Suppose that $\mathcal{Y} = (\Omega, \{\mathcal{S}_i\}_{0 \leq i \leq t})$ is another association scheme, with adjacency matrices $\{B_i\}_{0 \leq i \leq d}$. If for each k, $B_k = \sum_{\beta=1}^{r_k} A_{i_\beta}$ then $\mathcal{Y}$ is a **subscheme** of $\mathcal{X}$. In this situation, each eigenvalue μ_α of B_k is of the form $\sum_{i=1}^{r_k} \lambda_\alpha(i_\beta)$ for some character λ_α of $\mathcal{X}$ and if $P_{\mathcal{X}}$ is partitioned by grouping the columns corresponding to the $\{A_{i_\beta}\}_{\beta=1}^{r_k}$ together, there is a natural partition of the characters, in which all characters λ_m are in the same block if and only if the values $\sum_{\beta=1}^{r_k} \lambda_m(i_\beta)$ are the same for each k. In this case $\sum_{\beta=1}^{r_k} \lambda_m(i_\beta) = \mu_\alpha(k)$. This means that $P_{\mathcal{X}}$ may be partitioned into rectangular blocks, with row sums on the blocks constant, and such that the $\mathcal{P}$-matrix $P_{P_{\mathcal{X}}}$ of $\mathcal{Y}$ may be obtained from $P_{\mathcal{X}}$ by eliminating repeating rows, the entry in the kth column being the block sum corresponding to B_k.

Example 4.6 Let the $\mathcal{X}$ be the association scheme on $\{1, \ldots, 6\}$ with $\mathcal{P}$-matrix (or character table)

	A_0	A_1	A_2	A_3
λ_0	1	2	1	2
λ_1	1	2	-1	-2
λ_2	1	-1	1	-1
λ_3	1	-1	-1	1

Note that this gives enough information to determine the structure constants of the scheme arising from the products of the diagonalized matrices, for example

$$A_1 A_3 = diag(2, 2, -1, -1) diag(2, -2, -1, 1) = diag(4, -4, 1, -1) = 2A_2 + A_3.$$

If the blocks $\{B_k\}$ are given by $B_0 = A_0$, $B_1 = A_1$, $B_2 = A_2 + A_3$, to form a subscheme $\mathcal{Y}$, a rearrangement of the above matrix gives rise to the table with fused columns

	B_0	B_1	B_2
λ_0	1	2	3
λ_1	1	2	-3
λ_2	1	-1	0
λ_3	1	-1	0

which indicates that the $\mathcal{P}$-matrix of $\mathcal{Y}$ is

	B_0	B_1	B_2
μ_0	1	2	3
μ_1	1	2	-3
μ_2	1	-1	0

Definition 4.9 A **fusion** of a character table of an association scheme $\mathcal{X}$ is the character table of a subscheme $\mathcal{Y}$.

4.6.1 The Magic Rectangle Condition

The group normalized character table of an association scheme $\mathcal{X} = (\Omega, \{\mathcal{R}_i\}_{0 \le i \le d})$ has a fusion if and only if the following **magic rectangle condition** is satisfied. The group normalized character table may be regarded as the array $\chi_i(j)$ where the $\{\chi_1, \ldots, \chi_r\}$ are the basic characters with $\chi_i(j)$ the value of χ_i on $\mathcal{R}_j$. Thus $d_i = \chi_i(\mathcal{R}_0)$, $i = 1, \ldots, r$, and k_i is the valency of $\mathcal{R}_i$. Then there is a partition $\{B_i\}_{i=1}^f$ of the set $\{\mathcal{R}_i\}$ and a partition $\{\psi_i\}_{i=1}^f$ of the set $\{\chi_1, \ldots, \chi_r\}$. The character table of $\mathcal{X}$ can then be ordered so that it consists of rectangles, a typical rectangle consisting of the columns corresponding to the elements of $B_j = \{\mathcal{R}_{t_1}, \ldots, \mathcal{R}_{t_{r_j}}\}$ and rows corresponding to $\psi_i = \{\chi_{w_1}, \ldots, \chi_{w_{u_i}}\}$. If $\mathcal{R}_{t_i}$ is abbreviated by $\mathcal{R}_i$ and χ_{w_i} by χ_i, the magic rectangle condition states that for a given such rectangle, for each i the value

$$\tau_{ij} = \frac{\sum_{m=1}^{r_j} k_m \chi_i(m)}{[\sum_{m=1}^{r_j} k_m] d_i}$$

is constant and equal to the common value for each j of

$$\frac{\sum_{m=1}^{u_i} d_m \chi_m(j)}{[\sum_{m=1}^{u_i} d_m^2]}.$$

The fused table is an $f \times f$ table which has rows corresponding to the ψ_i and classes corresponding to B_j such that the value of ψ_i on B_j (with slight abuse of notation) is $\eta_i \tau_{ij}$ where

$$\eta_i = \sqrt{\sum_{m=1}^{u_i} d_m^2}.$$

The magic rectangle condition follows directly from the existence of a subscheme. It first appeared in [178] in the context of schemes arising from quasigroups.

4.6.2   Induction and Frobenius Reciprocity for Association Schemes

Induction and restriction of group class functions has been described in Chap. 2, Sect. 2.4.3. The definition can be extended to association schemes as follows. Let $\mathcal{X} = (\Omega, \{\mathcal{R}_i\}_{0 \leq i \leq r})$ be an association scheme. A **class function** on a scheme $\mathcal{X}$ is a function $\phi : \Omega \times \Omega \to \mathbb{C}$ which is constant on the $\mathcal{R}_i$. The inner product of two class functions φ, ψ is

$$\langle \varphi, \psi \rangle = \frac{1}{|\Omega \times \Omega|} \sum_{(s,t) \in \Omega \times \Omega} \varphi(s,t)\overline{\psi(s,t)}$$

or equivalently

$$\langle \varphi, \psi \rangle = \frac{1}{|\Omega \times \Omega|} \sum_{(s,t) \in \Omega \times \Omega} \varphi(s,t)\psi(t,s).$$

Let $\mathcal{X} = (\Omega, \{\mathcal{R}_i\}_{0 \leq i \leq r})$ be an association scheme, and let $\mathcal{Y} = (\Gamma, \{\mathcal{U}_i\}_{0 \leq i \leq u})$ be another association scheme such that Γ is a subset of Ω. Define $\mathcal{Y}$ to be a **refinement scheme** of $\mathcal{X}$ if for all $j = 1, \ldots, u$,

$$\mathcal{U}_j \subseteq \mathcal{R}_\alpha$$

for some α, or equivalently each relation $\mathcal{U}_j$ is a consequence of some relation $\mathcal{R}_\alpha$. Suppose λ is a class function on $\mathcal{Y}$, i.e a function from $\Gamma \times \Gamma$ to $\mathbb{C}$ which is constant on the $\mathcal{U}_i$. For any function ϕ and any set X, let $\phi(X) = \sum_{x \in X} \phi(x)$. Define the **induced class function** $\lambda_{\mathcal{Y}}^{\mathcal{X}}$ on $\mathcal{X}$ by

$$\frac{1}{|\Omega \times \Omega|} \lambda_{\mathcal{Y}}^{\mathcal{X}}(\mathcal{R}_i) = \frac{1}{|\Gamma \times \Gamma|} \lambda(\widehat{\mathcal{R}_i})$$

where

$$\widehat{\mathcal{R}_i} = \mathcal{R}_i \cap (\Gamma \times \Gamma) \text{ and } \lambda(\sum_{j \in J} \mathcal{U}_j) = \sum_{j \in J} \lambda(\mathcal{U}_j).$$

If $\widehat{\mathcal{R}_i}$ is empty then $\lambda(\widehat{\mathcal{R}_i})$ is taken to be 0.

The following was first shown for quasigroup class functions in [176], but subsequently a note [177] was published indicating the result for arbitrary association schemes. It may be seen that induction is an averaging process.

Theorem 4.4

(a) *Transitivity of induction. Let* $\mathcal{X} = (\Omega, \{\mathcal{R}_i\}_{0 \le i \le r})$ *be an association scheme, and let* $\mathcal{Y} = (\Gamma, \{\mathcal{U}_i\}_{0 \le i \le u})$ *be a refinement subscheme of* $\mathcal{X}$*. Further, let* $\mathcal{Z} = (\Lambda, \{\mathcal{V}_i\}_{0 \le i \le v})$ *be a refinement subscheme of* $\mathcal{Y}$*, and* ϕ *be a class function on* $\mathcal{Z}$*. Then*

$$(\phi_{\mathcal{Z}}^{\mathcal{Y}})_{\mathcal{Y}}^{\mathcal{X}} = \phi_{\mathcal{Z}}^{\mathcal{X}}.$$

(b) *Frobenius reciprocity. Let* ϕ *be a class function on* $\mathcal{Y}$ *and* ψ *be a class function on* $\mathcal{X}$*. The restriction* $\psi_{\mathcal{Y}}$ *of* ψ *to* $\mathcal{Y}$ *is defined in the obvious manner. Then*

$$\langle \phi, \psi_{\mathcal{Y}} \rangle_{\mathcal{Y}} = \langle \phi_{\mathcal{Y}}^{\mathcal{X}}, \psi \rangle_{\mathcal{X}}.$$

Proof

(a) From the definition of induction, for $\Delta \subseteq \Omega \times \Omega$,

$$\frac{(\phi_{\mathcal{Z}}^{\mathcal{Y}})_{\mathcal{Y}}^{\mathcal{X}}(\Delta)}{|\Omega \times \Omega|} = \frac{(\phi_{\mathcal{Z}}^{\mathcal{Y}})(\Delta \cap \Gamma \times \Gamma)}{|\Gamma \times \Gamma|} = \frac{\phi(\Delta \cap \Gamma \times \Gamma \cap \Lambda \times \Lambda)}{|\Lambda \times \Lambda|} = \frac{\phi(\Delta \cap \Lambda \times \Lambda)}{|\Lambda \times \Lambda|}$$

$$= \frac{\phi_{\mathcal{Z}}^{\mathcal{X}}(\Delta)}{|\Omega \times \Omega|}.$$

(b) Let ϕ and ψ be class functions on $\mathcal{Y}$ and $\mathcal{X}$ respectively. Then

$$\langle \psi, \phi_{\mathcal{Y}}^{\mathcal{X}} \rangle_{\mathcal{X}} = \frac{1}{|\Omega \times \Omega|} \sum_{i=1}^{r} \sum_{(z,t) \in \mathcal{R}_i} \psi(t, z) \phi_{\mathcal{Y}}^{\mathcal{X}}(z, t) \qquad (4.6.1)$$

Now let $\widehat{\mathcal{R}_i} = \mathcal{R}_i \cap (\Gamma \times \Gamma)$, thus $\widehat{\mathcal{R}_i}$ may be represented as $\mathcal{U}_{i1} \cup \mathcal{U}_{i2} \ldots \cup \mathcal{U}_{is_i}$. Then

$$\langle \psi, \phi_y^x \rangle_x = \frac{1}{|\Omega \times \Omega|} \sum_{i=1}^{r} \sum_{(z,t) \in \widehat{\mathcal{R}_i}} \psi(t,z) \frac{|\Omega \times \Omega|}{|\Gamma \times \Gamma|} \phi(z,t)$$

$$= |\Gamma \times \Gamma|^{-1} \sum_{i=1}^{r} \sum_{j=1}^{s_i} \sum_{(z,t) \in \mathcal{U}_{i1}} \psi y(t,z) \phi(z,t)$$

$$= \langle \phi, \psi y \rangle_y.$$

$\square$

4.6.3   Some Other Appearances of Association Schemes

The foregoing account has concentrated on the parts of association scheme theory which are closest to group theory. In [10] there is an account of other topics, for example the "Mckay observation" that connects representations of finite subgroups of $GL(2, \mathbb{C})$ with Dynkin diagrams, and the characterization of Askey-Wilson polynomials in terms of association schemes. Bannai and his school have calculated the character tables for many families of association schemes, mainly arising as Schurian schemes from permutation actions of finite classical groups. In particular they obtain the character tables of the unique family of nonassociative simple Moufang loops [11].

Interesting work within group theory was carried out in the context of double coset algebras by Frame in [103] and [104] and as in [271] the case is considered where the scheme is not commutative. It would be hard to indicate all the situations where association schemes arise in other areas. An especially good source of applications in the infinite case is [209]. The extensive theory of homogeneous spaces and its connection with special functions is just one example of how association scheme theory can be applied (see [268]). The development of harmonic analysis on finite groups is given later in Chap. 7. Mention must also be made of the work of the school of Kerber which is given an extensive treatment in [190]. An application to Tits' theory of buildings appears in [297]. The reader may also find interesting the work of Godsil and collaborators in explaining the relationship between factors and knot polynomials via association schemes (for example see [55]).

4.7 Quasigroup Characters

Quasigroups appear at the intersection of combinatorics and universal algebra, and may be regarded from one point of view as groups where the axioms have been weakened. Starting with group axioms, if the axiom of associativity is weakened various classes of loops are obtained. If further the identity axiom is not assumed, this leads to a quasigroup. The most visible appearance of non-associativity in mathematics has been in the theory of Lie algebras and Jordan algebras, which arise naturally in the geometry of mathematical physics, but more recently there have been suggestions that non-associative algebra of various forms is present in other geometrical contexts related to physics.

The approach to quasigroups along the lines of group theory can quickly lead to a plethora of complicated calculations. Proofs of results which are easy and transparent in group theory can become very complicated for loops and quasigroups, although increasingly computer software has been developed to help (for example [222]). Since character theory and representation theory have been used so successfully in group theory and applications, an extension to nonassociative objects offers intriguing possibilities. It may be noted that several of the mathematicians at the heart of finite group theory have had connections with nonassociative algebra.

Quasigroups briefly appeared in Chap. 1, Sect. 1.3.2. More formal definitions are given here, which highlight the connections with combinatorics and universal algebra.

Combinatorial definition: A quasigroup is a set Q together with a binary operation ($\circ$) such that $x \circ a = b$ and $a \circ y = b$ have unique solutions x and y for all $a, b \in Q$. A **latin square** is an $n \times n$ array whose entries are $\{u_1, u_2, \ldots, u_n\}$ such that each entry u_i appears exactly once in each row and column. Then Q is a finite quasigroup if and only if its multiplication table, the table whose $(i, j)^{th}$ entry is $u_i \circ u_j$, forms a latin square. A quasigroup Q with an identity element e is a **loop**, and an associative loop is a group.

Equational definition: a quasigroup $(Q, \cdot, /, \backslash)$ is a set Q equipped with three binary operations of multiplication, **right division** / and **left division** \, satisfying the identities:

$$y \backslash (y \cdot x) = x;$$
$$(x \cdot y)/y = x;$$
$$y \cdot (y \backslash x) = x;$$
$$(x/y) \cdot y = x;$$

Quasigroups $(Q_1, \circ)$ and $(Q_1, *)$ are **isotopic** if there exist a triple (U, V, W) of permutations on the set Q, such that

$$q_1 * q_2 = W(U(q_1) \circ V(q_2)).$$

The idea of isotopy can be expressed more combinatorially. Finite quasigroups Q_1, Q_2 are isotopic if and only if the latin square corresponding to Q_2 can be obtained from that corresponding to Q_1 by permuting rows and columns and renaming elements. Isotopy may be ignored in group theory, because two groups Q_1 and Q_2 are isotopic if and only if they are isomorphic. It is usual to write the operation of Q as juxtaposition ab. An important reference for loops and quasigroups is [34].

Group representations are usually approached via homomorphisms into the general linear group. However, a homomorphism from a quasigroup Q to a group cannot contain more information than that in the largest group which is a homomorphic image of Q (and this could be the trivial group). Thus if an effective character theory is to be produced, another path to the construction of characters is required. The work described in Chap. 1 indicates such a way, in that the original definition by Frobenius of group characters did not involve homomorphisms into a linear group. An extension of his approach leads to an association scheme arising from the permutation action of the multiplication group of a quasigroup Q, which produces a Gelfand pair, and the $\mathcal{P}$-matrix is renormalized to obtain a character theory for Q which coincides with the usual character theory if Q is a group. The papers [175? –179] set out such a character theory, the **combinatorial** character theory of Q (see also [183]). There are many other interesting generalizations of aspects of group representation theory to quasigroups in [263] and the reader is referred to this. In particular, Smith includes an interpretation of the Burnside ring via stochastic matrices. The discussion here is limited to the combinatorial character theory.

4.7.1   The Multiplication Group of a Quasigroup

Let Q be a quasigroup.

Definition 4.10 The **multiplication group** $Mlt(Q)$ of Q is the permutation group on the set Q generated by the maps $\{L(q), R(q)\}, q \in Q$ where $L(q) : x \to qx$ and $R(q) : x \to xq$.

It is seen that the above definition generalizes the definition in Example 4.1, and that for loops given above.

The **conjugacy classes** of Q are defined to be the orbits Δ_i of $Mlt(Q)$ acting on $Q \times Q$ with diagonal action. If Q is a loop the **inner mapping group** $I(Q)$ is the stabiliser of e in $Mlt(Q)$. A subloop K of a loop Q is **normal** if and only if $I(Q)K = K$.

Example 4.7 There is a commutative loop Q_6 of order 6, whose unbordered multiplication table is the following latin square.

$$
\begin{array}{cccccc}
1 & 2 & 3 & 4 & 5 & 6 \\
2 & 1 & 4 & 3 & 6 & 5 \\
3 & 4 & 5 & 6 & 1 & 2 \\
4 & 3 & 6 & 5 & 2 & 1 \\
5 & 6 & 1 & 2 & 4 & 3 \\
6 & 5 & 2 & 1 & 3 & 4
\end{array}
$$

The multiplication group is isomorphic to $A_4 \times C_2$. The inner mapping group is the subgroup with elements $\{e, (3, 4), (5, 6), (3, 4)(5, 6)\}$. Q_6 has the normal subloop $\{1, 2\}$ of order 2 (which since all loops of orders ≤ 4 are associative is isomorphic to C_2).

In the case where Q is a loop, the subsets C_i of Q defined by

$$C_i = \{q : (e, q) \in \Delta_i\} \tag{4.7.1}$$

are usually defined to be the conjugacy classes of Q, and it follows directly that the C_i coincide with the orbits of $I(Q)$. If Q is a group this coincides with the usual definition of conjugacy classes (see [175]). Compare also [259].

It is easily seen that the conjugacy classes of the loop Q_6 are

$$\{e\}, \{2\}, \{3, 4\}, \{5, 6\}.$$

As in Sect. 4.2 there is a basis $\{A_i\}$ of the centralizer ring $\mathbf{S}$ of the permutation action $(Mlt(Q), Q)$ associated to the orbits $\{\Delta_1, \Delta_2, \dots, \Delta_r\}$. Specifically, the $(j, k)^{th}$ entry of A_i is 1 if $(q_j, q_k) \in \Delta_i$ and 0 otherwise. The theory depends on the important fact that $\mathbf{S}$ is commutative, or in other words the pair $(Mlt(Q), H)$ where H is any point stabiliser in $Mlt(Q)$ is a Gelfand pair. As mentioned above, the fact that a Gelfand pair is associated to any quasigroup first appeared in [175] but does not seem to be widely known. This is proved below. First the loop case is addressed.

Proposition 4.2 *Let Q be a finite loop. The centralizer ring $\mathbf{S}$ of the permutation action $(Mlt(Q), Q)$ is commutative.*

Proof Assume that Q is a finite loop, $Q = \{e = q_1, q_2, \dots, q_n\}$. It is sufficient to show that the A_i as defined above commute.

To each Δ_i there is associated the conjugacy class $C_i = \{p_1, p_2, \dots, p_{r_i}\}$ of Q as in (4.7.1). As is described above, there is associated to each $g \in Mlt(Q)$ an $n \times n$ permutation matrix $T(g) : T(g)_{i,j} = 1$ if and only if $g(q_i) = q_j$, and further a permutation matrix $T(q)$ may be assigned to $q \in Q$ by $T(q) = T(R(q))$. It will be shown that

$$A_i = \sum_{j=1}^{r_i} T(p_j). \tag{4.7.2}$$

Consider the orbit Δ_i. The subset of elements of Δ_i with first coordinate e consists of $\{(e, p_j)\}_{j=1}^{r_i}$ corresponding to C_i. Then for a given q the subset $\{(q, qp_j)\}_{j=1}^{r_i}$ is exactly the subset of Δ_i of all elements whose first coordinate is q. Here the fact is used that the qp_j are distinct and that the valency of relation $\mathcal{R}_i$ (whose incidence matrix is A_i) is r_i. Thus the matrix A_i has entry 1 in the $(j, k)^{\text{th}}$ position if and only if there exists $p \in Q$ such that $q_j p = q_k$ and (4.7.2) follows. Since the A_i form a basis for the centralizer ring of $Mlt(Q)$, in particular they commute with $T(p)$ for all $p \in Q$ and thus A_i commutes with A_j for arbitrary i, j. $\square$

Lemma 4.7 *If Q is any finite quasigroup it is isotopic to a loop P with $Mlt(Q) \supseteq Mlt(P)$.*

Proof Consider the right maps $\{\mu_i\}_{i=1}^n$ and the left maps $\{\lambda_i\}_{i=1}^n$ of Q. The quasigroup Q' with right maps $\{\mu_i'\}_{i=1}^n = \{\mu_1^{-1}\mu_i\}_{i=1}^n$ and left maps $\{\lambda_i'\}_{i=1}^n = \{\lambda_{\mu_1^{-1}(i)}\}_{i=1}^n$ is isotopic to Q and clearly $Mlt(Q) \supseteq M(Q')$. Now the quasigroup Q'' with right maps $\{\mu_i''\}_{i=1}^n = \{\mu_{\lambda_1'^{-1}(i)}'\}_{i=1}^n$ and left maps $\{\lambda_i''\}_{i=1}^n = \{\lambda_1'^{-1}\lambda_i'\}_{i=1}^n$ has $\mu_1'' = \lambda_1'' = 1$, which implies that Q'' has an identity element and again $M(Q') \supseteq M(Q'')$. The lemma follows on taking $P = Q''$. $\square$

The above construction of a loop from a quasigroup may be thought of in terms of latin squares as follows. The multiplication table of a quasigroup whose elements are labelled as $\{1, 2, \ldots, n\}$ is a latin square

a_{11}	a_{12}	$\ldots$	a_{1n}
a_{21}	a_{22}	$\ldots$	a_{2n}
$\ldots$			
a_{n1}	a_{n2}	$\ldots$	a_{nn}

where $a_{ij} = i \cdot j$. If the rows are reordered so that $\{1, 2, \ldots, n\}$ appear in their natural order in the first column, the multiplication table of Q' is obtained, and then if the columns of the new table are rearranged so that $\{1, 2, \ldots, n\}$ appear in their natural order in the first row the multiplication table of Q'' is obtained, and it is clear that the element in the top left-hand corner is an identity element, i.e. Q'' is a loop.

Theorem 4.5 *If Q is a quasigroup with multiplication group $Mlt(Q)$, $q_0 \in Q$ and $I_{q_0}(Q) = \{g \in Mlt(Q) : g(q_0) = q_0\}$ then $(Mlt(Q), I_{q_0}(Q))$ form a Gelfand pair.*

Proof This follows from Proposition 4.2 and Lemma 4.7, noting that if $G \supseteq H$ are both permutation groups on the set Ω the orbits of G are unions of orbits of H and therefore the centralizer ring of G is contained in that of H. $\square$

It has now been shown that there is an association scheme attached to each quasigroup Q, and the $\mathcal{P}$-matrix P and the renormalized matrix Ψ of this scheme can be constructed as in Sect. 4.5.1.

Definition 4.11 The character table of the quasigroup Q is the matrix Ψ.

Example 4.8 Let M_{120} be the simple Moufang loop of order 120. This is the smallest member of the family of Paige loops, which is the unique family of simple non-associative Moufang loops. The $\mathcal{P}$-matrices of the corresponding association schemes are calculated in [11] and that of M_{120} is

$$\begin{bmatrix} 1 & 63 & 56 \\ 1 & -9 & 8 \\ 1 & 3 & -4 \end{bmatrix}$$

and the (group normalized) character table is

Class size →	1	63	56
χ_1	1	1	1
χ_2	$\sqrt{35}$	$-\sqrt{\frac{5}{7}}$	$\sqrt{\frac{5}{7}}$
χ_3	$\sqrt{84}$	$\frac{2}{\sqrt{21}}$	$-\sqrt{\frac{3}{7}}$

The reader is reminded that the renormalization between the two matrices is as follows. The entry in the (i, j)th position of the character table is $\sqrt{f_i}/n_j$ times the corresponding entry of the $\mathcal{P}$-matrix, where n_j is the size of the jth conjugacy class of Q and f_i is the degree of the corresponding irreducible subconstituent of the permutation representation of $Mlt(Q)$ acting on Q.

4.7.2 Class Functions on Quasigroups

In the case of a loop Q, a class function may be regarded as a function from Q to $\mathbb{C}$ which is constant on conjugacy classes (which are subsets of Q). For an arbitrary quasigroup Q, the conjugacy classes may only be regarded as subsets of $Q \times Q$, and a class function is defined as follows. Let $Cl(Q)$ denote the set of conjugacy classes of Q.

Definition 4.12 If Q is a quasigroup, a class function is a function $f : Q \times Q \to \mathbb{C}$ which is constant on the conjugacy classes of Q.

The character table of Q has rows indexed by the characters, and columns indexed by the conjugacy classes. A character χ is a class function in the above sense, using the convention that for (s, t) in $Q \times Q$, $\chi(s, t) = \frac{1}{|C_i|}\chi(C_i)$, where $(s, t) \in C_i$.

Definition 4.13 If φ and ψ are class functions on Q define the inner product $\langle , \rangle$ by

$$\langle \varphi, \psi \rangle = \sum_{C_i \in Cl(Q)} \varphi(C_i)\overline{\psi(C_i)} = \frac{1}{|Q \times Q|} \sum_{(s,t) \in Q \times Q} \varphi(s,t)\overline{\psi(s,t)}.$$

It also follows that

$$\langle \varphi, \psi \rangle = \frac{1}{|Q \times Q|} \sum_{(s,t) \in Q \times Q} \varphi(s,t)\psi(t,s),$$

using the fact that for any association scheme $p_i(j) = \overline{p_{i'}(j)}$. This inner product generalizes the inner product of group class functions.

4.7.2.1 Further Examples

Consider the commutative loop Q_6 of order 6 given in Example 4.7. Its character table is

Class size	1	1	2	2
χ_1	1	1	1	1
χ_2	1	1	ω	ω^2
χ_3	1	1	ω^2	ω
χ_4	$\sqrt{3}$	$-\sqrt{3}$	0	0

where $\omega = (-1+\sqrt{3}i)/2$. The **kernel** of a loop character χ of a loop Q is analogous to the kernel of a group character. It is the subset $\{x : \chi(x) = \chi(e)\}$. It may be seen that the kernels of the first three characters coincide as the unique normal subloop of order 2.

The Ward quasigroup corresponding to C_5 (see Chap. 1, Sect. 1.3.2) has character table

Class size	1	2	2
χ_1	1	1	1
χ_2	$\sqrt{2}$	$\frac{1}{\sqrt{2}}\alpha$	$\frac{1}{\sqrt{2}}\alpha^*$
χ_3	$\sqrt{2}$	$\frac{1}{\sqrt{2}}\alpha^*$	$\frac{1}{\sqrt{2}}\alpha$

where $\alpha = (-1+\sqrt{5})/2$, $\alpha^* = (-1-\sqrt{5})/2$.

The character table of the smallest non-associative Moufang loop (of order 12) is

Class size	1	2	3	3	3
χ_1	1	1	1	1	1
χ_2	1	1	1	-1	-1
χ_3	1	1	-1	1	-1
χ_4	1	1	-1	-1	1
χ_5	$2\sqrt{2}$	$-\sqrt{2}$	0	0	0

4.7.3 The Information in the Character Table of a Loop or Quasigroup

Let the $(i, j)^{th}$ entry in the character table of a quasigroup Q be $\chi_i(j)$. χ_i will be referred to as the ith **basic character** and $\chi_i(j)$ as its value on the jth class. The basic characters form a basis for the set of class functions. If Q is a loop the class functions may be regarded as functions from Q to $\mathbb{C}$ constant on the conjugacy classes regarded as subsets of Q, as indicated above, giving a closer analogy to the group case. However, in a loop Q there is not necessarily a two-sided inverse for an element $q \in Q$. Instead there is a **left inverse** q^λ which is the solution of $xq = e$, and a **right inverse** q^ρ which is the solution of $qx = e$.

The character tables exhibit some properties of group tables. There are row and column orthogonality relations.

Theorem 4.6 *If* $\{\chi_1, \chi_2, \ldots \chi_r\}$ *are the basic characters of a finite quasigroup Q then*

Proposition 4.3

(a) $\langle \chi_i, \chi_j \rangle = \delta_{ij}|Q|$.
(b) $\sum_{i=1}^r \chi_i(k)\chi_i(m) = \delta_{k,m}|Q|/|C_k|$.

It is easily seen that Theorem 4.6 follows from the orthogonality relations on the $\mathcal{P}$-matrix of the association scheme.

As for groups, the character table $\Psi(Q)$ of a finite quasigroup Q encodes a large amount of information about the quasigroup. In particular, it specifies the numbers k_i, m_i, p_{ij}, q_{ij} of the corresponding association scheme.

4.7.4   The Information in the Character Table of a Loop

The following definitions are parallel to those for groups.

Definition 4.14 A subloop of a loop Q is a subset $H \subset Q$ such that

(1) $e \in H$,
(2) for all $q \in H$, q^λ and q^ρ lie in H,
(3) for all q_1, q_2 in H, $q_1 q_2$ lies in H.

If $f : Q_1 \rightarrow Q_2$ is a homomorphism, the kernel is a normal subloop. Conversely, every normal subloop H of a loop Q is the kernel of the epimorphism

$$Q \rightarrow Q/H,$$

where the factor loop Q/H is defined on the cosets of H in an analogous manner to a factor group of a group. Note that if a loop Q has a non-normal subloop the cosets need not form a partition of Q, i.e. Lagrange's theorem need not hold.

Definition 4.15 The **center** of a loop Q is the subset $\{z : xz = zx, (xy)z = x(yz)$ for all $x, y \in Q\}$.

It necessarily follows that the center is an abelian group, and that it is normal in Q. Thus if Q has center Z_1 one can form the factor loop Q/Z_1 and then the **second center** Z_2 is defined by Z_2/Z_1 is the center of Q/Z_1, etc. This defines the **upper central series**.

In a loop Q the **commutator** of elements x and y is the element $v = (x, y)$ defined by

$$xy = (yx)v.$$

The **associator** of the elements x, y, z is the element $w = (x, y, z)$ defined by

$$(xy)z = (x(yz))w.$$

To define the lower central series, let $\gamma_1(Q) = Q'$, and define γ_{i+1}, $i \geq 2$ inductively by

$$\gamma_{i+1}(Q) = \langle (Q, \gamma_i(Q)), (Q, Q, \gamma_i(Q)), (Q, \gamma_i(Q), Q), (\gamma_i(Q), Q, Q) \rangle,$$

the subloop generated by the commutators and associators of the indicated types. Then the **lower central series** of Q is

$$Q = \gamma_0(Q) \supseteq \gamma_1(Q) \supseteq \cdots \supseteq \gamma_i(Q) \supseteq \cdots.$$

Lemma 4.8 *The character table of a loop Q contains the following information.*

(1) Each homomorphism of Q into an abelian group corresponds to a linear character of Q.

(2) The set of normal subloops. A subloop P of a loop Q is normal if and only if P is the intersection of the kernels of a set of basic characters.

(3) The center (the union of singleton classes).

(4) The upper and lower central series.

4.7.5   The Information in the Character Table of a Quasigroup

Modifications need to be made for quasigroups. For example, if Q_1, Q_2 are quasigroups a homomorphism $f : Q_1 \rightarrow Q_2$ does not necessarily provide a subquasigroup as a kernel. Instead there is a congruence. A **congruence** on a quasigroup Q is an equivalence relation which, as a subset of $Q \times Q$, is a subquasigroup of $Q \times Q$. The normal structure of a quasigroup is reflected in the lattice of congruences on Q under inclusion. This is determined by the character table of Q.

Theorem 4.7 ([175, 261, Th. 3.6]) *The congruence lattice of a finite quasigroup is determined by its character table.*

The proof of Theorem 4.7 is easy in the group or loop case since a subgroup or subloop is normal if and only if it is the intersection of the kernels of characters. The quasigroup case is less immediate, and uses the idea of [47, Prop. 3.1].

If Q and P are groups and if $f : Q \rightarrow P$ is a homomorphism from Q onto P and ϕ is an irreducible (i.e. basic) character of P it follows from the fact that ϕ is associated to a matrix representation that $\chi = f\phi$ (composition) is an irreducible character of Q. The statement that such a composition is a basic character for the quasigroup Q remains true for arbitrary quasigroups P and Q but the proof is not so direct. It appears in [178].

A linear character of a finite group G arises from a one-dimensional representation, and it is easily proved that any linear character of G is associated to a homomorphism into the abelian group G/G'. The statement of the corresponding result for quasigroups is analogous. A linear character of a quasigroup Q is defined to be a basic character ϕ such that $\phi(C_1) = 1$. Any non-empty quasigroup Q has a unique abelian group $Q^\natural$ which is called the **abelian replica** of Q, (analogous to G/G' above): it is the largest abelian group which is a homomorphic image of Q. Then a basic character of Q is linear if and only if it arises from a basic character of $Q^\natural$ by composition with the homomorphism $Q \rightarrow Q^\natural$. The proof is given in [179] and is considerably more involved than in the group case.

In contrast to the group case the entries in the character table of a quasigroup need not be algebraic integers, as is easily seen from the examples above, although the $\mathcal{P}$-matrix entries are algebraic integers. In the case of a group G the order of the

classes and the degree of the characters are integers dividing the order of G. In the
loop case, it is well-known that the order of a class need not divide the order of the
loop, and if the order of a class C of a quasigroup is defined to be $\frac{|C|}{|Q|}$ again this
need not divide $|Q|$. The examples in the previous section show that the degree of a
character $\chi\ (=\chi(C_1))$ need not be an integer.

4.7.6   Rank 2 Quasigroups

A non-empty, finite quasigroup Q has rank 2 if the classes are $C_1 = \{(x,x)\}$ and
$C_2 = Q \times Q - C_1$. Equivalently, the multiplication group G of Q acts 2-transitively
on Q. For example, any quasigroup of order 3 which is not an abelian group has
rank 2. It can easily be calculated from the orthogonality relations that the (group
normalized) character table of any rank 2 quasigroup Q of order n must be

$$
\begin{array}{c|cc}
 & C_1 & C_2 \\
\hline
\chi_0 & 1 & 1 \\
\chi_1 & \sqrt{n-1} & -(n-1)^{-1/2}
\end{array}
$$

In a well-defined sense, almost all quasigroups have rank 2. For a positive integer
n, let $l(n)$ denote the number of Latin squares of order n in the symbols $1, 2, \ldots, n$.
Let $\mathcal{P}$ be a certain property which a finite quasigroup may or may not possess. Let
$p(n)$ denote the number of Latin squares of order n that yield multiplication tables
of quasigroups possessing the property $\mathcal{P}$ when bordered on the left and on top with
$1, 2, \ldots, n$ in order. Then **almost all finite quasigroups** are said to **have property**
$\mathcal{P}$ if

$$\lim_{n\to\infty} \frac{p(n)}{l(n)} = 1, \tag{4.7.3}$$

while **hardly any finite quasigroups** are said to **have property** $\mathcal{P}$ if the limit
in (4.7.3) is 0. (Thus in the latter case, almost all finite quasigroups have the
complementary property $\neg\mathcal{P}$.) The following statement was conjectured in [261]:

Theorem 4.8 *Almost all finite quasigroups are rank 2 quasigroups.*

In fact, Theorem 4.8 follows from a stronger observation which C.E. Praeger
attributes to P.J. Cameron in [238]:

Theorem 4.9 *Almost all finite quasigroups Q have the symmetric group $Q!$ as their
multiplication group.*

Proof By [206], the smallest transitive permutation group containing a random
permutation is almost always the symmetric or alternating group. But by [133],
hardly any finite quasigroups have their multiplication group consisting entirely of
even permutations.   □

The above result means that the combinatorial character theory of almost all quasigroups contains very little information. However, for many interesting varieties of quasigroups and loops, it is rare to find an object with a trivial character table. For example every non-associative Moufang loop has a non-trivial character table. A group of order larger than 2 cannot the trivial character table.

4.7.7   Induction and Restriction for Quasigroup Characters and Frobenus Reciprocity

Let Q be a quasigroup with classes $\{\Delta_1, \ldots, \Delta_r\}$ with non-empty subquasigroup P. Let the classes of P (regarded as subsets of $P \times P$) be $\{\Gamma_1, \ldots, \Gamma_s\}$. Suppose that ϕ is a class function on P. Then the value of the induced class function ϕ^Q on the class Δ of Q is defined by

$$\phi_P^Q(\Delta)/|Q \times Q| = \phi(\Delta \cap P \times P)/|P \times P|.$$

It follows automatically that induced class functions may be defined for the case where a quasigroup P is a subquasigroup of a quasigroup Q. Theorem 4.4 then may be applied to obtain the following corollary.

Corollary 4.1   *The following hold*

(a) Transitivity of induction. Let H be a non-empty subquasigroup of P which is in turn a subquasigroup of Q, and ϕ be a class function on H. Then

$$(\phi_H^P)_P^Q = \phi_H^Q.$$

(b) Frobenius reciprocity. Let ϕ be a class function on P and ψ be a class function on Q. The restriction ψ_P of ψ to P is defined in the obvious manner. Then

$$\langle \phi, \psi_P \rangle_P = \langle \phi_P^Q, \psi \rangle_Q.$$

If Q and P are groups the results in (4.1) become the well-known results for group characters. However, whereas in the case of groups induction and restriction take a character into a character, for an arbitrary basic character χ of Q with subquasigroup P, the restriction χ_P is a linear combination of basic characters of P but the coefficients need not be integers. Similarly, if ψ is a basic character of P, ψ_P^Q need not be a linear combination of positive integral multiples of basic characters of Q. This means that care must be used in using induction to deduce the existence of basic characters. Another major technique in group character theory also fails to generalize. If ϕ and ψ are class functions on Q the product $\phi\psi$ is defined by

$$\phi\psi(q_1, q_2) = \phi(q_1, q_2)\psi(q_1, q_2).$$

In the group case $\phi\psi$ corresponds to the tensor product of the corresponding representations and is always a sum of irreducible characters with positive integral coefficients. For quasigroup characters the analogous statement is no longer true. For a quasigroup Q the **coefficient ring** is defined to be the ring over Z generated by $\langle \chi_i \chi_j, \chi_k \rangle$. This ring seems to be of interest and several conjectures about it are given in [170].

4.7.8 A Generalization of Artin's Theorem for Group Characters

The theorem of Artin for group characters may be stated as follows.

Theorem 4.10 *Any character χ of a finite group G may be expressed as*

$$\chi = \sum b_i \psi_i,$$

where the ψ_i are induced from cyclic subgroups and the b_i lie in $\mathbb{Q}$.

There follows a version of this theorem for quasigroups. A set of subquasigroups $\{P_i, i = 1, \ldots, N\}$ of a quasigroup Q is defined to be **protrusive** if for each conjugacy class C_i there exists a j such that $P_j \chi P_j \cap C_i \neq \emptyset$. For groups an obvious protrusive set is the set of all cyclic subgroups, and this generalizes to the set of singly generated subloops of a loop (although this is no longer the case for quasigroups). The proof of the following appears theorem appears in [176]).

Theorem 4.11 *Let $\{P_i, i = 1, \ldots N\}$ be a protrusive set of subquasigroups of Q. A basic character χ of Q satisfies*

$$\chi_i = \sum_j b_j \psi_{P_j}^Q$$

where ψ is a basic character of P_j and the b_j are algebraic numbers.

4.7.9 Methods of Calculation

As with group characters, a variety of different methods has been used to calculate quasigroup character tables. Below are some of the methods which have been applied.

(1) A direct calculation of the eigenvalues of the adjacency matrices. This is effective only in small cases.

(2) The calculation of the structure constants of the class algebra. The structure constants $\{a_{ij}^k\}$, which are non-zero integers, are defined by

$$\bar{C}_i \bar{C}_j = \sum a_{ij}^k \bar{C}_k.$$

A matrix representation of the class algebra can be obtained by representing $\bar{C}_i$ by the matrix whose (j, k)th entry is a_{ij}^k and the $\mathcal{P}$-matrix entries are again the eigenvalues. This method is similar to that by which Frobenius calculated the character table of $PSL(2, p)$ ([106]).

(3) If Q has a homomorphic image whose character table is known, a portion of the character table of Q may be calculated by using the composition with the homomorphism as described above.

(4) Under certain circumstances induction may be used to obtain basic characters, although care must be taken in the case where the coefficient ring is not integral.

(5) Fusion. It often happens that the character table of a quasigroup Q_1 is a fusion of the table of a quasigroup Q_2 of the same order. The magic rectangle condition introduced above becomes relevant. This method may be applied most easily when Q_1 is a loop with conjugacy classes which are unions of conjugacy classes of a group Q_2. The character table of Q_1 may then be obtained from that of Q_2, which can be calculated by the much easier methods of group theory. Fusion is also used to obtain the character tables of various families of loops arising from extensions of groups discussed in [170]. A much more extensive account of fusion in the group case appears in Chap. 9.

The initial development of loops and quasigroups was connected with the development of web geometry by Blascke and Bol. To any isotopy class of finite quasigroups there is associated a 3-web, and since character tables of isotopic loops coincide, it would be interesting if a direct construction of a character table from a 3-web could be found (see [20]).

Chapter 5
The 2-Characters of a Group and the Weak Cayley Table

Abstract Group character theory has many applications, but it is not easy to characterize the information contained in the character table of a group, or give concise information which in addition to that in the character table determines a group. The first part of this chapter examines the extra information which is obtained if the irreducible 2-characters of a group are known. It is shown that this is equivalent to the knowledge of the "weak Cayley table" (WCT) of the group G. A list of properties which are determined by the WCT and not by the character table is obtained, but non-isomorphic groups may have the same WCT. This work gives insight into why the problems in R. Brauer, Representations of finite groups, in "Lectures in Modern Mathematics, Vol. I," T.L. Saaty (ed.), Wiley, New York, NY, 1963, 133–175, have often been hard to answer.

There is a discussion of the work of Humphries on $\mathcal{W}(G)$ the group of weak Cayley table isomorphisms (WCTI's). This group contains the a subgroup $\mathcal{W}_0(G)$ generated by $Aut(G)$ and the anti-automorphism $g \to g^{-1}$. $\mathcal{W}(G)$ is said to be trivial if $\mathcal{W}(G) = \mathcal{W}_0(G)$. For some groups, such as the symmetric groups and dihedral groups $\mathcal{W}(G)$, is trivial, but there are many examples where $\mathcal{W}(G)$ is much larger than $\mathcal{W}_0(G)$. The discussion is not limited to the finite case.

The category of "weak morphisms" between groups is described. These are weaker than ordinary homomorphisms but preserve more structure than maps which preserve the character table. A crossed product condition on these morphisms is needed to make this category associative. This category has some interesting properties.

5.1 Introduction

A considerable amount of effort has been made to examine the information contained in the character table of a finite group, but there are many open problems, for example that of determining a list of properties which forces a table to be a group character table. A reason for some of the difficulties is that character tables

© Springer Nature Switzerland AG 2019

K. W. Johnson, *Group Matrices, Group Determinants and Representation Theory*, Lecture Notes in Mathematics 2233, https://doi.org/10.1007/978-3-030-28300-1_5

of objects such as loops and quasigroups can be constructed which, while having many properties of group character tables, are not character tables of groups.

The character theory of groups has enormous uses in both pure and applied group theory, but there remains the question of whether there is not a finer tool which, while not containing the full information on the group, can be used where character theory does not contain sufficient information. The 2-characters are a candidate for such a tool and this chapter looks at the information which they contain.

As mentioned in the Preface, Richard Brauer put forward a series of problems in [30] designed to illustrate the state of knowledge of representation theory. Among these is

Problem 5.1 Let G and G^* be two finite groups and assume that there exists a one-to-one mapping $K_j \rightarrow K_j^*$ of the set of conjugacy classes of G onto the set of conjugacy classes of G^*, and a one-to-one mapping of characters $\chi_i \rightarrow \chi_i^*$ of the irreducible characters of G to the irreducible characters of G^* such that

(a) $$\chi_i(K_j) = \chi_i^*(K_j^*)$$

(b) $$(K_j^{[m]})^* = (K_j^*)^{[m]} \text{ for all integers } m,$$

Are G and G^* isomorphic?

Here $K^{[m]}$ denotes the conjugacy class which consists of the mth powers of the elements of K.

Examples were quickly given by Dade [70] which show that the answer is negative, and there has been significant discussion about pairs of groups which satisfy the conditions of Problem 5.1 which have been given the name of **Brauer pairs**. There remains the question of whether there is a concise way of presenting information which added to that of the character table determines a group. The result on 3-characters given in Chap. 1, Theorem 1.9 provides an answer, but the information is not concise. Related to this is the question of how much more information is contained in the irreducible 1- and 2-characters of a group than in the character table. This is addressed in the first part of this chapter. It is also shown that the information contained in the irreducible 1- and 2-characters is precisely that in the weak Cayley table defined below.

Section 5.1 examines the information contained in the 1- and 2-characters of a group and introduces the weak Cayley table of a group (WCT). A list of properties is given which are not determined by the ordinary character table of a group but which are when the information on the 2-characters is included, but this list remains far from definitive.

In Sect. 5.2 the work of Humphries [151] is described on the group $\mathcal{W}(G)$ of weak Cayley table isomorphisms of a group. This group is in a sense a generalization of the automorphism group of a group. It necessarily contains the anti-homomorphisms. Kimmerle has discussed the analogous group of character table isomorphisms for finite groups (see [193] and the references there). The discussion in [151] is not restricted to finite groups. Often $\mathcal{W}(G)$ can be quite large, but for groups such as the symmetric groups and free groups it can be "trivial" in the sense explained below.

The category **Gwp** of weak Cayley morphisms of groups is the subject of Sect. 5.3. In **Gwp** the isomorphisms are weaker than group isomorphisms but produce isomorphisms of the corresponding weak Cayley tables, and hence of the corresponding character tables. In the Sect. 5.2 a weak Cayley table isomorphism is defined, but it was found that in the categorical context an extra assumption is necessary to obtain an associative category, involving a crossed product condition. One of the results described here is that there is a left adjoint to the forgetful functor from **Gwp** to **Gp**, the category of group homomorphisms.

5.2   The Information in the 1- and 2-Characters of a Group

As a prelude to discussing the natural extensions of the problems of Brauer there follows a summary of results on the information contained in the ordinary character table, which relies heavily on the thesis of Mattarei [217]. See also [219] for related work.

5.2.1   The Information Contained in the Ordinary Character Table

An exact statement of the data included in the character table of a group is given in the following definition.

Definition 5.1 Let G be a finite group with conjugacy classes $\{C_1, \ldots, C_r\}$ and ordinary irreducible characters $\{\chi_1, \ldots, \chi_r\}$. The character table $\mathcal{T}(G)$ of the finite group G is the $n \times n$ matrix of algebraic numbers whose rows are indexed by the χ_i and the columns by the C_j with (i, j)th entry $\chi_i(g)$, $g \in C_j$.

No additional information is assumed, for example the order of the elements in a given class is not assumed to be given.

Definition 5.2 Let the groups G and H have respectively conjugacy classes $\{C_1, \ldots, C_r\}$ and $\{D_1, \ldots, D_r\}$, and irreducible characters $\{\chi_1, \ldots, \chi_r\}$ (respectively $\{\psi_1, \ldots, \psi_r\}$). Then G and H **have the same character table** if there is a bijection

$$\alpha : \{C_1, \ldots, C_r\} \to \{D_1, \ldots, D_r\}$$

and a bijection

$$\beta : \{\chi_1, \ldots, \chi_r\} \to \{\psi_1, \ldots, \psi_r\}$$

such that $[\beta(\chi_i)]\alpha(h) = \chi_i(g)$ whenever $g \in C_j$, $h \in \beta(C_j)$ for all i, j.

The following is a list of some of the information contained in the character table of G.

1. The size of each conjugacy class (equivalent to the order of the centralizer of each element in the class).
2. The center.
3. The lattice of normal subgroups (including the order of each subgroup).
4. The character table of each factor group.
5. It can be decided whether G is nilpotent, supersoluble or soluble.
6. The terms of the upper central series may be found, and in particular if G is nilpotent its nilpotency class may be found.
7. The lower central series may be found, and in fact the conjugacy classes which contain elements of the form $[g_1, \ldots, g_i]$ may be determined.
8. It may be determined whether a normal subgroup N is nilpotent.
9. The Fitting subgroup (i.e. the largest nilpotent subgroup of G) may be determined.
10. If G is soluble, the Fitting series and the Fitting length may be determined.
11. The prime divisors of the orders of the elements in any given conjugacy class.
12. The chief series in the following sense. A chief series of a group is a maximal series of the form

$$\{e\} = K_0 < K_1 < \ldots < K_m = G$$

where each K_i is normal in G.

Theorem 5.1 (Kimmerle–Sandling [195]) *Assume that G and H have the same character table. Let*

$$\{e\} = K_0 < K_1 < \ldots < K_m = G \tag{5.2.1}$$

be a chief series of G. Then H has a chief series

$$\{e\} = L_0 < L_1 < \ldots < L_m = H \tag{5.2.2}$$

such that the chief factors K_i/K_{i-1} and L_i/L_{i-1} are isomorphic for each index i.

(13) It can be determined whether for a given set π of primes G has nilpotent Hall π-subgroups, and if the Hall π-subgroups are abelian they can be determined up to isomorphism.

The following is a list some of the properties which are not determined by the ordinary character table.

(i) The set of involutions.
(ii) The character table of a normal subgroup.

(iii) The derived length and the question of whether if a set of conjugacy classes which generate a normal subgroup N is given it can be determined whether N is abelian.

(iv) The power maps on conjugacy classes.

 (v) The Burnside ring.

The above list of properties is far from definitive.

5.2.2   The Weak Cayley Table of a Group

To address the question of the information in the 1- and 2-characters of a group there is a need for techniques to show that a given pair of groups have the same irreducible 1- and 2-characters. The obvious approach, which was first used, was to calculate the character tables of both groups and to try to set up a bijection between the elements which induces a bijection between the irreducible 1- and 2-characters. In the papers [173] and [174] this method was used to show that such a bijection occurred (a) between certain pairs of equidistributed permutation groups; and (b) between the non-abelian groups of order p^3 for p an odd prime. These calculations are significantly more complicated than those that show the corresponding pairs of groups have the same ordinary character table.

It was later discovered that there is a simpler method. The information in the irreducible 1- and 2-characters is the same as that in the weak Cayley table described below in Definition 5.3. Moreover, in order to show that a pair G_1, G_2 of groups have the same irreducible 1- and 2-characters, or equivalently the same weak Cayley table, it is sufficient to show that there is a map $\alpha : G_1 \rightarrow G_2$ such that $\alpha(gh)$ and $\alpha(g)\alpha(h)$ are conjugate in G_2 for all $g, h \in G_2$, i.e. the calculation of characters is unnecessary. In fact in many cases it appears that the easiest way to show that G_1 and G_2 have the same character table is to demonstrate such an α, since groups with identical weak Cayley tables necessarily have the same character tables. The concept of a Camina pair is often useful in demonstrating the existence of an α. The idea of a weak Cayley table or a weak Cayley table isomorphism does not depend on the finiteness of the group (see Sect. 5.3 below). The construction of such an α to show that pairs of 2-groups have the same irreducible 1- and 2-characters was first used in an unpublished preprint by Mckay and Sibley [213].

Definition 5.3 Suppose that $G = \{e = g_1, g_2, \cdots, g_n\}$. The **weak Cayley table** (or **WCT**) of G is the table whose rows and columns are indexed by the elements of G, the entry in the $(g, h)th$ position being the conjugacy class of gh.

Note that the row which corresponds to e determines the set of conjugacy classes of G.

For example let G be $\mathcal{S}_3$ and as before let $C_0 = \{e\}$, $C_1 = \{(123), (132)\}$ and $C_2 = \{(12), (23), (13)\}$ denote the conjugacy classes. The WCT of G is

		e	(123)	(132)	(12)	(23)	(13)
e		C_0	C_1	C_1	C_2	C_2	C_2
(123)		C_1	C_1	C_0	C_2	C_2	C_2
(132)		C_1	C_0	C_1	C_2	C_2	C_2
(12)		C_2	C_2	C_2	C_0	C_1	C_1
(23)		C_2	C_2	C_2	C_1	C_0	C_1
(13)		C_2	C_2	C_2	C_1	C_1	C_0

The reduced group matrix defined in Chap. 1 is related to the WCT. The reduced group matrix of $\mathcal{S}_3$ is the matrix

$$\begin{bmatrix} x_{C_0} & x_{C_1} & x_{C_1} & x_{C_2} & x_{C_2} & x_{C_2} \\ x_{C_1} & x_{C_0} & x_{C_1} & x_{C_2} & x_{C_2} & x_{C_2} \\ x_{C_1} & x_{C_1} & x_{C_0} & x_{C_2} & x_{C_2} & x_{C_2} \\ x_{C_2} & x_{C_2} & x_{C_2} & x_{C_0} & x_{C_1} & x_{C_1} \\ x_{C_2} & x_{C_2} & x_{C_2} & x_{C_1} & x_{C_0} & x_{C_1} \\ x_{C_2} & x_{C_2} & x_{C_2} & x_{C_1} & x_{C_1} & x_{C_0} \end{bmatrix}.$$

The determinant of the matrix consisting of the unbordered WCT with C_i replaced by x_{C_i} is up to sign the same as the reduced group determinant.

Proposition 5.1 *Let G be a finite group. The WCT can be constructed from the irreducible 1- and 2-characters of a group over $\mathbb{C}$. Conversely, if the WCT is given, the irreducible 1- and 2-characters can be calculated.*

Proof Assume the irreducible 1- and 2-characters are known for a group G. It is well-known that the conjugacy classes of G can be determined from the 1-characters over $\mathbb{C}$. Now suppose

$$\chi^{(2)}(g, h) = \chi(g)\chi(h) - \chi(gh)$$

is given. Since $\chi(g)\chi(h)$ can be read off from the 1-characters it follows that the function $(g, h) \rightarrow \chi(gh)$ is known for all $\chi \in Irr(G)$. But if $\chi(gh)$ is known for all irreducible χ this is sufficient to determine the conjugacy class of gh (This follows from the column orthogonality of the character table).

Conversely, suppose the WCT is given. The elements in each conjugacy class can be read off from the first row. If C_i and C_j are conjugacy classes the (C_i, C_j) block of the table whose rows correspond to elements in C_i and whose columns correspond to elements in C_j determines the integers a_{ij}^k in the product

$$\overline{C}_i\overline{C}_j = \sum a_{ij}^k \overline{C}_k. \tag{5.2.3}$$

It follows that the class algebra, which is isomorphic to the center of $\mathbb{C}G$, and hence the character table, can be constructed from the WCT. If χ is an irreducible character $\chi(gh)$, for any pair (g, h), may be read off from the WCT and thus $\chi^{(2)}(g, h)$ can be found.   □

If the characteristic of the field is 0 the WCT can be constructed from the knowledge of the irreducible 2-characters and the 1-dimensional characters. This is the case since $\chi^{(2)}(g, e) = \chi(g)(\chi(e) - 1)$ which implies that χ can be calculated from $\chi^{(2)}$ if $deg(\chi) \neq 1$.

The idea of a WCT is implicit in [213] and a proof of Proposition 5.1 and statements of the corollaries below can essentially be found there.

As in [173], groups G_1 and G_2 are said to have the same irreducible 1- and 2-characters if there is a bijection $\sigma : G_1 \rightarrow G_2$ which induces a bijection between the irreducible characters $\{\chi_1, \cdots, \chi_r\}$ of G and $\{\psi_1, \cdots, \psi_r\}$ of H

$$\chi_i(g) = \psi_i(\sigma(g)) \tag{5.2.4}$$

and such that

$$\chi_i^{(2)}(g_1, g_2) = \psi_i^{(2)}(\sigma(g_1), \sigma(g_2)) \tag{5.2.5}$$

for all i.

Definition 5.4 G_1 and G_2 **have the same WCT** if there is a bijection $\alpha : G_1 \rightarrow G_2$ such that

for all $g, h \in G_1, \alpha(gh)$ and $\alpha(g)\alpha(h)$ are conjugate in G_2.   (5.2.6)

The map α will then be called a **WCT isomorphism or WCTI**.

Remark 5.1 The definition of a WCT-isomorphism is modified in Sect. 5.4.

5.2.2.1 Properties of WCT-Isomorphisms

Proposition 5.2 *Suppose $\alpha : G \rightarrow H$ is a WCTI. Then*

(i) $\alpha(e) = e$.
(ii) *If g_1 and g_2 are conjugate in G then $\alpha(g_1)$ and $\alpha(g_2)$ are conjugate in H.*
(iii) $\alpha(g)^{-1} = \alpha(g^{-1})$ *for all $g \in G$.*

Proof First (ii) is addressed. Consider g_1, g_2 in G and suppose that $h_1 = \alpha(g_1)$ and $h_2 = \alpha(g_2)$. Then

$$\alpha(g_1 g_2) = \alpha(\alpha^{-1}(h_1)\alpha^{-1}(h_2)) = x^{-1}\alpha(\alpha^{-1}(h_1))\alpha(\alpha^{-1}(h_2))x$$

$$= x^{-1} h_1 h_2 x$$

for some $x \in G$. Similarly $\alpha(g_2 g_1) = y^{-1} h_2 h_1 y$ for some $y \in G$. Since $\alpha(g_1 g_2)$ and $\alpha(g_2 g_1)$ are conjugate and any pair of conjugate elements of G can be written as $g_1 g_2$ and $g_2 g_1$ (ii) follows. Then (ii) implies that $\alpha(e)$ lies in the *center* of H and also that $\alpha(e)\alpha(e) = \alpha(e \cdot e) = \alpha(e)$, whence $\alpha(e) = e$. Also

$$e = \alpha(e) = \alpha(gg^{-1}) = x^{-1}\alpha(g)(g^{-1})x$$

for some $x \in G$ which implies (iii). $\qquad\qquad\qquad\qquad\qquad\qquad\square$

Note that for a general WCTI α, $\alpha(ghk)$ and $\alpha(g)\alpha(h)\alpha(k)$ need not be conjugate.

Proposition 5.3 *Let α be a WCTI from G_1 to G_2. Let N be a normal subgroup of G_1. Then*

(a) $\alpha(N)$ is a normal subgroup of G_2.
(b) The image $\alpha(gN)$ of a coset of N in G_1 is a coset $\alpha(g)\alpha(N)$ in G_2.
(c) α induces a WCTI $\bar{\alpha} : G_1/N \to G_2/\alpha(N)$ via $\bar{\alpha}(gN) = \alpha(g)\alpha(N)$.

Proof It is clear that a WCTI α sends normal subgroups of G_1 to normal subgroups of G_2. Now let N be a normal subgroup of G_1 and $gN = hN$ for some $g, h \in G_1$. Since $h^{-1}g \in N$, it follows that $\alpha(h^{-1}g) \in \alpha(N)$. Now since $\alpha(h^{-1}g)$ and $\alpha(h^{-1})\alpha(g) = \alpha(h)^{-1}\alpha(g)$ are conjugate in G_2, they both lie in $\alpha(N)$ and $\alpha(g)\alpha(N) = \alpha(h)\alpha(N)$. Hence, for a normal subgroup N of G_1, the WCTI α induces a bijection

$$\bar{\alpha} : G_1/N \to G_2/\alpha(N) \qquad\qquad\qquad (5.2.7)$$

via $\bar{\alpha}(gN) = \alpha(g)\alpha(N)$, for all $g \in G_1$. It will be proved that $\bar{\alpha}$ is a WCTI.

To prove (b), let gN and hN be conjugate cosets of N, with $g, h \in G_1$. It follows that g is conjugate to hn in G_1, for some $n \in N$. Consequently $\alpha(g)$ is conjugate to $\alpha(hn)$ in G_2 and therefore is conjugate to $\alpha(h)\alpha(n)$. It follows that $\alpha(g)\alpha(N)$ and $\alpha(h)\alpha(n)\alpha(N) = \alpha(h)\alpha(N)$ are conjugate cosets of $\alpha(N)$. The same argument applied to the inverse map of α in place of α proves that (i) is satisfied.

It will now be shown that (c) holds. Suppose that $g, h \in G_1$. Then $\bar{\alpha}(ghN) = \alpha(ghn)\alpha(N)$, where $n \in N$. Since $\alpha(ghn)$ is conjugate to $\alpha(gh)\alpha(n)$, the coset $\alpha(ghn)\alpha(N)$ of $\alpha(N)$ is conjugate to

$$\alpha(gh)\alpha(n)\alpha(N) = \alpha(gh)\alpha(N).$$

Finally, since $\alpha(gh)$ and $\alpha(g)\alpha(h)$ are conjugate, it follows that $\alpha(gh)\alpha(N)$ is conjugate to

$$\alpha(g)\alpha(h)\alpha(N) = \bar{\alpha}(gN)\bar{\alpha}(hN)$$

which proves (ii). Hence α induces a WCTI $\bar{\alpha}$ between the quotient groups G_1/N and $G_2/\alpha(N)$, in a canonical way. $\qquad\qquad\qquad\qquad\qquad\square$

5.2.3 The 2-Character Table

The 2-character table of a group is defined as follows (this is a special case of the k-character table of a group for $k \geq 2$ which is defined in Chap. 6). It was first defined in [169]. **The 2-classes** of a group G are defined to be the orbits on $G \times G$ of $\langle G, C_2 \rangle$, where G acts by simultaneous conjugation and the generator of the C_2 takes (g, h) into (h, g) for all g, h in G. The columns of the 2-character table are indexed by the 2-classes. If $Irr(G) = \{\chi_i\}_{i=1}^{r}$, the rows are indexed by the extended 2-characters (called generalized 2-characters in [169]) which are the following functions constant on the 2-classes:

(a) $\chi_i^{(2,+)}$, $i = 1, \ldots, r$, defined by

$$\chi_i^{(2,+)}(g, h) = \chi_i(g)\chi_i(h) + \chi_i(gh).$$

(b) for each χ_i, χ_j with $i < j$ the function $\chi_i \circ \chi_j$ defined by

$$\chi_i \circ \chi_j(g, h) = \chi_i(g)\chi_j(h) + \chi_i(h)\chi_j(g),$$

and
(c) $\chi_i^{(2)}$ for $\chi_i(e) > 1$.

The entry in the sth row and tth column is the value of the sth extended 2-character on the tth orbit.

Example 5.1 For S_3 the ordinary character table is given in Chap. 1 as

Class size	1	2	3
Representative	e	a	b
χ_1	1	1	1
χ_2	1	1	-1
χ_3	2	-1	0

Here $a = (123)$ and $b = (12)$. A set of representatives of the 2-classes is

$$\{(e, e), (a, a), (b, b), (e, a), (e, b), (a, a^2), (b, ab), (a, b)\}.$$

In the table below these will be abbreviated to $\{r_1, \ldots, r_8\}$. The sizes of these classes are $1, 2, 3, 4, 6, 2, 6, 12$ in the order given. The information in the extended 2-characters can be presented in the following table. The norm of a 2-class function ψ is defined to be

$$\langle \psi, \psi \rangle_2 = \frac{1}{|G|^2} \sum_{g \in G \times G} \psi(g)\overline{\psi(g)}.$$

Class size →		1	2	3	4	6	2	6	12
Representative →		r_1	r_2	r_3	r_4	r_5	r_6	r_7	r_8
Character	Norm								
$\chi_1^{(2,+)}$	4	2	2	2	2	2	2	2	2
$\chi_1 \circ \chi_2$	2	2	2	-2	2	0	2	-2	0
$\chi_1 \circ \chi_3$	2	4	-2	0	1	2	-2	0	-1
$\chi_2^{(2,+)}$	4	2	2	2	2	-2	2	2	-2
$\chi_2 \circ \chi_3$	2	4	-2	0	1	-2	-2	0	1
$\chi_3^{(2,+)}$	3	6	0	2	-3	0	3	-1	0
$\chi_3^{(2)}$	1	2	2	-2	-1	0	-1	1	0
Orthogonal	6	0	-6	-4	0	0	6	2	0

The rows of the 2-character table are orthogonal under the inner product

$$\langle \phi, \psi \rangle_2 = \frac{1}{|G|^2} \sum_{\underline{g} \in G \times G} \phi(\underline{g}) \overline{\psi(\underline{g})}.$$

The orthogonal complement is a vector which in addition to the set of extended 2-characters forms an orthogonal basis for the set of 2-class functions.

The corollaries below follow directly.

Corollary 5.1 *Let G_1 and G_2 be finite groups. The following are equivalent.*

(i) G_1 and G_2 have the same WCT.
(ii) G_1 and G_2 have the same 2-character tables.

Proof That (ii) implies (i) follows directly from Proposition 5.1, since (ii) automatically implies that G_1 and G_2 have the same 1-, and 2-characters. Now assume (i) is satisfied. If χ_i and χ_j are known it follows that $\chi_i \circ \chi_j$ is known. Since for any χ, $\chi(gh)$ can be read off from the WCT then both $\chi_i^{(2)}$ and $\chi_i^{(2,+)}$ can also be determined. ☐

Corollary 5.2 *If G_1 and G_2 have the same WCT they have the same character tables.*

The information in the character table and in the WCT of a group G may be compared as follows. That in the character table is equivalent to the knowledge of the structure constants a_{ij}^k in (5.2.3) and gives the probability that the product of a random conjugate of g with a random conjugate of h is in a given class C for every $g, h \in G$ and for every conjugacy class C of G. If S is the ring generated by the class sums $\{\overline{C_i}\}$, such a collection of probabilities on $S \times S$ tells us which element of G is the identity (which of course is characterized by the property $e \cdot e = e$), and also which subsets of G are the conjugacy classes of G. The WCT of G specifies for each ordered pair $(g, h) \in G$ exactly which class C of G contains the product gh.

It is possible to use the class algebra to produce "character-free" statements and proofs of results on groups with the same character tables, see [184].

5.2.3.1 Connections with Brauer Pairs

First, the definition of a Brauer pair is restated as

Definition 5.5 The groups G_1 and G_2 form a *Brauer pair* via $\alpha : G_1 \to G_2$ if α is a character table isomorphism and if $\alpha(g)^m$ is a conjugate of $\alpha(g^m)$ in G_2 for every $g \in G_1$ and for every integer m.

In order to test for a Brauer pair it is only necessary to consider exponents m which are prime divisors of $|G_1|$, since for m prime to $|G_1|$ the conjugacy class containing g^m can be found from the algebraic conjugacy of character values. Shortly after Brauer's paper appeared Dade gave a counterexample to Brauer's conjecture by producing a Brauer pair which are not isomorphic (see [70]). The preprint of Mckay-Sibley [213] included the following corollary.

Corollary 5.3 *If G_1 and G_2 are 2-groups with the same WCT, they form a Brauer pair.*

Proof Let G_1, G_2 be 2-groups with the same WCT via the map α. Consequently $\alpha(g)^2$ is conjugate to $\alpha(g^2)$ in G_2. Since odd powers of classes in 2-groups can be determined from algebraic conjugacy of character values, it follows that all power maps in conjugacy classes are preserved by α. □

This was then used by Mckay and Sibley to provide a proof independent of computer calculations of the existence of Brauer pairs in 2-groups (the results of their calculations are given in [174]).

5.2.3.2 The WCT as a Frequency Square

A **frequency square** of order n and frequency vector $(\lambda_1, \cdots, \lambda_m)$ is an $n \times n$ array of m objects $a_1, \cdots, a_m$ such that a_i appears exactly λ_i times in each row and in each column (see [78], Ch. 12).

The following properties follow from the above definition of the WCT.

(W1) The unbordered WCT of a group G is a frequency square of order $n = |G|$ in which C_i appears $|C_i|$ times in each row and column.

(W2) The WCT is symmetrical about the diagonal.

(W3) The block condition. Corresponding to each pair (C_i, C_j) of conjugacy classes there is a block of the table whose rows are indexed by the elements of C_i and whose columns are indexed by the elements of C_j. In this block each entry C_k appears $a_{ij}^k |C_k|$ times where a_{ij}^k is an integer (see above).

(W4) $|C_i|$ divides n.
(W5)   It may be assumed that the first row and column correspond to a C_1 with $|C_1| = 1$, and that the first row is $(C_1, C_2, \ldots, C_r)$, while the first column is $(C_1, C_2, \ldots, C_r)^t$.

Problem 5.2 Find necessary and sufficient conditions for a frequency square to be the unbordered *WCT* of a group.

Example 5.2

$$
\begin{array}{cccccc}
1 & 2 & 2 & 3 & 3 & 3 \\
2 & 1 & 3 & 2 & 3 & 3 \\
2 & 3 & 1 & 3 & 2 & 3 \\
3 & 2 & 3 & 2 & 3 & 1 \\
3 & 3 & 2 & 3 & 1 & 2 \\
3 & 3 & 3 & 1 & 2 & 2
\end{array}
\tag{5.2.8}
$$

The frequency square (5.2.8) satisfies (W1), (W2) and (W4), but not (W3).

Pairs of groups which have the same character tables and different WCTs are easily obtained. The smallest example is that of Q_8 and D_4. It is well-known that they have the same character tables. In this case their irreducible representations are either linear or of degree 2. Now the irreducible factors of Θ_G of degree 2 can be read off from the irreducible 2-characters, and thus in this case the groups have the same WCT if and only if the factors of degree 2 corresponding to the 2-dimensional representations are pairwise equivalent. For this pair of groups there is only one such factor. That for Q_8 is already given in Chap. 7.3 as $\phi = \alpha^2 + \beta^2 + \gamma^2 + \delta^2$ and the factor for D_4 is $\phi_1 = \alpha_1^2 + \beta_1^2 - \gamma_1^2 - \delta_1^2$ where $\alpha_1 = x_1 - x_2, \alpha\beta_1 = x_3 - x_4, \gamma_1 = x_5 - x_6$ and $\delta_1 = x_7 - x_8$. It is easily seen that ϕ and ϕ_1 are not equivalent. (A simpler argument is available using the results in Theorem 5.2 below, since the two groups have different numbers of involutions). This shows directly that the WCT contains more information than the ordinary character table. It is more difficult to find examples of odd order with the same character tables and different WCTs, but examples are available of order 243.

5.2.4 Properties Determined by the Weak Cayley Table

The following theorem indicates some of the properties which are readily deduced from the WCT, but in general are not determined by the character table.

Theorem 5.2 *The following are determined by the WCT of a group G.*

  (i) *The set of involutions of G.*
 (ii) *For each element z of the center of a group the set $S(z) = \{g; g^2 = z\}$.*
(iii) *The set of 2-power elements, together with their orders.*

(iv) The Frobenius-Schur indicator of each character.

(v) The precise product of any central element z by any element g.

Proof By considering the diagonal of the WCT, (i) and (ii) follow immediately since for z central $\{z\}$ is a single element class. To prove (iii), note that from (i) the conjugacy classes which consist of involutions are known, and then it follows that since the conjugacy class of a square can be read off the elements of order 4 may be determined, and this procedure can be repeated.

The Frobenius-Schur indicator v_2 of a character χ may be defined as

$$v_2(\chi) = 1/|G| \sum_{g \in G} \chi(g^2). \tag{5.2.9}$$

If χ is an irreducible character, then $v_2(\chi)$ is used as follows (in the context of representations over $\mathbb{C}$):

(a) $v_2(\chi) = 0$ if χ is not real;

(b) $v_2(\chi) = 1$ if χ is afforded by a real representation;

(c) $v_2(\chi) = -1$ if χ is real but not afforded by any real representation. It may however be represented by a quaternionic representation.

For any character χ the values of $\chi^{(2)}(g, h)$ can be determined from the WCT, and in particular $\chi^{(2)}(g, g) = \chi(g)^2 - \chi(g^2)$ for arbitrary g. Since $\chi(g)$ is known from the ordinary character table, $\chi(g^2)$ may be found and consequently v_2 which shows (iv). The statement (v) follows because given $g \in G$ and $z \in Z(G)$ with product $gz = h$, it follows that $gh^{(-1)} = z$ and hence $h^{(-1)}$ and therefore h may be deduced from the WCT. □

The example of D_4 and Q_8 shows that the character table does not determine the number of involutions. It is a well-known fact that if the extra information of the Frobenius-Schur indicators is included then the number of involutions of a group (but not the actual set of involutions) can be obtained. The products of central elements can be deduced from the character table alone, the product being deduced from the product of character values divided by the corresponding degrees. Thus the isomorphism class of the center of the group can be deduced from the character table.

In the light of the above, (v) may be restated as "a WCTI preserves multiplication *by* central elements", while a character table isomorphism only preserves multiplication *of* central elements.

Some groups are determined by their WCT. In fact the groups in the class in the following theorem are determined by their character tables together with the 2-power maps (this information is contained in the WCT of a 2-group, see Corollary 5.3). The Frattini subgroup is the intersection of all maximal subgroups of G.

Theorem 5.3 *Let G_1 be a 2-group of class 2 with Frattini subgroup $\Phi(G_1) \leq Z(G_1)$. Let G_1 and G_2 form a Brauer pair via $\alpha : G_1 \to G_2$. Then G_1 and G_2 are isomorphic (although α need not be an isomorphism).*

The proof is given in [184].

5.2.5   Groups with the Same WCT

The source of the following work is motivated by results in the thesis of Mattarei [217] and the subsequent publications [216] and [218]. He constructed examples of groups with the same ordinary character table and different derived lengths, thus demonstrating that the derived length is not determined by the character table. It is shown here that his main results remain true on replacing the character table by the WCT, but with important restrictions. The proofs are for the most part quite different, using WCTIs. The work relies on the proofs of Mattarei that the pairs of groups constructed are not isomorphic.

In [163, p. 329] Ito states that 'it would be very desirable to obtain meaningful generalizations of the notions of Frobenius groups and Zassenhaus groups. Camina pairs form one such generalization. In [48] two hypotheses were introduced on a finite group G:

(F1) G has a proper, non-trivial normal subgroup H and a set of non-principal irreducible characters $\chi_1, \chi_2, \ldots, \chi_r, r \geq 1$, such that

(a) each χ_i vanishes on $G\backslash H$; and
(b) there are $\alpha_1, \alpha_2, \ldots \alpha_r \in \mathbb{N}$ such that $\sum_{i=1}^{r} \alpha_i \chi_i$ is constant on $H^{\#} = H\backslash\{e\}$.

(F2) G has a proper non-trivial normal subgroup H such that if $x \in G\backslash H$ and $y \in H$, then x and xy are conjugate.

These properties are satisfied by Frobenius groups [158] and Camina showed that (F1) and (F2) are equivalent in [48].

Definition 5.6   A group G satisfying either (F1) or (F2) (for some choice of H) is called a **Camina group** and the pair $(G; H)$ is called a **Camina pair**.

Camina pairs have been studied, for example in [60, 211], and [71].
There is a further class of groups which is less restrictive and which contains important examples. This is the class of Camina triples which were introduced in [184]. They these may be defined by either of the equivalent conditions given below.

Definition 5.7

(C1)   there are proper, non-trivial normal subgroups H, K of G where $K \subseteq H$, a set of non-principal irreducible characters $\chi_1, \chi_2, \ldots, \chi_r, r \geq 1$; and $\alpha_1, \alpha_2, \ldots, \alpha_r \in \mathbb{N}$ such that:

(a) each χ_i vanishes on $G\backslash H$,
(b) $\sum_{i=1}^{r} \alpha_i \chi_i$ is constant on $K^{\#} = K\backslash\{e\}$.

(c) $\sum_{i=1}^{r} \alpha_i \chi_i$ is zero on $H \backslash K$, and

(d) $\sum_{g \in K} \chi_i(g) = 0$ for all $i \leq r$.

(C2) G has proper, non-trivial normal subgroups H, K, with $K \subseteq H$ such that if $x \in G \backslash H$ and $y \in K$, then x and xy are conjugate.

A group satisfying either (C1) or (C2) will be called a **Camina triple group** and (G, H, K) will be called a **Camina triple**. Camina triples have also occurred in [152] and [153].

The following theorem provides a useful tool to determine that a pair of groups have the same WCT.

Theorem 5.4 *Suppose G is a group of odd order and that N is an abelian group which is a G-module. Suppose further that G_1 and G_2 are nonisomorphic groups which are extensions of N by G such that (G_1, N) and (G_2, N) are Camina pairs. Then G_1 and G_2 have the same weak Cayley tables.*

Proof From standard extension theory G_1 and G_2 may be written as structures on $N \times G$ with multiplication $\circ_i$, for $i = 1, 2$, where

$$(n_1, g_1) \circ_i (n_2, g_2) = (n_1 n_2^{g_1^{-1}} f_i(g_1, g_2), g_1 g_2) \qquad (5.2.10)$$

with $f_i : G \times G \rightarrow N$ and it will be assumed without loss of generality that $f_i(g, e) = f_i(e, g) = e$ for all $g \in G$. A bijection ϕ from $N \times G$ to itself is defined as follows. Divide $G - \{e\}$ arbitrarily into disjoint subsets S_1 and S_2 such that $S_1 \cup S_2 = G - \{e\}$ and $S_2 = \{g^{-1}; g \in S_1\}$. Then define

$$\phi(n, e) = (n, e) \qquad \text{for } n \in N,$$

$$\phi(n, g) = (n, g) \qquad \text{for } g \in S_1, \qquad (5.2.11)$$

$$\phi((n, g)^{\text{inv}(1)}) = (n, g)^{\text{inv}(2)} \qquad \text{for } g \in S_1,$$

where $(n, g)^{\text{inv}(i)}$ denotes the inverse of (n, g) in G_i, for $i = 1, 2$. It is claimed that the conjugacy classes of G_1 and G_2 regarded as subsets of $N \times G$ coincide and that ϕ is a WCTI between G_1 and G_2. Since the inverse of (n, g) in either group is of the form (m, g^{-1}) for some $m \in N$ it is easy to check that ϕ is a bijection. Using the assumption of the theorem, and by direct calculation, the conjugacy classes of each G_i consist of

(a) subsets of the form $\{(n^g, e); g \in G\}$,

(b) for a given conjugacy class $C \neq \{e\}$ of G, the set $\{(n, g); n \in N, g \in C\}$. It follows directly that ϕ preserves conjugacy classes. It will now be shown that $P_1 = (n_1, g_1) \circ_1 (n_2, g_2)$ and $P_2 = \phi(n_1, g_1) \circ_2 \phi(n_2, g_2)$ lie in the same conjugacy class. Three cases arise:

(i) $g_1 = g_2 = e$. Here $P_1 = P_2 = (n_1 n_2, e)$.

(ii) $g_1 \neq g_2^{-1}$. From (5.2.10) $P_1 = (m_1, g_1 g_2)$ and $P_2 = (m_2, g_1 g_2)$ where $m_1, m_2 \in N$ so that P_1 and P_2 lie in the same class of type (b).

(iii) The important case is when $g_2 = g_1^{-1} \neq e$. Without loss of generality it may be assumed that $g_1 \in S_1$, since xy and yx necessarily lie in the same conjugacy class. Note that in both extensions $(n, g) = (n, e) \circ (e, g)$ for all $n \in N, g \in G$. Consequently,

$$(n, g) \circ (m, g)^{-1} = (n, e) \circ (e, g) \circ ((m, e) \circ (e, g))^{-1}$$
$$= (n, e) \circ (e, g) \circ (e, g)^{-1} \circ (m, e)^{-1}$$
$$= (nm^{-1}, e),$$

independently of the specific extension in which it is computed. Now (n_2, g_2) can be written as $(m, g_1)^{\text{inv}(1)}$ for some $m \in N$. Hence

$$P_1 = (n_1, g_1) \circ_1 (m, g_1)^{\text{inv}(1)} = (n_1 m^{-1}, e)$$

and

$$P_2 = \phi(n, g_1) \circ_2 \phi((m, g_1)^{\text{inv}(1)}) = (n_1, g_1) \circ_2 (m, g_1)^{\text{inv}(2)} = (n_1 m^{-1}, e).$$

It follows that G_1 and G_2 have the same WCT. □

Since D_4 and Q_8 regarded as extensions of C_2 by $C_2 \times C_2$ satisfy the conditions of the theorem apart from the condition that G is of odd order, it is seen that a restriction of this type is necessary. The analogous result for character tables holds without this restriction [216].

There follows a restatement of Theorem 5.4 in a form directly applicable to the examples constructed in [216] and [218] (compare with Corollary 6 of [216]).

Corollary 5.4 *Suppose the extensions*

$$N_i \rightarrowtail G_i \longrightarrow K_i$$

are given, for $i = 1, 2$, where N_i is an abelian group and K_i is of odd order, such that

(i) *there exists a group isomorphism $\overline{\alpha} \colon K_1 \to K_2$;*
(ii) *there exists a K_1-module isomorphism $\tilde{\alpha} \colon N_1 \to N_2$ where the action of K_1 on N_2 is via $\overline{\alpha}$;*
(iii) *(G_1, N_1) and (G_2, N_2) are Camina pairs.*

Then G_1 and G_2 have the same WCT.

Corollary 5.5 *Under the assumption of Corollary 5.4, G_1 and G_2 have the same character tables.*

5.2.5.1 Some Examples

The examples of pairs of groups with the same character tables which have been constructed to answer questions related to the information contained in character tables often, but not always, satisfy the conditions of the above corollaries.

(I) The pair of groups of order p^3 for p odd were shown in [174] to have the same WCT by an argument using character tables. This result follows directly from the fact that they are Camina pairs with respect to the center (of order p) which satisfy Corollary 5.4.

(II) In [173] the first examples of pairs of non-isomorphic groups with the same 1- and 2-characters are produced. These groups are subgroups of groups of permutations on a finite field which are equidistributed, i.e. there is a 1 : 1 correspondence between them which preserves cycle type. These are not Camina pairs. For the details see [173] (see also [194]).

(III) Now the results which motivated the introduction of the WCT are introduced. In [216] groups G, H are constructed with the following properties. Both have order $p^{3(q-1)}q$ where p and q are odd primes, $q \geq 5$ and the multiplicative order of p is $q - 1$ modulo q. Also both have a normal Sylow p-subgroup with an elementary abelian center of order p^{q-1}, such that all of its quotients modulo a maximal subgroup of its center are extraspecial p-groups (such p-groups P are sometimes called **semi-extraspecial**, they satisfy $P' = Z(P) = \Phi(P)$, and (P, P') is a Camina pair). The Sylow p-subgroup P of G (resp. $\bar{P}$ of H) has a cyclic complement Q (resp. $\bar{Q}$) which acts irreducibly on P' (resp. $\bar{P}'$) and on A/P' (resp. $X/\bar{P}'$), where $A = [P, Q]$ (resp. $X = [\bar{P}, \bar{Q}]$). It follows that $G' = A$ and $H' = X$. The important difference is that A is abelian (in fact, P is a Sylow p-subgroup of $SL(3, p^{q-1})$, and A is one of its maximal abelian subgroups), while X is not (in fact $X' = \bar{P}'$). Therefore, G is metabelian and H is of derived length 3.

A proof is given in [216] that (G, P') and $(H, \bar{P}')$ are Camina pairs, so that Corollary 5.4 applies using $G_1 = G, N_1 = P', G_2 = H, N_2 = \bar{P}'$ to conclude that G and H have the same WCT.

Corollary 5.6 *The pair G, H of groups given in [216] have the same WCT.*

More examples of groups with different derived lengths and the same WCT are described in [184].

5.2.5.2 Modified Problems of Brauer

The problems of Brauer in [30] have natural extensions to properties determined by the WCT. Two of these may be answered as follows.

Corollary 5.7 *The derived length of a group cannot be deduced from its WCT.*

Remark 5.2 Humphries and Rode have recently shown that the derived length of a group is determined by the WCT together with the "2-S-ring".

This work will be presented in Chap. 10.

Corollary 5.8 *If a set of conjugacy classes of a group which make up a normal subgroup N is given, it cannot be determined from the weak Cayley table of G whether N is abelian.*

Proof As is indicated in the discussion at p. 92 of [216], if Corollary 5.8 were false it would follow that the derived length could be determined from the WCT, contradicting Corollary 5.7.   □

A further useful result is:

Corollary 5.9 *If p is odd, any pair of extra-special p-groups of the same order have the same WCT.*

Proof It is easy to check that these groups satisfy the conditions of Theorem 5.4 where N is the center of each group.   □

More results of this nature are given in [184].

5.3   Weak Cayley Isomorphisms and Weak Cayley Table Groups

The following is an account of the work of Humphries in [151]. A WCTI between groups G and H has been defined in the previous section. In the case where G and H coincide the WCTIs form a group $\mathcal{W}(G)$ called the **Weak Cayley table group** of G. This group contains a subgroup $\mathcal{W}_0(G)$ generated by the automorphism group $Aut(G)$ and the antiautomorphism $\mathfrak{I} : g \rightarrow g^{-1}$. If $\mathcal{W}(G) = \mathcal{W}_0(G)$ then $\mathcal{W}(G)$ is said to be **trivial**. It follows easily that if $\alpha : G \rightarrow H$ is a WCTI then $\mathcal{W}(G) = \mathcal{W}(H)$.

The definition of a WCTI makes sense for infinite groups. While there does not appear to be an easy way to define an arbitrary pair of infinite groups to "have the same character table" it appears much easier to define "having the same WCTs". In [193] the group of automorphisms of the character table of a finite group is discussed, but it would appear to be difficult to extend this to infinite groups.

In the following, $x \smile y$ is used to denote that x is conjugate to y.

5.3.1   Finite Groups with Trivial Weak Cayley Groups

The following is a generalization of Hölder's Theorem on the outer automorphisms of $\mathcal{S}_n$.

Theorem 5.5 *For $n > 1$ the group $\mathcal{W}(\mathcal{S}_n)$ is trivial.*

The proof depends on the following lemma.

Lemma 5.1 *Let G be a group containing a non-trivial conjugacy class C such that*

(i) $G = \langle C \rangle$.

   If $f \in \mathcal{W}(G)$ then there exists $f' \in \mathcal{W}_0(G)$ such that the following three statements are true:

(ii) $ff'(C) = C$ *and* $ff'|_C = id_C$;

(iii) $ff'(C^{(2)}) = C^{(2)}$ *and* $ff'|_{C^{(2)}} = id_{C^{(2)}}$ *where $C^{(2)}$ is the set of elements of G of length 2 relative to the generating set C;*

(iv) $ff'(C^{(3)}) = C^{(3)}$ *and* $ff'|_{C^{(3)}} = id_{C^{(3)}}$ *where $C^{(3)}$ is the set of elements of G of length 3 relative to the generating set C;*

(v) *for all $x, y \in G$ with $x \backsim y$ and $x \neq y$ there exists $c \in C \cup C^{(2)}$ such that $xc \nsim yc$.*

Then $\mathcal{W}(G)$ is trivial.

Proof Let $\alpha = ff'$ and λ be the length function for G relative to the generating set C. Let $x \in G$ be a shortest word such that $y = \alpha(x) \neq x$. Then $\lambda(x) > 2$ by (ii) and (iii), and x can be expressed as $x = x'c$ where $c \in C$ and $\lambda(x') < \lambda(x)$. Further, $\alpha(x') = x'$ and hence $y = \alpha(x'c) \backsim \alpha(x')\alpha(c) = x'c = x$. Thus by (v) there exists $d \in C \cup C^{(2)}$ such that $yd \nsim xd$. But this leads to a contradiction since it follows from (iv) that

$$yd \backsim \alpha(xd) = \alpha(x'cd) \backsim \alpha(x')\alpha(cd) = x'cd = xd.$$

and the lemma is proved.   □

The remainder of the proof of Theorem 5.5 consists of showing that Lemma 5.1 can be applied in the case where $G = \mathcal{S}_n$ and C is the set of all transpositions (ij) in $\mathcal{S}_n$. For the details the reader is referred to [151].

Recall that D_n denotes the dihedral group of order $2n$.

Theorem 5.6 *For $n > 1$ the group $\mathcal{W}(D_n)$ is trivial.*

Proof If $D_n = \langle a, b | a^n = b^2 = e, bab = a^{-1} \rangle$ the non-trivial conjugacy classes are

(i) $\{a^k, a^{-k}\}, 0 < k < \frac{n}{2}$.

(ii) If n is odd, the subset $\mathcal{R}$ of involutions $\{a^k b; 0 \leq k \leq n\}$

(iii) If n is even, the subsets of $\mathcal{R}$ of the form $\{a^k b; k = 1, 3, \ldots, n - 1\}$ and $\{a^k b; k = 0, 2, \ldots, n - 2\}$

(iv) If n is even $\{a^{n/2}\}$.

Now Aut(D_n) acts transitively on $\mathcal{R}$ so that if $f \in \mathcal{W}(D_n)$ after possible composition with an automorphism it can be assumed that $f(b) = b$.

Since f preserves classes it follows that for all k, $f(a^k) = a^{\alpha(k)}$ with α a bijection from $\mathbb{Z}/n\,\mathbb{Z}$ to itself, and $f(a^k b) = a^{\beta(k)}b$ where β is also a bijection from $\mathbb{Z}/n\,\mathbb{Z}$ to itself. Then

$$a^{\alpha(k+m)} = f(a^{k+m}) = f(a^k a^m) = a^{\alpha(k)}a^{\alpha(m)} = a^{\alpha(k)+\alpha(m)};$$

$$a^{\alpha(k)} = f(a^k) = f(b.ba^k) \sim b.ba^{\beta(k)} = a^{\beta(k)};$$

$$a^{\alpha(m-k)} = f(a^{m-k}) = f(ba^k.ba^m) \sim ba^{\beta(k)}.ba^{\beta(m)} = a^{\beta(m)-\beta(k)}.$$

The following relations may be deduced between α and β.

$$\alpha(k+m) = \pm(\alpha(k)+\alpha(m)); \tag{5.3.1}$$

$$\alpha(k) = \pm\beta(k); \tag{5.3.2}$$

$$\alpha(m-k) = \pm(\beta(m)+\beta(-k); \tag{5.3.3}$$

for all $k, m \in \mathbb{Z}/n\,\mathbb{Z}$.

It will be shown that both α and β are automorphisms. Suppose $\alpha(1) = r$. Then (5.3.1) together with the fact that f is a bijection implies that $(r, n) = 1$ and hence $\alpha(-1) = -r$ which further implies that $\alpha(2) = \pm 2r$. Since $\alpha(3) = \pm(\pm 2r + r)$ it follows since α is a bijection that $\alpha(2) = 2r$. Then $\alpha(-2) = -2r$ and hence $\alpha(3) = \pm 3r$. This argument can be continued to show that $\alpha(k) = kr$ for all k and hence α is an automorphism.

The argument that β is an automorphism is similar since it can be deduced from (5.3.1)–(5.3.3) that $\beta(k+m) = \pm(\beta(k)+\beta(m))$ and moreover $\beta(1) = \pm r$. If $\beta(1) = -r$ then f can be composed with $\mathcal{I}(D_n)$ so that $\alpha(1) = -r$ and $\beta(1)$ remains unchanged.

The proof is then completed by a direct verification that f is an automorphism.

$\square$

5.3.2   Groups with Non-Trivial $\mathcal{W}(G)$

Many examples occur when (G, N) is a Camina pair.

Example 5.3 Let $G = A_4$ and $N = V_4 = \{e, (1, 2)(3, 4), (1, 3)(2, 4), (1, 4)(2, 3)\}$. The conjugacy classes are $C_1 = \{e\}$, $C_2 = N - \{e\}$, $C_3 = (1, 2, 3)N$, $C_4 = (1, 3, 2)N$. It is easily seen that (G, N) is a Camina pair.

The WCT may be described as follows. The product of any pair of distinct elements of N is in C_2, the products ab for all $a, b \in (1, 2, 3)N = C_3$ all lie in $(1, 3, 2)N = C_4$. The product of any pair of elements in $(1, 3, 2)N$ lie in

$(1, 2, 3)\dot{N} = C_3$. There remains the products of an element a in $(1, 2, 3)N$ with an element b in $(1, 3, 2)N$. The WCT can be summarized as

$$
\begin{array}{c|ccc}
 & b \in N & b \in C_3 & b \in C_4 \\
\hline
a \in N & B_{11} & B_{12} & B_{13} \\
a \in C_3 & B_{21} & B_{22} & B_{23} \\
a \in C_4 & B_{31} & B_{32} & B_{33}
\end{array} \;,
$$

where

$$
B_{11} = \begin{matrix} C_1 & C_2 & C_2 & C_2 \\ C_2 & C_1 & C_2 & C_2 \\ C_2 & C_2 & C_1 & C_2 \\ C_2 & C_2 & C_2 & C_1 \end{matrix}, \quad
B_{12} = \begin{matrix} C_3 & C_3 & C_3 & C_3 \\ C_3 & C_3 & C_3 & C_3 \\ C_3 & C_2 & C_3 & C_3 \\ C_3 & C_3 & C_3 & C_3 \end{matrix},
$$

$$
B_{13} = \begin{matrix} C_4 & C_4 & C_4 & C_4 \\ C_4 & C_4 & C_4 & C_4 \\ C_4 & C_4 & C_4 & C_4 \\ C_4 & C_4 & C_4 & C_4 \end{matrix}, \quad
B_{23} = \begin{matrix} C_1 & C_2 & C_2 & C_2 \\ C_2 & C_2 & C_1 & C_2 \\ C_2 & C_2 & C_2 & C_1 \\ C_2 & C_1 & C_2 & C_2 \end{matrix},
$$

$$
B_{32} = \begin{matrix} C_1 & C_2 & C_2 & C_2 \\ C_2 & C_2 & C_2 & C_1 \\ C_2 & C_1 & C_2 & C_2 \\ C_2 & C_2 & C_1 & C_2 \end{matrix}, \quad B_{21} = B_{33} = B_{12}, \; B_{22} = B_{33} = B_{13} = B_{22}.
$$

It is clear that the WCT remains unchanged if C_2 and C_3 are subjected to arbitrary permutations $\sigma : C_2 \to C_2$ and $\tau : C_3 \to C_3$, and since for any WCTI f, $f(a) = f(a^{-1})^{-1}$ any two such permutations can be completed in only one way to a WCTI. It follows that the number of such WCTIs is $3!4! = 144$. But $\mathfrak{I}$ is not included in the above set and hence $|\mathcal{W}(G)| = 288$. Since $\mathrm{Aut}(A_4) = \mathrm{Aut}(S_4) = S_4$, $|\mathcal{W}_0(G)| = 48$.

This process may be repeated for the Frobenius groups Fr_{20} and Fr_{21}. (Fr_{20}, N) is a Camina pair where N is the normal subgroup of order 5. Then $|\mathcal{W}(Fr_{20})| = 2(4!(5!)^2) = 691,200$. Again, (Fr_{21}, N) is a Camina pair where N is the normal subgroup of order 7 and $|\mathcal{W}(Fr_{21})| = 2(6!7!) = 7,257,600$.

Another set of examples is the non-abelian p-groups of order p^3 (p odd). These are Camina pairs on the center Z (of order p). Suppose the generators of G are a and b. If f is any bijection on each of the cosets $a^i b^j Z$ for $i, j \le \frac{p-1}{2}$ it can be extended to a WCTI $\hat{f}$ on G by taking $\hat{f}(z) = z$ for $z \in Z$ and $\hat{f}(x^{-1}) = (\hat{f}(x))^{-1}$ for all $x \notin Z$. If such a map is not the identity it cannot be an automorphism and hence these groups also have non-trivial WCTI's. Note that since there is a WCTI between the pair of non-isomorphic non-abelian groups of order p^3 for p odd they have isomorphic weak Cayley table groups.

5.3.3   Simple Finite Groups

If G and H are finite simple groups then (using the classification of finite simple groups) Chen proved in [56] that if G and H have the same character table then $G \cong H$. It clearly follows that if $f : G \to H$ is a WCTI then $G \cong H$.

The following observation is also due to Humphries. For a group G define the subgroup $U(G)$ to be the subgroup generated by all elements of the form

$$f(ab)f(b)^{-1}f(a)^{-1}$$

for all $f \in W(G)/W_0(G)$ and all $a, b \in G$. The following hold.

Proposition 5.4 *The subgroup U is characteristic in G and $U \subset G'$.*

Corollary 5.10 *If G is a simple group then either $W(G)$ is trivial or $U(G) = G$.*

For the proofs the reader is referred to [151].

Problem 5.3 Is there a finite simple group G with $W(G)$ non-trivial?

5.3.4   Infinite Groups

One question which naturally arises is whether the free group $\mathcal{F}_n$ of rank n can have non-trivial WCTIs. The answer is negative, except possibly in the case where $n = 3$.

Theorem 5.7 (Humphries) *If $\mathcal{F}_n$ is the free group of rank n then $W(\mathcal{F}_n)$ is trivial for $n \neq 3$.*

This theorem is proved after a technical discussion of WCTIs between groups G and H of the form $G = G_1 * G_2$, $H = H_1 * H_2$ where $*$ denotes free product. Some intermediate results are the following.

Theorem 5.8 *If H_1 is indecomposable (i.e. cannot be represented as a non-trivial free product), is not infinite cyclic, and H_1 is not isomorphic to a subgroup of G_1 or G_2 then there are no WCTIs $f : G_1 * G_2 \to H_1 * H_2$.*

Corollary 5.11 *There are no WCTIs $f : G*G \to H*H$ where G, H are nontrivial indecomposable groups such that H is not infinite cyclic and not isomorphic to a subgroup of G.*

The group G is **cohopfian** if it is not isomorphic to a proper subgroup of itself. A subgroup $K \subseteq H$ is **conjugacy closed** in H if whenever $r, s \in K$ are conjugate in H, then they are conjugate in K.

Theorem 5.9 *Let G_1, G_2, H_1, H_2 be nontrivial with H_1, H_2 co-Hopfian and freely indecomposable. Suppose that $G_1 \ncong H_1$ and $G_2 \ncong H_2$. Then there are no WCTs*

$f : G_1 * G_2 \rightarrow H$, where H has a conjugacy closed subgroup isomorphic to $H_1 * H_2$.

Theorem 5.10 *Let H_1, H_2 be non-abelian infinite groups. Then $\mathcal{W}(H_1 * H_2)$ is trivial.*

The following result shows that a free group cannot have the same WCT as a non-free group on the same number of generators.

Theorem 5.11 *Let $f : \mathcal{F}_n \rightarrow G$ be a WCTI where G is generated by a set with n elements. Then $G \cong \mathcal{F}_n$.*

Humphries also discusses the example of the integral Heisenberg group (the group of all upper triangular 3×3 matrices with 1 on the diagonal). This group has a subgroup of non-trivial WCTIs which is isomorphic to $\mathbb{Z}^\infty$. One proof relies on the fact that, in a similar way to the case where the groups of order p^3 were discussed above, a Camina pair is present with N the derived group.

5.4 The Category of Weak Cayley Morphisms

The work in this section is due to J.D.H Smith and the author and appeared in [180]. The definition of a weak Cayley table function $f : G \rightarrow H$ in Sect. 5.2 is a function which sends conjugate elements to conjugate elements and such that $f(g_1 g_2)$ is conjugate to $f(g_1)f(g_2)$ for all $g_1, g_2 \in G$ (the case considered is usually where f is a bijection). Since such functions are weaker than homomorphisms, but retain more structure than the character table, it was thought interesting to examine them from a categorical point of view. On further investigation, it appears to be more suitable to add into the definition the specification of the conjugating element x such that $f(g_1)f(g_2) = f(g_1 g_2)^x$ and a crossed product condition made precise below. In this section a weak Cayley table function together with an appropriate specification of these conjugating elements is defined to be a **weak Cayley table morphism**, or more briefly just a **weak morphism** (Definition 5.8). The crossed-product condition included in the definition is in the sense of [67]—compare Remark 5.3(d) below. Theorem 5.12 then shows that under composition, the weak morphisms form a category **Gwp**. Without the crossed-product condition, the weak morphisms would only form a "non-associative category" under composition.

Since group homomorphisms are weak morphisms, there is a forgetful functor U to **Gwp** from the category **Gp** of group homomorphisms. Theorem 5.13 constructs a left adjoint to this functor. On the other hand, forgetting the specific choice of conjugating elements embodied in a weak morphism leads to a projection functor P from **Gwp** to the category **Set** of sets. Two weak morphisms are said to be **homotopic** if they project under P to the same weak Cayley table function. The monoid $P_1^{-1}\{id_G\}$ of weak morphisms that are homotopic to the identity morphism on a group G is then examined. As shown by Proposition 5.8(d), this monoid may contain non-invertible elements. Problem 5.4 asks whether each weak Cayley table bijection is the image under the projection functor P of an isomorphism in the category **Gwp**. An invertible element of the monoid $P_1^{-1}\{\mathrm{id}_G\}$ yields a left quasigroup structure on the set G. The left multiplication maps of these various left quasigroups form a group which is defined to be the **perturbation group** Γ_G of the group G. Theorem 5.14 analyzes the structure of this group. In turn, the perturbation group Γ_G appears in the structure of the group U_G of units of the monoid $P_1^{-1}\{\mathrm{id}_G\}$, as described by Theorem 5.15. Note that for each weak morphism $\alpha : G \to H$, composition in the category **Gwp** affords a left action of U_G and a right action of U_H on the homotopy class of α.

Throughout this section, notational conventions and definitions not otherwise explained follow the usage of [264].

Definition 5.8 Let G and H be groups. Then a **weak (Cayley table) morphism** $\alpha : G \to H$ consists of a triple $(\alpha_1, \alpha_2, \alpha_3)$ of functions

$$\alpha_1 : G \to H;\ g \mapsto g^{\alpha_1}$$

and

$$\alpha_i : G^2 \to H;\ (g_1, g_2) \mapsto \alpha_i(g_1, g_2)$$

for $i = 2, 3$ such that:

(1) α_1 maps the identity of G to the identity of H; and
(2) $g_1^{\alpha_1} g_2^{\alpha_1} \alpha_2(g_1, g_2) = \alpha_2(g_1, g_2)(g_1 g_2)^{\alpha_1}$,
(3) $g_1^{\alpha_1} \alpha_3(g_1, g_2) = \alpha_3(g_1, g_2)(g_2^{-1} g_1 g_2)^{\alpha_1}$,
(4) $\alpha_3(g_1, g_2 g_3) = \alpha_3(g_1, g_2)\alpha_3(g_2^{-1} g_1 g_2, g_3)$

for all g_1, g_2, g_3 in G.

In the above notation of Sect. 5.2, if f is a WCTI, then α_1 corresponds to f, $\alpha_2(g_1, g_2)$ corresponds to the choice of a conjugating element x mentioned above and the element $\alpha_3(g_1, g_2)$ is the conjugating element which ensures that f takes conjugacy classes into conjugacy classes.

Remark 5.3 Denote projection on the second factor by

$$\pi_2 : G^2 \to G;\ (g_1, g_2) \mapsto g_2.$$

(a) If $f : G \to H$ is a group homomorphism, then the triple $(f, e_H, \pi_2 f)$ is a weak morphism, the second component being the constant map $e_H : G^2 \to H$ whose value is the identity element of H.
(b) As above let $\mathcal{J} : G \to G;\ g \mapsto g^{-1}$ be the inversion map on a group G. Then

$$(\mathcal{I}, \pi_2, \pi_2) : G \to G$$

is a weak morphism.

(c) Condition (1) of Definition 5.8 is not redundant. Consider $G = S_3$. Let $\alpha_1 : G \to G$ be the constant map with value (123). Let $\alpha_2 : G^2 \to G$ be the constant map with value (23) and let $\alpha_3 : G^2 \to G$ be the constant map with value (e). Then $(\alpha_1, \alpha_2, \alpha_3)$ satisfies Conditions (2)–(4) of Definition 5.8, but not Condition (1).

(d) In Definition 5.8, consider the conjugation action of G on itself. Then in the language of [67], Condition (4) says that α_3 is a crossed product.

Proposition 5.5 *Let $\alpha : G \to H$ be a weak morphism. Then for all g in G, the following hold:*

(1) $[g^{\alpha_1}, \alpha_2(1, g)] = [g^{\alpha_1}, \alpha_2(g, 1)] = 1$;
(2) $(g^{-1})^{\alpha_1} = (g^{\alpha_1})^{-1}$;
(3) The map $G \to H$; $x \mapsto \alpha_3(1, x)$ may be chosen as an arbitrary homomorphism;
(4) $\alpha_3(g, 1) = 1$.

Proof Statement (1) follows by specialization of the arguments g_1, g_2 in Condition (2) of Definition 5.8. Statement (2) follows by Condition (1) on setting $g_1 = g$ and $g_2 = g^{-1}$ in Condition (2)—compare [184, p. 398]. Statement (3) is apparent upon specialization of Condition (3), the homomorphic property being required for consistency with Condition (4). Finally, Statement (4) follows on setting $g_1 = g$ and $g_2 = g_3 = 1$ in Condition (4) of Definition 5.8. □

Remark 5.4 In view of Proposition 5.5(3), a normalization $\alpha_3(1, g) = 1$ for g in G could be chosen as part of the requirements of Definition 5.8. However, this would preclude the convenient use of the second projection π_2 in contexts such as Remark 5.3 *(a)* and *(b)*.

5.4.1 The Category of Weak Morphisms

Proposition 5.6 *Let $\alpha : G \to H$ and $\beta : H \to K$ be weak morphisms. Then there is a composite weak morphism $\alpha\beta : G \to K$ with components $(\alpha\beta)_1 = \alpha_1\beta_1$,*

$$(\alpha\beta)_2(g_1, g_2) = \beta_2(g_1^{\alpha_1}, g_2^{\alpha_1})\beta_3(g_1^{\alpha_1}g_2^{\alpha_1}, \alpha_2(g_1, g_2)), \tag{5.4.1}$$

and

$$(\alpha\beta)_3(g_1, g_2) = \beta_3(g_1^{\alpha_1}, \alpha_3(g_1, g_2)) \tag{5.4.2}$$

for g_1, g_2 in G.

Proof It must be verified that $\alpha\beta$ satisfies the conditions of Definition 5.8. Condition (1) is immediate. To verify Condition (3), note that Condition (3) on α implies

$$(g_2^{-1}g_1g_2)^{\alpha_1} = \alpha_3(g_1, g_2)^{-1}g_1^{\alpha_1}\alpha_3(g_1, g_2) \tag{5.4.3}$$

for g_1, g_2 in G. Condition (3) on β applied to 5.4.3 then yields

$$(g_2^{-1}g_1g_2)^{\alpha_1\beta_1} = \beta_3(g_1^{\alpha_1}, \alpha_3(g_1, g_2))^{-1} g_1^{\alpha_1\beta_1} \beta_3(g_1^{\alpha_1}, \alpha_3(g_1, g_2))$$

as required, given $(\alpha\beta)_3$ expressed by (5.4.2). To verify Condition (2) on $\alpha\beta$, note that Condition (2) on α implies

$$(g_1g_2)^{\alpha_1} = \alpha_2(g_1, g_2)^{-1} g_1^{\alpha_1} g_2^{\alpha_1} \alpha_2(g_1, g_2) \qquad (5.4.4)$$

for g_1, g_2 in G. Applying Condition (3) on β to (5.4.4) gives

$$(g_1g_2)^{\alpha_1\beta_1} = \beta_3(g_1^{\alpha_1} g_2^{\alpha_1}, \alpha_2(g_1, g_2))^{-1} (g_1^{\alpha_1} g_2^{\alpha_1})^{\beta_1} \beta_3(g_1^{\alpha_1} g_2^{\alpha_1}, \alpha_2(g_1, g_2)).$$

Expansion of the middle term $(g_1^{\alpha_1} g_2^{\alpha_1})^{\beta_1}$ of the right hand side of this equation using Condition (2) on β then yields the required Condition (2) on $\alpha\beta$, with $(\alpha\beta)_2$ being specified by (5.4.1). Finally, for g_i in G, Eq. (5.4.2) and Definition 5.8 give

$$(\alpha\beta)_3(g_1, g_2g_3) = \beta_3(g_1^{\alpha_1}, \alpha_3(g_1, g_2g_3))$$

$$= \beta_3(g_1^{\alpha_1}, \alpha_3(g_1, g_2)\alpha_3(g_2^{-1}g_1g_2, g_3))$$

$$= \beta_3(g_1^{\alpha_1}, \alpha_3(g_1, g_2))\beta_3(\alpha_3(g_1, g_2)^{-1} g_1^{\alpha_1}\alpha_3(g_1, g_2), \alpha_3(g_2^{-1}g_1g_2, g_3))$$

$$= \beta_3(g_1^{\alpha_1}, \alpha_3(g_1, g_2))\beta_3((g_2^{-1}g_1g_2)^{\alpha_1}, \alpha_3(g_2^{-1}g_1g_2, g_3))$$

$$= (\alpha\beta)_3(g_1, g_2)(\alpha\beta)_3(g_2^{-1}g_1g_2, g_3),$$

verifying Condition (4) on $\alpha\beta$.   □

Proposition 5.7 *For weak morphisms* $\alpha : G \to H$, $\beta : H \to K$ *and* $\gamma : K \to L$, *the associative law* $(\alpha\beta)\gamma = \alpha(\beta\gamma)$ *holds.*

Proof The equality $(\alpha\beta.\gamma)_1 = (\alpha.\beta\gamma)_1$ represents the associativity

$$(\alpha_1\beta_1)\gamma_1 = \alpha_1(\beta_1\gamma_1)$$

of functional composition, while $(\alpha\beta.\gamma)_3 = (\alpha.\beta\gamma)_3$ follows easily from (5.4.2). Now for elements x, y, z of G, the compositions (5.4.1) and (5.4.2) reduce $(\alpha\beta.\gamma)_2(x, y)$ to the product of $\gamma_2(x^{\alpha_1\beta_1}, y^{\alpha_1\beta_1})$ with

$$\gamma_3(x^{\alpha_1\beta_1} y^{\alpha_1\beta_1}, \beta_2(x^{\alpha_1}, y^{\alpha_1})\beta_3(x^{\alpha_1} y^{\alpha_1}, \alpha_2(x, y))) \qquad (5.4.5)$$

and $(\alpha.\beta\gamma)_2(x, y)$ to the product of $\gamma_2(x^{\alpha_1\beta_1}, y^{\alpha_1\beta_1})$ with

$$\gamma_3(x^{\alpha_1\beta_1} y^{\alpha_1\beta_1}, \beta_2(x^{\alpha_1}, y^{\alpha_1}))\gamma_3((x^{\alpha_1} y^{\alpha_1})^{\beta_1}, \beta_3(x^{\alpha_1} y^{\alpha_1}, \alpha_2(x, y))). \qquad (5.4.6)$$

Since β satisfies Definition 5.8 (2), the latter term of (5.4.6) may be rewritten as

$$\gamma_3(\beta_2(x^{\alpha_1}, y^{\alpha_1})^{-1} x^{\alpha_1 \beta_1} y^{\alpha_1 \beta_1} \beta_2(x^{\alpha_1}, y^{\alpha_1}), \beta_3(x^{\alpha_1} y^{\alpha_1}, \alpha_2(x, y))).$$

The equality between (5.4.5) and (5.4.6) then follows since γ satisfies Definition 5.8(4). □

Theorem 5.12 *There is a locally small category* **Gwp** *whose object class is the class of all groups, such that for groups G and H, the morphism class* **Gwp**(G, H) *is the set of all weak morphisms from G to H. The identity morphism at a group G is the weak morphism $\iota_G = (id_G, e_G, \pi_2)$, while the composition of weak morphisms is given by Proposition 5.6.*

Proof Consider a weak morphism $\alpha : G \to H$. By (5.4.1),

$$(\alpha \iota_H)_2(g_1, g_2) = e_H(g_1, g_2)\pi_2(g_1^{\alpha_1} g_2^{\alpha_1}, \alpha_2(g_1, g_2)) = \alpha_2(g_1, g_2).$$

By (5.4.2), it follows that

$$(\alpha \iota_H)_3(g_1, g_2) = \pi_2(g_1^{\alpha_1}, \alpha_3(g_1, g_2)) = \alpha_3(g_1, g_2).$$

Thus $\alpha \iota_H = \alpha$. Again by (5.4.1),

$$(\iota_G \alpha)_2(g_1, g_2) = \alpha_2(g_1, g_2)\alpha_3(g_1 g_2, e_G(g_1, g_2)) = \alpha_2(g_1, g_2),$$

the latter equation holding by Statement (4) of Proposition 5.5. By (5.4.2),

$$(\iota_G \alpha)_3(g_1, g_2) = \alpha_3(g_1, \pi_2(g_1, g_2)) = \alpha_3(g_1, g_2).$$

Thus $\iota_G \alpha = \alpha$. The partial associativity of the composition in **Gwp** is given by Proposition 5.7. □

Corollary 5.12 *There is a forgetful functor $U : $ **Gp** $\to$ **Gwp** *from the category of (homomorphisms between) groups, with morphism part $U : (f : G \to H) \mapsto (f, e_H, \pi_2 f)$.*

Proof Compare Remark 5.3(a). Verification of the functoriality is straightforward. □

5.4.2 The Adjunction

Let G be a group. Let W be the free group on the disjoint union $G + G^2 + G^2$ of the set G with two copies of G^2. Let $\eta_1' : G \to W$ insert the generators from G. For $i = 2, 3$, let $\eta_i' : G^2 \to W$ insert the generators from the $(i - 1)$-th copy of G^2. Let GF be the quotient of W obtained by imposing the relations

1. $1_G^{\eta_1'} = 1_W$;
2. For all $g_1, g_2 \in G$,

$$g_1^{\eta_1'} g_2^{\eta_1'} \eta_2'(g_1, g_2) = \eta_2'(g_1, g_2)(g_1 g_2)^{\eta_1'};$$

3. For all $g_1, g_2 \in G$,

$$g_1^{\eta_1'} \eta_3'(g_1, g_2) = \eta_3'(g_1, g_2)(g_2^{-1} g_1 g_2)^{\eta_1'};$$

4. For all $g_i \in G$,

$$\eta_3'(g_1, g_2 g_3) = \eta_3'(g_1, g_2)\eta'(g_2^{-1} g_1 g_2, g_3);$$

corresponding to the respective conditions of Definition 5.8. Let η_i for $1 \le i \le 3$ denote the composite of η_i' with the projection from W to GF. Note that η_G or

$$(\eta_1, \eta_2, \eta_3) : G \to GFU \tag{5.4.7}$$

is a weak morphism.

Theorem 5.13 *The forgetful functor* $U : \mathbf{Gp} \to \mathbf{Gwp}$ *has a left adjoint* $F : \mathbf{Gwp} \to \mathbf{Gp}$.

Proof Let $\alpha : G \to H$ be a weak morphism. There is a unique homomorphism from W to H defined by $g_1^{\eta_1'} \mapsto g_1^{\alpha_1}$ and $\eta_i'(g_1, g_2) \mapsto \alpha_i(g_1, g_2)$ for $i = 2, 3$ and g_1, g_2 in G. This homomorphism factorizes through a unique homomorphism $\bar{\alpha} : GF \to H$.

It can now be verified that $(\eta_1, \eta_2, \eta_3)(\bar{\alpha}, e_H, \pi_2 \bar{\alpha}) = (\alpha_1, \alpha_2, \alpha_3)$. For g in G, one has

$$g^{\eta_1 \bar{\alpha}} = g^{\alpha_1} \tag{5.4.8}$$

by the definition of $\bar{\alpha}$. For g_1, g_2 in G, (5.4.1) yields

$$(\eta \bar{\alpha}^U)_2(g_1, g_2) = \eta_2(g_1, g_2)^{\bar{\alpha}} = \alpha_2(g_1, g_2), \tag{5.4.9}$$

while (5.4.2) gives

$$(\eta \bar{\alpha}^U)_3(g_1, g_2) = \eta_3(g_1, g_2)^{\bar{\alpha}} = \alpha_3(g_1, g_2). \tag{5.4.10}$$

On the other hand, the final equations in the lines (5.4.8)–(5.4.10) specify the homomorphism $\bar{\alpha} : GF \to H$ uniquely.   □

Corollary 5.13 *Let* $\prod_{i \in I} H_i$ *be the product (in* $\mathbf{Gp}$*) of a family of groups, equipped with projections* $p_i : \prod_{j \in I} H_j \to H_i$ *for each* i *in* I. *Then the group* $\prod_{i \in I} H_i$, *equipped with projections $p_i^U : \prod_{j \in I} H_j \to H_i$ for each i in I, is the product in* **Gwp** *of the family of groups.*

Proof The right adjoint $U :$ **Gp** $\to$ **Gwp** creates products.   □

Corollary 5.14 *For a group G, the weak morphism 5.4.7 is the component at G of the unit of the adjunction of Theorem 5.13.*

As a dual to Corollary 5.14, note that the component at a group G of the counit of the adjunction of Theorem 5.13 is the group homomorphism $\varepsilon_G : GF \to G$ given by

$$g_1^{\eta_1} \mapsto g_1, \quad \eta_2(g_1, g_2) \mapsto e, \quad \eta_3(g_1, g_2) \mapsto g_2$$

for g_1, g_2 in G.

5.4.3 Homotopy

Let $P :$ **Gwp** $\to$ **Set** be the functor to the category of sets projecting each weak morphism $\alpha : G \to H$ to its first component $\alpha_1 : G \to H$. The image of the functor P, a subcategory **Gwp**^{P} of **Set**, is called the **category of weak Cayley table functions**. The basic open problem concerning the relation between weak Cayley table bijections and the categorical considerations of this section is the following.

Problem 5.4 Is each weak Cayley table bijection the projection under P of a weak isomorphism?

The following concept may help to put Problem 5.4 into context. Two parallel weak morphisms α, $\beta : G \to H$ are said to be **homotopic** if $\alpha_1 = \beta_1$. The proposition below shows that the homotopy class of an isomorphism may contain weak morphisms which are not isomorphisms. In other words, there may be non-invertible weak morphisms that project under P to a weak Cayley table bijection.

Proposition 5.8 *Let G be a group.*

(a) *The homotopy class $P_1^{-1}\{id_G\}$ of the identity $\iota_G = (id_G, e_G, \pi_2)$ forms a monoid.*

(b) *For each α in $P_1^{-1}\{id_G\}$, one has*

$$\alpha_2(x, y) \in C_G(xy)$$

and

$$\alpha_3(x, y) \in C_G(x)y$$

for all x, y in G.

(c) For elements α, β of $P_1^{-1}\{id_G\}$, one has $\alpha\beta = \iota_G$ if and only if

$$\beta_3(x, \alpha_3(x, y)) = y \qquad (5.4.11)$$

and

$$\beta_2(x, y) = \beta_3(xy, \alpha_2(x, y))^{-1} \qquad (5.4.12)$$

for all x, y in G.
(d) If G is nontrivial, then the monoid $P_1^{-1}\{id_G\}$ is not a group.

Proof

(a) is an immediate consequence of the functoriality of P.
(b) follows from Definition 5.8(2), (3).
(c) follows from (5.4.1) and (5.4.2), along with the definition of ι_G.
(d) For α in $P_1^{-1}\{id_G\}$ to be invertible, (5.4.11) shows that

$$\widehat{\alpha}_x : G \to G; \, y \mapsto \alpha_3(x, y) \qquad (5.4.13)$$

must biject for each x in G. On the other hand, Proposition 5.5(3) shows that the homomorphism $\widehat{\alpha}_1 : G \to G$ may be chosen arbitrarily, and in particular need not biject if G is nontrivial.

$\square$

Equation (5.4.11) shows that each invertible weak morphism α homotopic to the identity map on a group G yields a left quasigroup structure $(G, \alpha_3, (\alpha^{-1})_3)$ on the underlying set G of the group. The maps (5.4.13) are the left multiplications in the left quasigroup. The following definition gives a different description of these maps. For a set X, let $S(X)$ denote the group of permutations of X. For a subgroup H of a group G, let $H \setminus G$ denote the set $\{Hx \mid x \in G\}$ of right cosets of H.

Definition 5.9 Let G be a group. Then a *perturbation* of G is a map $\theta : x \mapsto \theta_x$ with domain G, such that

(1) $\theta_x \in \prod_{X \in C_G(x) \setminus G} S(X)$ and
(2) $(yz)\theta_x = y\theta_x . z\theta_{y^{-1}xy}$

for all x, y, z in G. Such a map θ is said to *perturb* G.

Remark 5.5 Let θ perturb a group G with elements x and y.

(a) The map θ_x restricts to an automorphism of $C_G(x)$. In particular, a perturbation of an abelian group A is just an indexed collection of automorphisms of A.
(b) By Definition 5.9(2), knowledge of θ_x implies knowledge of $\theta_{y^{-1}xy}$. Thus a perturbation is specified completely by its values on a set of representatives for the conjugacy classes of G. These various values, in turn, are independent of each other.

Proposition 5.9 *The set Γ_G of all perturbations of a group G forms a group under the multiplication $(\theta, \varphi) \mapsto (x \mapsto \theta_x \varphi_x)$.*

Proof Let θ and φ be perturbations. Then for x, y, z in G, one has

$$(yz)\theta_x \varphi_x = (y\theta_x . z\theta_{y^{-1}xy})\varphi_x = y\theta_x \varphi_x . z\theta_{y^{-1}xy}\varphi_{v^{-1}xv}$$

with $v = y\theta_x$. However, $v^{-1}xv = y^{-1}xy$ by Definition 5.9(1) for θ. $\qquad\square$

The structure of the group Γ_G of perturbations of a group G is described as follows.

Theorem 5.14 *Let G be a group, and let $\{g_i \mid 0 \le i < s\}$ be a set of representatives for the conjugacy classes of G, with $g_0 = e$. For each $0 \le i < s$, let n_i be the cardinality of the conjugacy class of g_i, and let $Aut(C_G(g_i))$ act diagonally on the power $C_G(g_i)^{n_i-1}$. Then the group Γ_G of perturbations of G is isomorphic to the product*

$$\prod_{0 \le i < s} C_G(g_i)^{n_i-1} \rtimes Aut(C_G(g_i)) \qquad (5.4.14)$$

of split extensions.

Proof Consider a particular representative $g \in \{g_i \mid 0 \le i < s\}$, with conjugacy class of cardinality m. Let $\{x_1, \ldots, x_m\}$ be a set of representatives of the right cosets of $C_G(g)$ in G, with $x_1 = e$. For a perturbation θ, denote the restriction of θ_g to $C_G(g)$ by $\overline{\theta}_g$. By Remark 5.5(a), $\overline{\theta}_g$ is an automorphism of $C_G(g)$. For perturbations θ and φ, suppose that $x_i \theta_g = c_i x_i$ and $x_i \varphi_g = d_i x_i$ with c_i, d_i in $C_G(g)$. The permutation θ_g of G is specified completely by the m-tuple $(\overline{\theta}_g, c_2, \ldots, c_m)$, since for $x = cx_i \in C_G(g)x_i$, one has $x\theta_g = (cx_i)\theta_g = c\theta_g . x_i \theta_{c^{-1}gc} = c\overline{\theta}_g . c_i x_i$. Moreover, $x_i(\theta_g \varphi_g) = (c_i x_i)\varphi_g = (c_i \overline{\varphi}_g . d_i)x_i$, so that Γ_G maps homomorphically to the product (5.4.14).

Conversely, consider an element of (5.4.14) whose component at g is $(\overline{\theta}_g, c_2, \ldots, c_m)$. For an element $x = cx_i \in C_G(g)x_i$ of G, define $x\theta_g = c\overline{\theta}_g . c_i x_i$. These specifications, for the various conjugacy class representatives g, completely specify a unique perturbation θ in accordance with Remark 5.5(b). $\qquad\square$

Proposition 5.10 *Let G be a group. If α is an invertible element of the monoid $P_1^{-1}\{id_G\}$, then $\widehat{\alpha}$ given by its values (5.4.13) is a perturbation of G.*

Proof Satisfaction of Definition 5.9(1) by $\widehat{\alpha}$ follows from Proposition 5.8 and its proof. Condition (2) of Definition 5.9 for $\widehat{\alpha}$ is an immediate consequence from Condition (4) of Definition 5.8 for α. $\qquad\square$

Theorem 5.15 *Let G be a group, and let U_G be the group of units $P_1^{-1}\{id_G\}^*$ of the homotopy class $P_1^{-1}\{id_G\}$.*

(a) U_G contains a normal subgroup

$$U_2 = \{(id_G, \alpha_2, \pi_2) \mid \forall x, y \in G, \ \alpha_2(x, y) \in C_G(xy)\}$$

isomorphic to $\prod_{(x,y)\in G^2} C_G(xy)$.

(b) U_G contains a subgroup

$$U_3 = \{\alpha = (id_G, e_G, \alpha_3) \mid \widehat{\alpha} \in \Gamma_G\}$$

isomorphic to the perturbation group Γ_G.

(c) U_G is the semidirect product of the subgroup U_2 by the subgroup U_3. The action of U_3 on U_2 is given by

$$(id_G, e_G, \beta_3)^{-1}(id_G, \alpha_2, \pi_2)(id_G, e_G, \beta_3) = (id_G, \alpha'_2, \pi_2) \qquad (5.4.15)$$

with $\alpha'_2 : (x, y) \mapsto \alpha_2(x, y)\widehat{\beta}_{xy}$.

Proof

(a) The isomorphism is given by

$$(id_G, \alpha_2, \pi_2) \mapsto ((x, y) \mapsto \alpha_2(x, y)^{-1}).$$

Then for $\alpha \in U_2$, $\beta \in U$, and $x, y \in G$, one has

$$(\alpha\beta)_3(x, y) = \beta_3(x, \pi_2(x, y)) = y\widehat{\beta}_x$$

and

$$(\beta^{-1}.\alpha\beta)_3(x, y) = (\alpha\beta)_3(x, \beta_3^{-1}(x, y)) = y\widehat{\beta}_x^{-1}\widehat{\beta}_x = y,$$

so U_2 is normal in U_G. If now $\beta \in U_3$, then

$$(\alpha\beta)_2(x, y) = e_G(x, y)\beta_3(xy, \alpha_2(x, y)) = \alpha_2(x, y)\widehat{\beta}_{xy}$$

and

$$(\beta^{-1}.\alpha\beta)_2(x, y) = (\alpha\beta)_2(x, y).(\alpha\beta)_3(xy, \beta_2^{-1}(x, y))$$
$$= \alpha_2(x, y)\widehat{\beta}_{xy}.id_G\widehat{\beta}_{xy} = \alpha_2(x, y)\widehat{\beta}_{xy},$$

the last equality holding by Remark 5.5(a). Thus (5.4.15) is verified.

(b) The isomorphism is given by the map $(id_G, e_G, \alpha_3) \mapsto \widehat{\alpha}$ and its inverse $\widehat{\alpha} \mapsto (id_G, e_G, (x, y) \mapsto y\widehat{\alpha}_x)$.

(c) Certainly $U_2 \cap U_3$ is trivial. Consider a general element $\alpha = (id_G, \alpha_2, \alpha_3)$ of U_G. Then $\alpha = (id_G, e_G, \alpha_3)(id_G, \alpha_2, \pi_2)$. Thus $U_G = U_3.U_2$.

□

Remark 5.5(a) yields the following special case of Theorem 5.15.

Corollary 5.15 *For a finite abelian group A of order n, the group of units U_A is isomorphic to the semidirect product $A^{n^2} \rtimes Aut(A)^n$.*

Chapter 6
The Extended k-Characters

Abstract The subject of this chapter is the k-characters $\chi^{(k)}$ of a finite group G, and their extensions to more general objects. These characters are constant on certain subsets of G^k, the k-classes. Here work of Vazirani is presented which provides a set of "extended k-characters" for arbitrary k. These connect with various aspects of the representation theory of the symmetric groups and the general linear groups.

Immanent k-characters are defined for arbitrary k and any irreducible representation λ of S_n. They coincide with the usual k-characters if λ is the sign character and in the cases $k = 2$ and $k = 3$ they had appeared with other names. There are connections with the representation theory of wreath products, with invariant theory and Schur functions. There are orthogonality relations and the Littlewood-Richardson coefficients appear in the decomposition of products of extended k-characters.

6.1 Introduction

For any character χ of the finite group G the k-characters $\chi^{(k)}$, defined in Chap. 1, are constant on certain subsets of G^k, the k-classes, which are defined below. In [169] it was shown that an inner product $\langle,\rangle_k$ can be defined such that if $\{\chi_i\}_{i=1}^r$ are the irreducible characters of G then $\langle \chi_i^{(k)}, \chi_j^{(k)} \rangle_k = 0$ for $i \neq j$. A 2-character table for a group is also defined there, with rows consisting of "generalized 2-characters" and this is also described in Chap. 5. The rows of this table are orthogonal with respect to $\langle,\rangle_2$, but they do not span the space of 2-class functions. Later in [149] a 3-character table was defined in terms of "generalized 3-characters" which also form an orthogonal set.

Sometime during the 1990s Kerber made a remark to the author to the effect that the k-characters should be related to the wreath product of G with S_k. Vazirani discovered this independently and in [284], gave a general description of higher characters and class functions which include "immanent characters" and orthogonality relations and this chapter describes the relationship between her ideas and the k-character work. Here in order to avoid confusion with the use of the term

© Springer Nature Switzerland AG 2019   211
K. W. Johnson, *Group Matrices, Group Determinants*
and Representation Theory, Lecture Notes in Mathematics 2233,
https://doi.org/10.1007/978-3-030-28300-1_6

"generalized character" to describe a linear combination of irreducible characters with integer coefficients, the term extended k-characters will be used to denote the higher characters which are defined in the next section.

In Sect. 6.2 there are given the definitions of the k-classes and the extended k-characters, which include immanent characters. In Sect. 6.2.3 the immanent characters are related to the representations of the wreath product of G with S_k and also to the Schur functions which appear in the representation theory of S_k and $GL(n, \mathbb{C})$. Section 6.3 shows that the immanent characters may be regarded as a multilinearization of Schur functions, and in Sect. 6.4 they are interpreted as certain traces and invariants. In Sect. 6.5 an orthogonal set of k-class functions is presented which spans the same space as the extended k-characters and proofs of orthogonality and other results are given.

6.2 k-Classes and Extended k-Characters

The ordinary orthogonality relations between distinct group characters are given in Chap. 1, Corollary 1.3. Here they are expressed slightly differently by defining the inner product of class functions ψ_1, ψ_2 to be

$$\langle \psi_1, \psi_2 \rangle = \frac{1}{|G|} \sum_{g \in G} \psi_1(g)\overline{\psi_2(g)},$$

where $\overline{\alpha}$ denotes complex conjugate of α. If χ_i and χ_j are irreducible characters of G,

$$\langle \chi_i, \chi_j \rangle = \delta_{ij}.$$

These are usually referred to as the row orthogonality relations. Let

$$\{C_1 = \{e\}, C_2, \dots, C_r\}$$

be the set of conjugacy classes of G. There are the following column orthogonality relations

$$\sum_{i=1}^{r} \chi_i(g)\overline{\chi_i(h)} = \frac{|C_j|}{|G|} \text{ if both } g \text{ and } h \text{ lie in } C_j, 0 \text{ otherwise.}$$

These relations have been presented in Chap. 4 in the more general situation of an association scheme. As is well-known, they provide an important tool in group theory.

The 2-classes of a group and the extended 2-characters of a group have been defined in the previous chapter. Here the generalization to arbitrary k-classes and generalized k-characters is described.

6.2.1 k-Classes

The k-**classes** of a group G are defined to be the orbits on G^k (the direct product of k copies of G) under the action of simultaneous conjugation by G and the action of $\mathcal{S}_k$ by permuting the factors. More precisely the action of an element $g \in G$ is

$$(g_{i_1}, g_{i_2}, \ldots, g_{i_n})^g = (g^{-1}g_{i_1}g, g^{-1}g_{i_2}g, \ldots, g^{-1}g_{i_n}g)$$

and the action of $\sigma \in \mathcal{S}_k$ is

$$(g_{i_1}, g_{i_2}, \ldots, g_{i_n})^\sigma = (g_{i_{\sigma^{-1}(1)}}, g_{i_{\sigma^{-1}(2)}}, \ldots, g_{i_{\sigma^{-1}(n)}}). \tag{6.2.1}$$

A k-**class function**: $G^k \to \mathbb{C}$ is a function which is constant on the k-classes of G.

Remark 6.1 A k-class algebra has been defined by Humphries and Rode, with basis elements the sums of elements in each k-class. This algebra is an S-ring over G^k. Whereas the information in the 1-classes is exactly equivalent to that in the character table, Humphries and Rode have shown that this is no longer true if $k = 2$. These results will be presented in Chap. 10.

6.2.2 Relevant Representation Theory of $\mathcal{S}_n$ over $\mathbb{C}$

A reference for the following is [112]. Consider the expression of $\sigma \in \mathcal{S}_n$ as a product of disjoint cycles (including cycles of length 1)

$$(a_1, a_2, \ldots, a_{\mu_1})(b_1, b_2, \ldots, b_{\mu_2}) \ldots (d_1, d_2, \ldots, d_{\mu_m})$$

where

$$\mu_1 \geq \mu_2 \geq \mu_3 \geq \ldots \geq \mu_m.$$

Then $\mu = (\mu_1, \mu_2, \mu_3, \ldots, \mu_m)$ is a partition of n, called the **shape** of σ, denoted by $sh(\sigma)$. In general $\mu \vdash n$ will be written if μ is a partition of n, and the **length** $\ell(\mu)$ is defined to be m. The conjugacy class containing σ will also be denoted by μ. Let $z_\mu = |C_{\mathcal{S}_n}(\sigma)|$. Then the size of this conjugacy class is $\frac{n!}{z_\mu}$. Each irreducible character of $\mathcal{S}_n$ can also be associated to a partition $\lambda \vdash n$, and this character will be denoted by χ_λ or by λ if no confusion arises. To any partition

$\mu = (\mu_1, \mu_2, \mu_3, \ldots, \mu_m)$ there corresponds a **Young diagram** with μ_i boxes in the ith row. For example if $\mu = (4, 3, 2, 1, 1)$ the corresponding Young diagram is

Young diagrams can be used to obtain the irreducible representations of $\mathcal{S}_n$. A **tableau** is obtained from a Young diagram by arbitrarily inserting the integers $\{1, \ldots, n\}$ into the boxes. Consider the partition $\mu = (3, 2, 2, 1)$. A tableau corresponding to its Young diagram is

$$
\begin{array}{|c|c|c|}
\hline
1 & 2 & 3 \\
\hline
\end{array}
$$

The tableau gives rise to two subgroups of $\mathcal{S}_n$, which in the example are

$$P_\mu = \mathcal{S}_{\{1,2,3\}} \times \mathcal{S}_{\{4,5\}} \times \mathcal{S}_{\{6,7\}} \times \mathcal{S}_{\{8\}}$$

and

$$Q_\mu = \mathcal{S}_{\{1,4,6,8\}} \times \mathcal{S}_{\{2,5,7\}} \times \mathcal{S}_{\{3\}}.$$

In general P_μ consists of the permutations which preserve each row and Q_μ those which preserve each column of the tableau. Here $\mathcal{S}_X$ denotes the symmetric group on X so that $\mathcal{S}_X \subseteq \mathcal{S}_Y$ if $X \subseteq Y$. The elements a_μ and b_μ of $\mathbb{C}\mathcal{S}_n$ are defined by

$$a_\mu = \sum_{g \in P_\mu} g \quad \text{and } b_\mu = \sum_{g \in Q_\mu} sgn(g)g \;,$$

and the **Young symmetrizer** c_μ is defined to be $a_\mu b_\mu$. The module $\mathbb{C}\mathcal{S}_n c_\mu$ gives the irreducible representation V_μ of $\mathcal{S}_n$ with character χ_λ (this construction, up to equivalence of representations, is independent of the way in which the tableau is numbered).

There is a cruder decomposition of $\mathbb{C}\mathcal{S}_n$ using the central idempotent

$$e_\mu = \frac{1}{n!}\mu(e) \sum_{\sigma \in \mathcal{S}_n} \mu(\sigma)\sigma$$

which gives rise to the module $E_\mu = \mathbb{C}S_n e_\mu$ giving the representation with character $\chi_\mu(e)\chi_\mu$. The further decomposition

$$e_\mu = e_\mu^{(1)} + e_\mu^{(2)} + \ldots + e_\mu^{(\chi_\mu(e))}$$

decomposes E_μ into isomorphic copies of V_μ.

From the point of view of Chap. 1, the process described above may be explained as follows. The group matrix X of S_n is first transformed to $P^{-1}XP$ where the columns of P are the eigenvectors of the reduced matrix X^{red}. The transformed matrix is a block diagonal matrix, the ith block being of size m_i^2 which is the group matrix corresponding to E_λ, and the further reduction of each such block gives m_i blocks each of size m_i which are irreducible group matrices corresponding to the character μ. The group determinant then factors into the determinants of the blocks.

The definition of the k-characters, $k = 1, 2, \ldots$ corresponding to a character χ of an arbitrary group G is given in Chap. 1 and is shown there to be equivalent to the following.

Recall that for a cycle $\sigma = (j_1, j_2, \ldots, j_t)$ of S_k

$$\chi_\sigma(g_{i_1}, g_{i_2}, \cdots, g_{i_k}) = \chi(g_{i_{j_1}} g_{i_{j_2}} \cdots g_{i_{j_t}}).$$

and that for any $\tau \in S_k$, $\tau = \sigma_1 \sigma_2 \ldots \sigma_s$ expressed as a product of disjoint cycles, (including cycles of length 1)

$$\chi_\tau(g_{i_1}, g_{i_2}, \cdots, g_{i_k}) = \chi_{\sigma_1}(g_{i_1}, g_{i_2}, \cdots, g_{i_k}) \chi_{\sigma_2}(g_{i_1}, g_{i_2}, \cdots, g_{i_k})$$
$$\cdots \chi_{\sigma_s}(g_{i_1}, g_{i_2}, \cdots, g_{i_k}).$$

Then

$$\chi^{(k)}(g_{i_1}, g_{i_2}, \cdots, g_{i_k}) = \sum_{\tau \in S_k} sgn(\tau) \chi_\tau(g_{i_1}, g_{i_2}, \cdots, g_{i_k}).$$

The generalization below is due to Vazirani.

Definition 6.1 For any character χ of G and any irreducible character λ of S_k the **immanent k-character** $\chi_\lambda^{(k)}$ is defined by

$$\chi_\lambda^{(k)}(g_{i_1}, g_{i_2}, \cdots, g_{i_k}) = \sum_{\tau \in S_k} \lambda(\tau) \chi_\tau(g_{i_1}, g_{i_2}, \cdots, g_{i_k}).$$

It will be shown later that $\chi_\lambda^{(k)} = 0$ if $\chi(e) < \ell(\lambda)$.

Remark 6.2 If f is any trace-like function on an algebra the $f_\lambda^{(k)}$ can be defined. It may be interesting to examine the properties of these functions.

If $k = 2$ there are only two partitions $\lambda_1 = 2$ and $\lambda_2 = 1 + 1$. These correspond to the extended 2-characters $\chi^{(2,+)}$ and $\chi^{(2)}$ given in Chap. 5.

Example 6.1 For $k = 3$ there are three irreducible characters of $\mathcal{S}_3$: These are χ_1, χ_2, χ_3 in Example 5.1, and they correspond to partitions as follows: $\lambda_1 = $ id $(= \chi_1)$, $\lambda_2 = $ sgn $(= \chi_2)$, and $\lambda_3 = \chi_3$, where $\lambda_1 = 3$, $\lambda_2 = 1 + 1 + 1$ and $\lambda_3 = 2 + 1$. Then for an arbitrary character χ of a group G,

$$\chi_{\lambda_1}^{(3)}(g, h, k) = \chi(g)\chi(h)\chi(k) + \chi(g)\chi(hk) + \chi(h)\chi(gk) + \chi(k)\chi(gh) + \chi(ghk)$$
$$+ \chi(gkh),$$

$$\chi_{\lambda_2}^{(3)}(g, h, k) = \chi(g)\chi(h)\chi(k) - \chi(g)\chi(hk) - \chi(h)\chi(gk) - \chi(k)\chi(gh) + \chi(ghk)$$
$$+ \chi(gkh)$$

(the usual 3-character) and

$$\chi_{\lambda_3}^{(3)}(g, h, k) = 2\chi(g)\chi(h)\chi(k) - \chi(ghk) - \chi(gkh).$$

Definition 6.2 For any two functions $\phi : G^{r_1} \to \mathbb{C}$ and $\psi : G^{r_2} \to \mathbb{C}$ where $r_1 + r_2 = k$ define

$$(\phi \circ \psi)^{(k)}(g_1, g_2, \cdots, g_k)$$

$$= \frac{1}{r_1! r_2!} \sum_{\sigma \in \mathcal{S}_k} \phi(g_{\sigma(1)}, g_{\sigma(2)}, \cdots, g_{\sigma(r_1)}) \psi(g_{\sigma(r_1+1)}, g_{\sigma(r_1+2)}, \cdots, g_{\sigma(r_1+r_2)}).$$

This product, up to a scalar, multilinearizes the tensor product of the functions $(\phi.\psi)(g, h) = \phi(g) \circ \psi(h)$.

If $\chi(e) = 1$, i.e. χ is a homomorphism, there is only one non-zero immanent k-character corresponding to $\lambda_1 = (k)$ (the trivial character), namely

$$\chi_{\lambda_1}^{(k)} = \chi(g_1)\chi(g_2) \cdots \chi(g_k),$$

and this may be written as

$$(\chi \circ \chi \circ \chi \ldots \circ \chi) \ (k \text{ factors}).$$

For arbitrary k a set of extended basic k-characters of G may be built up from the $\{\chi_\lambda^{(r)}\}$, $r \le k$ using the operation $\circ$.

Example 6.2 Let $G = \mathcal{S}_3$ and $k = 3$. Let the irreducible characters be χ_1, χ_2, χ_3 (as given in the table in Example 5.1). The following is a set of basic 3-characters.

(a) $(\chi_i)_{\lambda_1}^{(3)}$, $i = 1, \ldots, 3$.
(b) $\chi_1 \circ \chi_2 \circ \chi_3$.
(c) $\chi_i \circ \chi_i \circ \chi_3$, $i = 1, 2$.

(d) $\chi_1 \circ \chi_1 \circ \chi_2$.

(e) $\chi_2 \circ \chi_2 \circ \chi_1$.

(f) $\chi_i \circ \chi_3^{(2)}$, $i = 1, 2$.

(g) $\chi_i \circ \chi_3^{(2,+)}$, $i = 1, 2$.

(h) $(\chi_3)_{\lambda_j}^{(3)}$, $j = 2, 3$.

It is possible to define a set of extended basic k-characters for all k but this becomes more complicated as k increases in size. The result of Vazirani to obtain an orthogonal set of extended basic k-class functions is much more easily described and is given below.

6.2.3 The Connection with Wreath Products

The wreath product $G \, Wr \, P$ of an arbitrary group G with a permutation group P acting on the set $\Omega = \{1, 2, \ldots, k\}$ is defined to be the split extension $B \rtimes P$ where the base group B is $G^{(k)}$, the direct product of k copies of G, with the action of an element $\pi \in P$ on B given by

$$(g_{i_1}, g_{i_2}, \ldots, g_{i_k})^\pi = (g_{i_{\pi^{-1}(1)}}, g_{i_{\pi^{-1}(2)}}, \ldots, g_{i_{\pi^{-1}(k)}}).$$

Now let G_k denote the wreath product $G \, Wr \, \mathcal{S}_k$. Denote the diagonal $\Delta(G)$ to be the copy of G inside B consisting of elements of the form $(g, g, \ldots, g)$. The conjugacy classes $\{D_i\}$ of G_k contained in B do not coincide with the k-classes of G but each D_i is a union of k-classes, since the conjugacy classes are the orbits of the action of B and $\mathcal{S}_k$ on B and the k-classes are the orbits of the action of $\Delta(G)$ and $\mathcal{S}_k$ acting on B.

Example 6.3 The 2-classes of $G = \mathcal{S}_3$ are listed in Example 5.1. The classes with representatives (a, a^2) and (a, a) fuse to a conjugacy class in G_2 of size 4, and the classes with representatives (b, b) and (b, ab) fuse to a conjugacy class of size 9. The remaining 2-classes are conjugacy classes in G_2.

Remark 6.3 In the above context, the class sums $\overline{D_i} = \sum_{b \in D_i} b$ form an S-ring over B (see Chap. 4).

The operations of restriction and induction characters have already occurred in this book and the corresponding operations on representations are given in [112]. Let χ be a character of an arbitrary group G of degree m with corresponding module V. The immanent k-character $\chi_\lambda^{(k)}$ corresponding to a representation λ of $\mathcal{S}_k$ is constant on the k-classes of G. The representation of G corresponding to V extends naturally to an irreducible representation $V^{\otimes k}$ of $B = G^k$ with character denoted by $\chi^{\otimes k}$. The action of $\mathcal{S}_k$ on $V^{\otimes k}$ via

$$(w_1 \otimes w_2 \otimes \cdots \otimes w_k)^\sigma = (w_{\sigma^{-1}(1)} \otimes w_{\sigma^{-1}(2)} \otimes \cdots \otimes w_{\sigma^{-1}(k)}))$$

makes $V^{\otimes k}$ into an irreducible module for G_k. Let $\widetilde{\chi}$ denote the character of this action. Then $\mathrm{Res}_B^{G_k}\widetilde{\chi} = \chi^{\otimes k}$ and $\mathrm{Res}_{\Delta(G)}^{G_k}\widetilde{\chi} = \chi^k$, where χ^k is the repeated product of χ in the character ring of G.

Example 6.4 Let $G = S_3$ with generators $a = (1,2,3)$, $b = (12)$ and consider G_2. The following table is the character table of G_2. The generator of S_2 is denoted by c.

Class size	1	4	9	4	6	12	6	12	18
Rep.	(e,e)	(a,a)	(b,b)	(e,a)	(e,b)	(a,b)	$(\alpha,\alpha)c$	$(e,\alpha)c$	$(\alpha,\beta)c$
Char.									
ϖ_1	1	1	1	1	1	1	1	1	1
ϖ_2	1	1	1	1	-1	-1	-1	-1	1
ϖ_3	1	1	1	1	-1	-1	1	1	-1
ϖ_4	1	1	1	1	1	1	-1	1	-1
ϖ_5	2	2	-2	2	0	0	0	0	0
ϖ_6	4	-2	0	1	-2	1	0	0	0
ϖ_7	4	-2	0	1	2	-1	0	0	0
ϖ_8	4	1	0	-2	0	0	-2	1	0
ϖ_9	4	1	0	-2	0	0	2	-1	0

Let χ be the irreducible character of G of degree 2. Then $\chi \otimes \chi$ is constant on the classes of G^2 inside B and takes on the vector of values $(4, 1, 0, -1, 2, 0)$ on these, i.e. the same set of values as either ϖ_8 or ϖ_9. The induced character $\widetilde{\chi}$ takes on the values $(8, 2, 0, -2, 4, 0)$ on these classes and is 0 on the classes outside B. Thus $\widetilde{\chi} = \varpi_8 + \varpi_9$.

Some relevant work on the representation theory of $GL(m, V)$ is now introduced (see [112]).

The **Schur function** $s_\lambda(x_1, \ldots, x_r)$ corresponding to the partition $\lambda \vdash k$ is defined as follows. Let $\lambda = (\lambda_1, \lambda_2, \ldots, \lambda_r)$ with $\lambda_1 \geq \lambda_2 \geq \ldots \geq \lambda_r$. Let $\delta = (k-1, k-2, \ldots, 1, 0)$. Let

$$v_\lambda(x_1, \ldots, x_r) = \det \begin{bmatrix} x_1^{\lambda_1} & x_2^{\lambda_1} & \ldots & x_r^{\lambda_1} \\ x_1^{\lambda_2} & x_2^{\lambda_2} & \ldots & x_r^{\lambda_2} \\ \ldots & \ldots & \ldots & \ldots \\ x_1^{\lambda_r} & x_2^{\lambda_r} & \ldots & x_r^{\lambda_r} \end{bmatrix}$$

and let $H = \prod_{1 \leq i < j \leq r} (x_j - x_i)$ be the Vandermonde determinant. Then

$$s_\lambda(x_1, \ldots, x_r) = \frac{v_{(\lambda+\delta)}(x_1, \ldots, x_r)}{H}.$$

Since v_λ and H are skew symmetric functions it follows that s_λ is a symmetric function.

Assume as above that V is a module for the finite group G. The representation of $GL(m, V)$ on $V^{\otimes k}$ splits into irreducible representations of the form $S_\lambda V$, where each $S_\lambda V$ is associated to the Schur function s_λ. More specifically, the character value at an element h of $GL(V)$ with respect to the representation $S_\lambda V$ is $s_\lambda(\zeta_1, \ldots, \zeta_m)$ where $\zeta_1, \ldots, \zeta_m$ are the eigenvalues of the matrix h. An alternative formula for the Schur function s_λ in terms of the power sum symmetric functions $p_\mu = p_{\mu_1} \cdots p_{\mu_l}$ where

$$p_{\mu_i} = x_1^{\mu_i} + \ldots + x_r^{\mu_i}$$

is

$$s_\lambda = \sum_{\mu \vdash k} \frac{1}{z_\mu} \lambda(\mu) p_\mu. \tag{6.2.2}$$

There is also the formula

$$S_\lambda V = \mathrm{Hom}_G(\mathrm{Res}_{G^k}^{G_k} \mathrm{Ind}_{S_k}^{G_k} \lambda, V^{\otimes k}).$$

$S_\lambda V$ is a module for G, but it is not necessarily irreducible.

Lemma 6.1 $S_\lambda V$ *as a representation of* G *has character*

$$\chi_{S_\lambda V}(g) = \frac{1}{r!} \chi_\lambda^{(k)}(g, g, \ldots, g).$$

Proof If $\{\zeta_1, \ldots, \zeta_m\}$ are the eigenvalues of g under the action on V, then as above

$$\chi_{S_\lambda V}(g) = s_\lambda(\zeta_1, \ldots, \zeta_m)$$

(see [112]). Since g^j has eigenvalues $\{\zeta_1^j, \ldots, \zeta_m^j\}$,

$$\chi(g^j) = \zeta_1^j + \zeta_2^j + \cdots + \zeta_m^j = p_j(\zeta_1, \ldots, \zeta_m).$$

Then from the expansion of the Schur functions in terms of the power sum symmetric functions (6.2.2) gives:

$$\chi_{S_\lambda V}(g) = s_\lambda((\zeta_1, \ldots, \zeta_m)$$

$$= \sum_{\mu \vdash r} \frac{1}{z_\mu} \lambda(\mu) p_\mu(\zeta_1, \ldots, \zeta_m)$$

$$= \frac{1}{k!} \sum_{\mu \vdash k} \frac{k!}{z_\mu} \lambda(\mu) \chi(g^{\mu_1}) \chi(g^{\mu_2}) \cdots \chi(g^{\mu_l})$$

$$= \frac{1}{k!} \chi_\lambda^{(k)}(g, g, \ldots, g).$$

$\square$

6.3 χ_λ as a Multilinearization

Let M_{s_λ} be the multilinearization of s_λ.

Lemma 6.2 *For $g \in G$, $\chi_{S_\lambda V}(g) = \chi_\lambda(g, g, \ldots, g)$ and*

$$\chi_\lambda(g_1, g_2, \ldots, g_k) = M_{s_\lambda}(g_1, g_2, \ldots, g_k). \tag{6.3.1}$$

Proof The first statement follows from Lemma 6.1. As above let $p_\lambda = p_{\lambda_1} \cdots p_{\lambda_l}$ be the power sum symmetric function. If Z is an $m \times m$ matrix with eigenvalues $\{\eta_1, \eta_2, \ldots, \eta_m\}$, and $\lambda \vdash k$, then let

$$p_\lambda(Z) = p_\lambda(\eta_1, \eta_2, \ldots, \eta_m) = Tr(Z^{\lambda_1}) Tr(Z^{\lambda_2}) \cdots Tr(Z^{\lambda_l}).$$

The multilinearization is

$$M_{p_\lambda}(Y_1, \ldots, Y_k) = \frac{1}{k!} \sum_{\sigma \in \mathcal{S}_k} Tr\left(Y_{\sigma(1)} Y_{\sigma(2)} \cdots Y_{\sigma(\lambda_1)}\right) \cdots Tr\left(Y_{\sigma(r-\lambda_l+1)} \cdots Y_{\sigma(k)}\right).$$

If $\tau \in \mathcal{S}_k$ is represented in cycle notation by

$$\tau = (\sigma(1)\sigma(2) \ldots \sigma(\lambda_1)) (\sigma(\lambda_1+1) \ldots \sigma(\lambda_1+\lambda_2)) \cdots (\sigma(r-\lambda_l+1) \ldots \sigma(k)),$$

then the inner summand of the above is $Tr_\tau(Y_1, \ldots, Y_k)$. As σ ranges over $\mathcal{S}_k$, τ ranges over all permutations of shape λ in $\mathcal{S}_k$ and this is a z_λ-to-1 correspondence. Hence

$$M_{p_\lambda}(Y_1, \ldots, Y_k) = \frac{1}{k!} z_\lambda \sum_{\tau : sh(\tau) = \lambda} Tr_\tau(Y_1, \ldots, Y_k).$$

Now using (6.2.2) the multilinearization of s_λ is

$$M_{s_\lambda}(Y_1, \ldots, Y_k) = \frac{1}{k!} \sum_{\mu \vdash k} \frac{1}{z_\mu} \lambda(\mu) \, z_\mu \sum_{\tau : sh(\tau) = \mu} Tr_\tau(Y_1, \ldots, Y_k)$$

$$= \frac{1}{k!} \sum_{\sigma \in \mathcal{S}_k} \lambda(\sigma) Tr_\sigma(Y_1, \ldots, Y_k). \tag{6.3.2}$$

If $\rho : G \to GL_k(\mathbb{C})$ is a representation with character χ and $Y_i = \rho(g_i)$, (6.3.2) becomes equivalent to the statement that the multilinearization of the Schur function evaluated at the Y_i is exactly $\frac{1}{k!}\chi_\lambda^{(k)}(g_1, g_2, \ldots, g_k)$. $\qquad\square$

Corollary 6.1 *If $\lambda \vdash k$ has $\ell(\lambda) > \chi(1)$, then $\chi_\lambda^{(k)}(g_1, g_2, \ldots, g_k) = 0$.*

Proof From standard theory, if the length of λ is greater than $n = \dim V$ then L_λ does not occur in $V^{\otimes r}$ (see [207] or [112]). Hence $\chi_{S_\lambda V} = s_\lambda = 0$ and so its multilinearization is also 0. $\qquad\square$

In particular, if $\deg(\chi) = k$ then $\chi_{sgn}^{(r)} = 0$ for $r \geq k + 1$. This result was known to Frobenius and is related to the characterization of characters in Chap. 10.

The following theorem gives a formula for the immanent k-characters of a sum of characters, in terms of the Littlewood-Richardson coefficients.

Theorem 6.1 *Let $\chi, \psi \in Irr(G)$, $\lambda \vdash k$. Then*

$$(\chi + \psi)_\lambda^{(k)} = \sum_{k_1+k_2=k,\, \mu\vdash k_1,\, \nu\vdash k_2} N_{\mu\nu\lambda}(\chi_\mu \circ \psi_\nu), \qquad (6.3.3)$$

*where $N_{\mu\nu\lambda}$ are the **Littlewood-Richardson coefficients**.*

Proof The Littlewood-Richardson coefficients appear when representations of $S_{k_1} \times S_{k_2}$ are induced to $S_{k_1+k_2}$. More precisely

$$\text{Ind}_{S_{k_1} \times S_{k_2}}^{S_{k_1+k_2}} \mu \otimes \nu = \sum N_{\mu\nu\lambda}\lambda.$$

Using Schur Weyl duality, as in [112], it follows that over $GL(V) \times GL(W)$

$$S_\lambda(V \oplus W) = \bigoplus N_{\mu\nu\lambda} S_\mu V \otimes S_\nu W.$$

This means that $\chi_{S_\lambda(V\oplus W)} = \sum N_{\mu\nu\lambda}\chi_{S_\mu V} \cdot \chi_{S_\nu W}$. Applying Lemma 6.1 and (6.3.1) and the remark above, it follows directly that (6.3.3) holds. $\qquad\square$

Corollary 6.2 *As above let $\chi^{(k)}$ denote the higher character $\chi_{(1^k)}$, i.e. the usual k-character. Suppose χ and ψ are characters of G. Then*

$$(\chi + \psi)^{(k)} = \sum_{a+b=k} \chi^{(a)} \circ \psi^{(b)}.$$

Proof It is known (see [111]) that $N_{\mu\nu(1^k)} = 1$ exactly when $\mu = (1^a)$, $\nu = (1^b)$ and 0 otherwise. Equivalently if $\wedge$ denotes the usual exterior product $\wedge^r(V \oplus W) = \bigoplus_{a+b=r} \wedge^a V \otimes \wedge^b W$. $\qquad\square$

Example 6.5 For characters χ and ψ of G,

$$(\chi + \psi)^{(3)} = \chi^{(3)} + \psi^{(3)} + \chi^{(2)} \circ \psi + \chi \circ \psi^{(2)}.$$

Remark 6.4 It appears to be difficult to obtain a similar formula for the tensor product character $\chi \cdot \psi$. In the case $k = 2$ there is the formula

$$(\chi \cdot \psi)^{(2)}(g, h) = \chi(g)\chi(h)\psi(g)\psi(h) - \chi(gh)\psi(gh)$$

$$= \frac{1}{2}(\chi^{(2)}(g, h)\psi^{(2,+)}(g, h) + \psi^{(2)}(g, h)\chi^{(2,+)}(g, h)),$$

but for $k > 2$ such a formula for $(\chi \cdot \psi)^{(k)}$ does not appear to be available.

6.4 χ_λ as a Trace and an Invariant

In Lemma 6.1, it was shown that $\frac{\lambda(1)}{r!}\chi_\lambda^{(k)}(g, \ldots, g)$ is the trace of the diagonal action of g on $e_\lambda V^{\otimes k}$ and therefore is equal to the trace of the action of $(g, g, \ldots, g)e_\lambda$ on $V^{\otimes k}$. If $\tilde{\chi}$ is defined as Sect. 6.2.3 and its definition has been extended to $\mathbb{C}G_k$, consider $\tilde{\chi}((g_1, g_2, \ldots, g_k)e_\lambda)$. Although this does not give a representation of G_k, $\mathrm{Hom}(\mathrm{Ind}_{S_k}^{G_k} E_\lambda, V^{\otimes k})$ is a representation of $\mathrm{Hom}(\mathrm{Ind}_{S_k}^{G_k} E_\lambda,$ $\mathrm{Ind}_{S_k}^{G_k} E_\lambda)$, and in effect this computes the trace of this action.

Lemma 6.3 *For $\sigma \in S_k$ and $\underline{g} = (g_1, g_2, \ldots, g_k) \in G^r$, $\tilde{\chi}(\underline{g}\sigma) = \chi_{\sigma^{-1}}(\underline{g})$.*

Proof Let $\sigma \in S_k$ and express σ^{-1} as a product of disjoint cycles by

$$\sigma^{-1} = (j_1, j_2, \ldots, j_{t_1})(m_1, m_2, \ldots, m_{t_2}) \ldots (u_1, u_2, \ldots, u_{t_w}).$$

Take $\{v_i\}_1^n$, a basis for V and write $[a_{ij}^{(h)}] = A^{(h)}$ for the matrix of g_h with respect to this basis. Then

$$\underline{g}\sigma(v_{i_1} \otimes v_{i_2} \otimes \ldots \otimes v_{i_k}) = \underline{g}(v_{i_{\sigma^{-1}(1)}} \otimes v_{i_{\sigma^{-1}(2)}} \otimes \ldots \otimes v_{i_{\sigma^{-1}(k)}})$$

$$= \sum_{j=1}^n a_{j i_{\sigma^{-1}(1)}}^{(1)} v_j \otimes \sum_{j=1}^n a_{j i_{\sigma^{-1}(1)}}^{(2)} v_j \otimes \ldots \otimes \sum_{j=1}^n a_{j i_{\sigma^{-1}(k)}}^{(k)} v_j.$$

In order to calculate the trace of this action the coefficient of $v_{i_1} \otimes v_{i_2} \otimes \ldots \otimes v_{i_k}$ in the above formula is needed. This is

$$a_{i_1 i_{\sigma^{-1}(1)}}^{(1)} a_{i_2 i_{\sigma^{-1}(2)}}^{(2)} a_{i_3 i_{\sigma^{-1}(3)}}^{(3)} \ldots a_{i_k i_{\sigma^{-1}(k)}}^{(k)}.$$

Corresponding to the cycles $(j_1, j_2, \ldots, j_{t_1}), (m_1, m_2, \ldots, m_{t_2}) \ldots$ of σ^{-1} the order of this product can be rearranged as

$$(a^{(j_1)}_{i_{j_1} i_{j_2}} a^{(j_2)}_{i_{j_2} i_{j_3}} a^{(j_3)}_{i_{j_3} i_{j_4}} \cdots a^{(j_{t_1})}_{i_{j_{t_1}} i_1})(a^{(m_1)}_{i_{m_1} i_{m_2}} a^{(m_2)}_{i_{m_2} i_{m_3}} a^{(m_3)}_{i_{m_3} i_{m_4}} \cdots a^{m_{t_2}}_{i_{m_{t_2}} i_{m_1}}) \ldots$$

$$(a^{(u_1)}_{i_{u_1} i_{u_2}} a^{(u_2)}_{i_{u_2} i_{u_3}} a^{(u_3)}_{i_{u_3} i_{u_4}} \cdots a^{(u_{tw})}_{u_{j_{tw}} i_{u_1}}).$$

Then

$$\widetilde{\chi}(\underline{g}\sigma) = \sum_{i_{j_1}=1}^{n} \cdots \sum_{i_{u_{tw}}=1}^{n} (a^{(j_1)}_{i_{j_1} i_{j_2}} a^{(j_2)}_{i_{j_2} i_{j_3}} a^{(j_3)}_{i_{j_3} i_{j_4}} \cdots a^{(j_{t_1})}_{i_{j_{t_1}} i_1})(a^{(m_1)}_{i_{m_1} i_{m_2}} a^{(m_2)}_{i_{m_2} i_{m_3}} a^{(m_3)}_{i_{m_3} i_{m_4}} \cdots a^{m_{t_2}}_{i_{m_{t_2}} i_{m_1}})$$

$$\ldots (a^{(u_1)}_{i_{u_1} i_{u_2}} a^{(u_2)}_{i_{u_2} i_{u_3}} a^{(u_3)}_{i_{u_3} i_{u_4}} \cdots a^{(u_{tw})}_{i_{u_{tw}} i_{u_1}}).$$

The right-hand side is a product of factors of the form

$$\sum_{i_{j_1}=1}^{n} \sum_{i_{j_2}=1}^{n} \cdots \sum_{i_{j_{t_1}}=1}^{n} (a^{(j_1)}_{i_{j_1} i_{j_2}} a^{(j_2)}_{i_{j_2} i_{j_3}} a^{(j_3)}_{i_{j_3} i_{j_4}} \cdots a^{(j_{t_1})}_{i_{j_{t_1}} i_1}),$$

$$\sum_{i_{m_1}=1}^{n} \sum_{i_{m_2}=1}^{n} \cdots \sum_{i_{m_{t_2}}=1}^{n} (a^{(m_1)}_{i_{m_1} i_{m_2}} a^{(m_2)}_{i_{m_2} i_{m_3}} a^{(m_3)}_{i_{m_3} i_{m_4}} \cdots a^{m_{t_2}}_{i_{m_{t_2}} i_{m_1}})$$

$$\ldots \sum_{i_{u_1}=1}^{n} \sum_{i_{u_2}=1}^{n} \cdots \sum_{i_{u_{tr}}=1}^{n} (a^{(u_1)}_{i_{u_1} i_{u_2}} a^{(u_2)}_{i_{u_2} i_{u_3}} a^{(u_3)}_{i_{u_3} i_{u_4}} \cdots a^{(u_{tw})}_{i_{u_{tw}} i_{u_1}}).$$

Now

$$\sum_{i_{j_1}=1}^{n} \sum_{i_{j_2}=1}^{n} \cdots \sum_{i_{j_{t_1}}=1}^{n} (a^{(j_1)}_{i_{j_1} i_{j_2}} a^{(j_2)}_{i_{j_2} i_{j_3}} a^{(j_3)}_{i_{j_3} i_{j_4}} \cdots a^{(j_{t_1})}_{i_{j_{t_1}} i_1}) = Tr(A^{(j_1)} A^{(j_2)} \ldots A^{(j_{t_1})})$$

with similar formulae for the other factors, so that the complete product is

$$Tr(A^{(j_1)} A^{(j_2)} \ldots A^{(j_{t_1})}) Tr(A^{(m_1)} A^{(m_2)} \ldots A^{(m_{t_1})}) \ldots Tr(A^{(u_1)} A^{(u_2)} \ldots A^{(u_{t_1})})$$

$$= \chi_{\sigma^{-1}}(\underline{g}).$$

$\square$

Theorem 6.2 $\tilde{\chi}(\underline{g}e_\lambda) = \frac{\lambda(1)}{k!}\chi_\lambda^{(k)}(\underline{g}).$

Proof From Lemma 6.3, it follows

$$\tilde{\chi}(\underline{g}e_\lambda) = \frac{\lambda(1)}{k!}\sum_{\sigma \in S_k}\lambda(\sigma^{-1})\tilde{\chi}(\underline{g}\sigma)$$

$$= \frac{\lambda(1)}{k!}\sum_{\sigma \in S_k}\lambda(\sigma^{-1})\chi_{\sigma^{-1}}(\underline{g}) = \frac{\lambda(1)}{k!}\sum_{\sigma \in S_k}\lambda(\sigma)\chi_\sigma(\underline{g})$$

$$= \frac{\lambda(1)}{k!}\chi_\lambda^{(r)}(\underline{g}).$$

In other words, $\chi_\lambda^{(r)}(\underline{g}) = \frac{r!}{\lambda(1)}\tilde{\chi}(e_\lambda \underline{g}e_\lambda)$ (since $e_\lambda^2 = e_\lambda$), and $\tilde{\chi}(e_\lambda \underline{g}e_\lambda)$ computes the trace of an element of

$$e_\lambda\mathbb{C}[G_k]e_\lambda \simeq Hom_{G_r}(\text{Ind}_{S_k}^{G_k}E_\lambda, \text{Ind}_{S_k}^{G_k}E_\lambda) \text{ on } e_\lambda V^{\otimes k} \simeq Hom(\text{Ind}_{S_k}^{G_k}E_\lambda, V^{\otimes k}).$$

$\square$

6.4.1 χ_λ as an Invariant

If G is taken as $GL(V)$ in the above then $S_\lambda V$ is irreducible, and $\chi_\lambda^{(k)}$ may be regarded as a multilinear invariant (with respect to the $\Delta(G)$-action map of $End(V)^{\otimes k}$). Sections 6.3 and 6.4 above can be described in the language of invariants used in Procesi's work [239], which is outlined as follows.

The following identifications hold:

$$(End(V)^{\otimes k})^* \simeq ((V^* \otimes V)^{\otimes k})^* \simeq (V^{\otimes k})^* \otimes V^{\otimes k} \simeq End\left(V^{\otimes k}\right).$$

A G-invariant map $End(V^k) \to \mathbb{C}$ may be viewed as an element of $End((V)^{\otimes k})$ which commutes with the action of $GL(V)$.

Now S_r and $GL(V)$ act naturally on $V^{\otimes k}$. If GL_k is defined as $GL(V)WrS_k$ this extends to a natural action of GL_k on $V^{\otimes k}$ which is centralized by the diagonal action of $G = GL(V) \simeq GL(m, \mathbb{C})$. From standard theory all operators commuting with the G-action can be realised as elements of $\mathbb{C}[S_k]$, and therefore those commuting with both the G-action and the S_k-action are linear combinations of the e_λ.

In Theorem 1.2 [239], Procesi shows that under the above identifications $\sigma \in S_k$ corresponds to an element he calls $\lambda_\sigma \in End(V^{\otimes k})$ via

$$\lambda_\sigma : (w_1 \otimes w_2 \otimes \ldots \otimes w_k) \to (w_{\sigma^{-1}(1)} \otimes w_{\sigma^{-1}(2)} \otimes \ldots \otimes w_{\sigma^{-1}(k)}).$$

which in turn corresponds to an invariant $\mu_\sigma \in (End(V)^{\otimes k})^*$, and this is $\chi_{\sigma^{-1}}$. Thus any multilinear invariant of k matrices of size $m \times m$ is a linear combination of the χ_σ.

Using the fact that the center of $\mathbb{C}(\mathcal{S}_k)$ is spanned by the e_λ, it also follows that the invariants that commute with both the G-action and the $\mathcal{S}_k$-action are linear combinations in the $\chi_\lambda^{(k)}$. Corollary 6.1 is essentially equivalent to Theorem 4.3 in ([239]) that all "trace identities are linear combinations of the $\chi_\lambda^{(k)}$ for which $\ell(\lambda) > m$".

6.5 k-Class Functions

Theorem 6.2 states that $\chi_\lambda^{(k)}(\underline{g}) = \frac{k!}{\lambda(1)}\widetilde{\chi}(\underline{g}e_\lambda)$ where $\widetilde{\chi}$ is the character of $V^{\otimes k}$ as a representation of the wreath product G_k. Here the situation is generalized to any character of G_k and the corresponding functions are examined.

Recall that the action of $\sigma \in \mathcal{S}_k$ on $B = G^k$ is

$$(g_1, g_2, \dots, g_k)^\sigma = (g_{\sigma^{-1}(1)}, g_{\sigma^{-1}(2)}, \dots, g_{\sigma^{-1}(k)})$$

and the action of $\Delta(G)$ on B is by conjugation.

Definition 6.3 For any pair ψ_1, ψ_2 of k-class functions on G define $\langle, \rangle_k$ by

$$\langle \psi_1, \psi_2 \rangle_k = \frac{1}{|G|^k} \sum_{\underline{g} \in G^k} \psi_1(\underline{g})\overline{\psi_2(\underline{g})}.$$

Definition 6.4 For any character ϖ of G_k and any partition λ define $\psi_\lambda^\varpi : G^k \to \mathbb{C}$ by

$$\psi_\lambda^\varpi(\underline{g}) = \varpi(\underline{g}e_\lambda).$$

where ϖ has been extended implicitly to $\mathbb{C}G_k$.

Lemma 6.4 ψ_λ^ϖ is a k-class function.

Proof Now ϖ is a class function on G_k, and e_λ and $\Delta(G)$ commute with each element of $\mathbb{C}(\mathcal{S}_k)$. Thus

$$\psi_\lambda^\varpi(\sigma^{-1}\underline{g}\sigma) = \varpi(\sigma^{-1}\underline{g}\sigma e_\lambda) = \varpi(\sigma^{-1}(\underline{g}e_\lambda)\sigma) = \varpi(\underline{g}e_\lambda)$$

$$= \psi_\lambda^\varpi(\underline{g})$$

$$\psi_\lambda^\varpi(\underline{a}^{-1}\underline{g}\underline{a}) = \varpi(\underline{a}^{-1}\underline{g}\underline{a}e_\lambda) = \varpi(\underline{a}^{-1}(\underline{g}e_\lambda)\underline{a}) = \varpi(\underline{g}e_\lambda)$$

$$= \psi_\lambda^\varpi(\underline{g}).$$

$\square$

The character of any $e_\lambda \mathbb{C} e_\lambda$-module will yield a k-class function in this manner, which will not in general be a class function of G^k.

6.5.1   Orthogonality

If ϖ is an irreducible character of G_k and λ is an irreducible character of $\mathcal{S}_k$ which is a homomorphic image of G_k then since λ can be lifted to a character of G_k the product $\lambda \cdot \varpi$ can be regarded as a character of G_k.

Theorem 6.3 *Let* $\varpi, \phi \in Irr(G_k)$, $\lambda, \mu \in Irr(\mathcal{S}_k)$. *If* $\langle \lambda \cdot \varpi, \mu \cdot \phi \rangle = 0$ *then*

$$\langle \psi_\lambda^\varpi, \psi_\mu^\phi \rangle_k = 0.$$

Proof The orthogonality relations for characters of G_k imply that for any $\tau \in \mathcal{S}_k \subset G_k$,

$$\sum_{w \in G_k} \lambda \cdot \varpi(w) \overline{\mu \cdot \phi(\tau w)} = 0.$$

Scaling, summing over τ and reordering summations yields

$$0 = \frac{\lambda(1)}{k!} \frac{\mu(1)}{k!} \sum_{\tau \in \mathcal{S}_k} \sum_{\underline{g}\sigma \in G_k} \lambda \cdot \varpi(\underline{g}\sigma) \overline{\mu \cdot \phi(\tau \underline{g}\sigma)}$$

$$= \frac{\lambda(1)}{k!} \frac{\mu(1)}{k!} \sum_{\underline{g} \in G^k} \sum_{\sigma \in \mathcal{S}_k} \lambda(\sigma) \varpi(\underline{g}\sigma) \sum_{\tau \in \mathcal{S}_k} \overline{\mu(\tau\sigma)\phi(\tau\underline{g}\sigma)}$$

$$= \sum_{\underline{g} \in G^k} \varpi\Big(\sum_{\sigma \in \mathcal{S}_k} \frac{\lambda(1)}{k!} \lambda(\sigma) \underline{g}\sigma \Big) \overline{\phi\Big(\sum_{\tau \in \mathcal{S}_k} \frac{\mu(1)}{k!} \mu(\tau\sigma)\tau\underline{g}\sigma \Big)} = \sum_{\underline{g} \in G^k} \varpi(\underline{g}e_\lambda) \overline{\phi(\underline{g}e_\lambda\mu)}$$

$$= \sum_{\underline{g} \in G^k} \varpi(\underline{g}e_\lambda) \overline{\phi(\underline{g}e_\mu)} = \langle \psi_\lambda^\varpi, \psi_\mu^\phi \rangle.$$

$\square$

Theorem 6.4 *If* $\lambda \cdot \varpi$ *is a sum of distinct irreducible characters, then*

$$\langle \psi_\lambda^\varpi, \psi_\lambda^\varpi \rangle_k = \frac{\lambda(1)}{\varpi(1)} \langle \lambda, Res_{\mathcal{S}_k}^{G_r} \varpi \rangle_{\mathcal{S}_k} = \frac{1}{\varpi(1)} \varpi(e_\lambda).$$

Proof Because irreducible characters occur in $\lambda \cdot \chi$ with multiplicity 1, the orthogonality relations for characters of G_k give

$$\frac{1}{|G_k|} \sum_{w \in G_k} \lambda \cdot \varpi(w) \overline{\lambda \cdot \varpi(\tau w)} = \frac{1}{\lambda \cdot \varpi(1)} \overline{\lambda \cdot \varpi(\tau)}.$$

Then, using the work above, it follows that

$$\frac{1}{|G^k|} \sum_{g \in G^k} \psi_\lambda^\varpi(g) \overline{\psi_\lambda^\varpi(g)} = \frac{\lambda(1)}{\varpi(1)} \frac{1}{k!} \sum_{\tau \in S_k} \lambda(\tau) \overline{\varpi(\tau)} = \frac{1}{\varpi(1)} \varpi(e_\lambda)$$

$$= \frac{\lambda(1)}{\varpi(1)} \langle \lambda, \mathrm{Res}_{S_k}^{G_k} \varpi \rangle_{S_k}.$$

$\square$

Corollary 6.3 *For fixed λ and k the extended higher characters $\chi_\lambda^{(k)}$ for $\chi \in Irr(G)$ are orthogonal with respect to $\langle, \rangle_k$. If $\chi_\lambda(e) = n$ the norm of $\frac{1}{r!}\chi_\lambda^{(k)}$ is $\frac{1}{n^k} s_\lambda(1, \ldots, 1)$ if it is non-zero and $\lambda \cdot \tilde\chi$ consists of distinct characters.*

Proof As above, if $\chi_i \in Irr(G)$ is the character of a representation V_i, let $\tilde\chi_i \in Irr(G_k)$ be the irreducible character of the associated action of G_k on $V_i^{\otimes k}$. Then

$$\langle \mathrm{Res}_{G^k}^{G_k} \lambda \cdot \tilde\chi_i, \mathrm{Res}_{G^k}^{G_k} \mu \cdot \tilde\chi_j \,_{G^k}\rangle_{G^k} = \lambda(1)\mu(1)\langle \chi_i^{\otimes k}, \chi_j^{\otimes k} \rangle_{G^k} = 0.$$

Thus it must also follow that $\langle \lambda \cdot \tilde\chi_i, \mu \cdot \tilde\chi_j \rangle_{G_k} = 0$ and the result is a consequence of Theorem 6.3. Observe that the action of $\sigma \in S_k$ on $V^{\otimes k}$: $\tilde\chi(\sigma) = n^{\ell(sh(\sigma))}$ is known where $n = \chi(1)$. Hence, if $\lambda \cdot \tilde\chi$ is a sum of distinct irreducible characters, then Theorems 6.2 and 6.4 give the norm of $\frac{1}{k!}\chi_\lambda^{(k)}$ as

$$\frac{1}{\widetilde{\chi(1)}} \frac{1}{k!} \sum_{\tau \in S_k} \lambda(\tau) n^{\ell(sh(\tau))} = \frac{1}{n^k} s_\lambda(1, 1, \ldots, 1).$$

$\square$

Now in the situation where $\varpi \in Irr(G_k)$ has $\mathrm{Res}_{G^k}^{G_k} \varpi$ irreducible, i.e. of the form $\varphi_1^{\mu_1} \otimes \cdots \otimes \varphi_\ell^{\mu_\ell}$ for $\varphi_i \in Irr(G)$, $\mu \vdash k$, then the ψ_λ^ϖ are symmetrized products of the $\varphi_{i\lambda}^{(\mu_i)}$ (in a similar fashion to the multilinearizations in the examples of Chap. 1.6, 3.2.2), and these include the functions $\chi_i \circ \chi_j$ appearing above.

6.5.2 The Number of k-Classes and Extended k-Functions

Lemma 6.5 *The dimension of the space of all k-class functions on a finite group G is*

$$\frac{1}{|G|}\sum_{g\in G}\sum_{\mu\vdash k}\frac{1}{z_\mu}|C_G(g^{\mu_1})|\cdots|C_G(g^{\mu_l})|.$$

Proof It is clear that this dimension is the number of k-classes. By the Frobenius-Burnside lemma, which states that if H is any permutation group acting on a set Ω,

$$|\Omega/H| = \frac{1}{|H|}\sum_{h\in H}|\Omega^h| \qquad (6.5.1)$$

where $|\Omega/H|$ is the number of orbits of H and Ω^h denotes the fixed point set of h. Note that the right-hand side of (6.5.1) is $\langle 1, \varsigma(h)\rangle_H$ where $\varsigma(h) = Fix(h)$ is the permutation character of the natural action of H on Ω. In the situation here $H = \Delta(G)S_k$ and $\Omega = G^k$. The permutation action of G on itself by conjugation has character $\varsigma(g) = |C_G(g)|$, and $\Delta(G)$ acts on G^k under the diagonal action with character ς^k. The number of orbits required is the multiplicity of the trivial character in $Sym^k(\mathbb{C}(\mathbb{G})) = S_{(k)}(\mathbb{C}(\mathbb{G})) = S_1(\mathbb{C}(\mathbb{G}))$. Now the character of G on $S_1(\mathbb{C}(\mathbb{G}))$ is just $\psi_1^{(k)}|_{\Delta(G)}$. Hence the number of generalized k-classes is

$$\langle 1, \psi_1^{(k)}\rangle_{\Delta(G)} = \frac{1}{|G|}\sum_{g\in G}\sum_{\mu\vdash k}\frac{1}{z_\mu}\psi(g^{\mu_1})\psi(g^{\mu_2})\cdots\psi(g^{\mu_l})$$

$$= \frac{1}{|G|}\sum_{g\in G}\sum_{\mu\vdash k}\frac{1}{z_\mu}|C_G(g^{\mu_1})|\cdots|C_G(g^{\mu_l})|.$$

$\square$

Theorem 6.5 *Let $\mathcal{B} = \{\psi_1^\varpi : \varpi(e_1) \neq 0\}$. Then $\mathcal{B}$ is an orthogonal basis for the span $\{\psi_\lambda^\chi : \lambda \in Irr(S_k), \varpi \in Irr(G_k)\}$. In particular, the extended higher characters are all linear combinations of the ψ_1^ϖ.*

Proof Theorem 6.3 shows that the ψ_1^ϖ are orthogonal. Take any $\psi_\lambda^\varpi : g \mapsto \varpi(g\sigma)$. Recall (as in the proof of Theorem 6.3) that λ may be lifted up to a character of G_k, also denoted λ, which will have G^k in its kernel. Suppose that $\lambda \cdot \varpi =$

$\sum_{\varpi_i \in Irr(G_k)} m_i \varpi_i$. Then

$$\varpi(ge_\lambda) = \varpi(g \sum_{\sigma \in S_k} \lambda(\sigma)\sigma) = \sum_{\sigma \in S_k} \lambda(\sigma)\varpi(g\sigma) = \sum_{\sigma \in S_k} \lambda(g\sigma)\varpi(g\sigma)$$

$$= \sum_{\sigma \in S_k} \lambda \cdot \varpi(g\sigma) = \lambda \cdot \varpi(g \sum_{\sigma \in S_k} \sigma) = \lambda \cdot \varpi(ge_1)$$

$$= \sum_{\varpi_i \in Irr(G_k)} m_i \varpi_i(ge_1).$$

Hence $\psi_\lambda^\varpi = \sum_{\chi_i \in Irr(G_k)} m\psi_1^\varpi$. By the remark below, $\psi_1^\varpi = 0$ if $\psi_1^\varpi \notin \mathcal{B}$, thus $\psi_\lambda^\varpi \in \text{span}(\mathcal{B})$.   □

Remark The set $\{\varpi \in Irr(G_k) : \varpi(e_\lambda) \neq 0\}$ contains exactly the support of $Ind_{S_k}^{G_k} L_\lambda$, since

$$\varpi(e_\lambda) = \frac{\lambda(1)}{k!} \sum_{\sigma \in S_k} \lambda(\sigma)\varpi(\sigma) = \lambda(1) < Res_{S_k}^{G_k} L_\lambda \varpi, \lambda_{S_k} = \lambda(1)\langle \varpi, Ind_{S_k}^{G_k} \lambda \rangle.$$

Let U be the representation of G_k with character ϖ. U and $Hom_{G_k}(Ind_{S_k}^{G_k} 1, U)$ are isomorphic as $Hom_{G_k}(Ind_{S_k}^{G_k} 1, Ind_{S_k}^{G_k} 1)$-modules and ψ_1^ϖ is the trace of this action. When $\varpi(e_1) = 0$ then $Hom_{G_r}(Ind_{S_k}^{G_k} 1, U) = 0$, and so its trace ψ_1^χ is also 0.

Remark 6.5 Theorem 6.5 is also valid when the trivial character $\lambda = (r)$ is replaced by the sign character $\lambda = (1^r)$.

Remark 6.6 In the case where G is abelian, the number of k-classes is equal to the number of basic extended k-characters. This is seen because all the irreducible characters of G are linear, and hence for any irreducible character χ of G, $\chi_\lambda^k = 0$ if $\lambda \neq (r)$.

Chapter 7
Fourier Analysis on Groups, Random Walks and Markov Chains

Abstract In Chap. 1 it is explained that if p is a probability on a finite group G the group matrix $X_G(p)$ is a transition matrix for a random walk on G. If f is an arbitrary function on G the process of transforming $X_G(f)$ into a block diagonal matrix is equivalent to the obtaining the Fourier transform of f. This chapter explains the connections with harmonic analysis and the group matrix. Most of the discussion is on probability theory and random walks.

The fusion of characters discussed in Chap. 4 becomes relevant, and also the idea of fission of characters is introduced, especially those fissions which preserve diaonalizability of the corresponding group matrix. As an example of how the group matrix and group determinant can be used as tools, their application to random walks which become uniform after a finite number of steps is examined.

7.1 Introduction

In the article [208] (incorporated and extended in the book [209]) Mackey gave an expansive account from a historical point of view of many ways in which, often with hindsight, group representation theory can be seen to be behind the results in what is usually understood in a broad sense as harmonic analysis. Fourier analysis in the classical setting of differential equations may be regarded as the decomposition of a function on $\mathbb{C}$ into a linear combination of the irreducible characters of $\mathbb{C}$, and for any group G there is a corresponding decomposition. This is in turn equivalent to the diagonalization or partial diagonalization of some form of a group matrix or equivalently the factorization of the corresponding determinant. Although this may not be so clear in the setting of infinite groups, a translation may be made where the x_g are regarded as functions with finite support.

The following are direct quotes from [209].

> For some time I had become increasingly impressed by the extent and interrelatedness of the applications of the theory of unitary group representations to physics, probability theory and number theory... also quantum physics and the theory of automorphic functions.

© Springer Nature Switzerland AG 2019
K. W. Johnson, *Group Matrices, Group Determinants and Representation Theory*, Lecture Notes in Mathematics 2233, https://doi.org/10.1007/978-3-030-28300-1_7

Large branches of these subjects may be looked upon as nearly identical with certain branches of group representations. Moreover, one obtains a clearer view of many known relationships between the subjects in question by looking at them in this way.

Classical Fourier analysis can be illuminatingly regarded as a chapter in the representation theory of compact commutative groups.

The author's philosophy is that harmonic analysis "is" the decomposition of induced representations (provided one includes the representations of the "inertial subgroups" defined by properly ergodic actions).

Key: the operation of translation of functions on G is carried over to the operation of multiplication by a fixed function for the functions on $\widehat{G}$. More significantly, it converts differentiation into multiplication and therefore converts differential equations into algebraic equations.

In the above, if G is abelian $\widehat{G}$ is the dual of G, i.e. the functions from G to $\mathbb{C}$ with suitable topology. If G is nonabelian then modifications need to be made (see Appendix A).

The discussion in this chapter is mainly restricted to the interaction of the techniques developed in earlier chapters with probability, but it is likely that the ideas also have relevance to more general harmonic analysis. Some examples where the group matrix and group determinant appear, such as in control theory and the theory of wavelets, will be described in Chap. 10. It has already been remarked that the (full) group matrix may be regarded as a transition matrix for the Markov chain coming from a random walk on a group, and in this chapter other group matrices will appear for more general Markov chains where a symmetry group is present. The reduced group matrix and the work in Chap. 4 is relevant, because the theory is much simpler if diagonalization is possible.

Given that there is a very extensive literature on the material addressed here, especially on random walks and Markov chains, written from many points of view, some justification should be set out for yet another presentation by an author outside the field. The motivation for this chapter is that there is an algebraic underpinning of a significant part of probability theory which has a direct connection via group representations to the various forms of the group matrix, group determinant and other concepts which have appeared in preceding chapters.

It is hoped that the presentation here will be helpful to probabilists. It may also be possible that group theory can benefit from the probabilistic perspective. It will become clear that the object is to suggest how techniques related to the group matrix and group determinant may offer advantages over the existing uses of representation theory in various applications.

It appears to be impossible to give an account of the possible applications without describing examples, but it also seems that the intuition of an experienced probabilist is needed to bring out the essence of these examples. Thus apologies are in order. The examples below tend to be taken directly out of existing publications, especially the two books mentioned below, and an effort has been made to refer to the sources as accurately as possible.

The starting point and basic reference for the discussion here is the book by Diaconis [82]. In the preface to the book the following sentence appears. "As the reader will see, there are endless places where "someone should develop a theory that makes sense of this" or try it out, or at least state an honest theorem, or...". The later book [54] develops aspects of material in [82], but, probably because there is such a variety of ideas in [82], there appears to be much room for further expansion.

As mentioned above, the focus here is on random walks where a finite symmetry group is present, which often leads to the use of the representation theory of the group. The literature in this case is reasonably compact. There are the two books mentioned above and a number of papers, a high proportion of which seem to be either coauthored or inspired by Diaconis. For applications the reader should consult the bibliographies in above two books, as well as that in [205].

By contrast there are many presentations in book form of random walks on infinite objects, with a bewildering variation in definitions. It would have been ideal if a summary could have been made of how many of the techniques relating to finite groups have arisen first in the infinite setting, but this is beyond this book. At best, an indication has been given to the descriptions in [82] and other places. An exception is made by including an account of spherical functions in the infinite case in Appendix A in order to give insight into the finite case and with a view to widening applications of the techniques developed here.

The following comment of Diaconis appears in [82].

> Natural mixing schemes can sometimes be represented as random walks on groups or homogeneous spaces. Then, representation theory allows a useful Fourier analysis. If the walks are invariant under conjugation, only the characters are needed. If the walks are bi-invariant under a subgroup giving a Gelfand pair, the spherical functions are needed.

In both these cases the essential fact from the point of view in this book is that the corresponding group matrix is diagonalizable. When this is no longer the case, methods arising from group matrices and group determinants can offer advantages over standard techniques. In some situations, given a representation $\rho : G \to GL(n, \mathbb{C})$ the coefficient functions ϕ_{ij} which take an element g to the matrix coefficient $\rho(g)_{ij}$ are important (and they can give rise to special functions (see Appendix A)). These functions depend on the choice of the representing matrices. Since the k-characters and the corresponding $s_k(x)$ occuring in Chap. 1.6 are independent of such a choice, they provide a more intrinsic set of data for ρ.

A random walk on a finite group G may be described as follows. The states of the walk are the elements of G. To each element g of G there is associated a probability $p(g) \in [0, 1]$ such that $\sum_{g \in G} p(g) = 1$. If at stage i the walk has reached the element h then it moves to the element gh with probability $p(g)$. It follows that the probability of a move from h to k is $p(kh^{-1})$. This may be regarded as a Markov chain (see below) with transition matrix K which is the transpose of $X_G(p)$. In the following the discussion will be given using $X_G(p)$ itself since minor modifications are needed to adapt the work to the transpose. In the study of random walks the powers K^r are important, and these are most easily obtained by diagonalization or partial diagonalization. As mentioned above, if the probabilities which are used are constant on conjugacy classes, then $X_G(p)$ is obtained from the reduced group matrix of Frobenius and is similar to a diagonal matrix. If arbitrary probabilities are used the techniques for diagonalization of the group matrix discussed in Chap. 2 become relevant.

In the more general situation of a random walk acting on a finite set Ω which has a group G of symmetries (for example Ω could be the vertices of a cube) this is also a Markov chain whose transition matrix is essentially the group matrix $X_G^\upsilon(p)$ of the corresponding permutation representation υ, where p is a probability which is constant on the right cosets of the point stabilizer K. If (G, K) is a Gelfand pair and p is constant on the double cosets of K then $X_G^\upsilon(p)$ may be diagonalized to a matrix $\widehat{X_G^\upsilon(p)}$, the non-zero diagonal elements giving rise to spherical functions. The Fourier transform of p can be regarded as $\widehat{X_G^\upsilon(p)}$, and the inverse transform is the reverse process of obtaining $X_G^\upsilon(p)$ from $\widehat{X_G^\upsilon(p)}$. In both the finite and infinite cases the Fourier transform is often described in terms of special functions, in the finite case often where there is a distance regular graph whose vertex set is Ω. The analysis in the finite case may be described in terms of association schemes. In the case where (G, K) is not Gelfand methods to diagonalize X_G^υ become relevant. The existence of regular subgroups produces connections with the theory of S-rings given in Chap. 4, and the paper [270] brings in these ideas.

The chapter is laid out as follows. Section 7.2 provides a summary of the techniques of Fourier analysis coming from group representation theory which are used in applications, together with some suggestions for alternative techniques which arise from considering the group determinant and the group matrix. Section 7.3 illustrates the close connection between the algebraic ideas in Chap. 4 and probability. It describes a result of Smith to the effect that induction of characters in association schemes may be regarded as conditional expectation. In Sect. 7.4 the discrete Fourier transform on $\mathbb{Z}_n$ is introduced. A brief account of the Fast Fourier Transform is given describing the method in terms of the group matrix and group determinant of a cyclic group. Then a sketch is given of how a fast version of the Fourier transform for arbitrary finite groups could be implemented using group matrix ideas.

In Sect. 7.5 basic aspects of Markov chains are set out. After a brief historical review relevant definitions are given and some standard examples are described. Then the discussion is restricted to random walks on groups. The cut-off phenomenon is introduced as an important concept. Next a question of Vishnevetskiy and Zhmud on random walks on groups which converge in a finite number of steps is studied in some detail, since it illustrates the use of the group matrix and presents situations where the cut-off is exact. Suggestions are then made of alternative techniques to examine random walks on groups in general.

For random walks on groups associated to a probability which is constant on conjugacy classes the basic results of association schemes given in Chap. 4 can be applied. Results on "fission" of the class algebra of a group are given, since the probabilities which are constant on the fissioned classes produce diagonalized group matrices and the analysis of the associated random walks proceeds in a similar manner.

In Chap. 6 Markov chains on objects with a finite symmetry group are discussed. The difficult cases are where the centralizer ring of the corresponding action is not commutative, and in this case even if the probablities are constant on the double cosets the corresponding group matrix is not diagonalizable. Questions of how to fission the classes of an association scheme are raised.

7.2 Fourier Analysis on Groups

If G is an arbitrary topological group, Fourier analysis usually assumes that a function is in $L^2(G)$, the set of square integrable functions from G to $\mathbb{C}$. This notation will be used here, although the condition of square integrability is redundant in the finite case.

Definition 7.1 Let G be an arbitrary finite group and f lie in $L^2(G)$. Let $\rho : G \to GL(m, \mathbb{C})$ be a representation so that for each $t \in G$, $\rho(t)$ is an $m \times m$ matrix. The **Fourier transform** of f with respect to ρ is defined to be the $m \times m$ matrix

$$\mathcal{F}_\rho(f) = \widehat{f}_\rho = \sum_{t \in G} f(t)\rho(t). \tag{7.2.1}$$

It may be seen that $\widehat{f}_\rho$ is $X_G^\rho(\underline{x})$, where $\underline{x} = (f(g_1), f(g_2), \ldots, f(g_n))$ for some ordering $(g_1, \ldots g_n)$ of G and $X_G^\rho(\underline{x})$ denotes the group matrix of G with respect to ρ as defined in Chap. 2. The notation $X_G^\rho(f)$ will be used interchangeably with $\widehat{f}_\rho$.

For f_1 and f_2 in $L^2(G)$ their **convolution** $f * g$ has already been defined in Chap. 1 (1.3.6), and is equivalent to

$$(f_1 * f_2)(s) = \sum_{tu=s} f_1(t) f_2(u).$$

Lemma 7.1 *For f_1 and f_2 in $L^2(G)$*

$$\mathcal{F}_\rho(f_1 * f_2) = \mathcal{F}_\rho(f_1).\mathcal{F}_\rho(f_2).$$

where the operation on the right-hand side is matrix multiplication.

Proof The statement of the Lemma is

$$X_G^\rho(f_1)X_G^\rho(f_2) = X_G^\rho(f_1 * f_2)$$

which is equivalent to the condition that $X_G^\rho(\underline{x})X_G^\rho(\underline{y}) = X_G^\rho(\underline{x} * \underline{y})$, see Theorem 2.1, Chap. 2. □

Lemma 7.2 *For f in $L^2(G)$ and $h \in G$,*

$$f(h) = \frac{1}{|G|} \sum_{\rho_i \in Irr'(G)} d_i Tr(\widehat{f}_{\rho_i} \rho_i(h^{-1})). \tag{7.2.2}$$

where $d_i = \deg(\rho_i)$.

Proof Since both sides are linear in f it is sufficient to prove the result for the function δ_g defined by $\delta_g(h) = \delta_{gh}$. In this case $(\widehat{\delta_g})_{\rho_i} = \rho_i(g)$ and the right-hand side is

$$\frac{1}{|G|} \sum_{\rho_i \in Irr'(G)} d_i Tr(\rho_i(g)\rho_i(h^{-1})) = \frac{1}{|G|} \sum_{\chi_i \in Irr(G)} d_i \chi_i(gh^{-1}) = \frac{1}{|G|} \pi(gh^{-1}),$$

where π is the character of the regular representation. Thus the function on the right-hand side is 0 except when $g = h$ in which case it is 1. This is exactly $\delta_g(h)$. □

Definition 7.2 The inner product of functions f_1, f_2 in $L^2(G)$ is

$$\langle f_1, f_2 \rangle = \frac{1}{|G|} \sum_{g \in G} f_1(g) f_2(g^{-1}).$$

It is seen that this definition, apart from a factor of $|G|$, is an extension of that for class functions given in Chap. 4.

Without loss of generality it can be assumed that each ρ_i is unitary.

Lemma 7.3 *If f_1, f_2 lie in $L^2(G)$ then*

$$\langle f_1, f_2 \rangle = \frac{1}{|G|} \sum_{\rho_i \in Irr'(G)} d_i Tr(\widehat{f_1}_{\rho_i} \overline{\widehat{f_2}_{\rho_i}}). \tag{7.2.3}$$

Proof Again, by linearity it is sufficient to prove the result for $f_1 = \delta_g$. Then the left-hand side of (7.2.3) is equal to $\frac{1}{|G|} f_2(g^{-1})$ and the right-hand side is

$$\frac{1}{|G|} \sum_{\rho_i \in Irr'(G)} d_i Tr(\rho_i(g) \overline{\widehat{f_2}_{\rho_i}})$$

and they are equal by Lemma 7.2. □

Lemma 7.2 gives the procedure for Fourier inversion in an arbitrary finite group, and Lemma 7.3 gives what is usually known as Plancherel's result. Define $\|f\|$ to be $\sqrt{\langle f, f \rangle}$, and for a matrix M define $\|M\|$ to be $\sqrt{Tr(M)}$. Then it is a consequence of (7.2.3) that

$$\|f\|^2 = \frac{1}{|G|} \sum_{\rho_i \in Irr'(G)} d_i \|\widehat{f}_{\rho_i}\|^2. \tag{7.2.4}$$

The Fourier transform and Fourier inversion may be explained in terms of the group matrix as follows.

Take a group G. Then as is indicated in Chap. 1, the full group matrix

$$X_G = \sum_{g \in G} \pi(g) x_g$$

where π is the right regular representation.

If $f : G \to \mathbb{C}$ then as mentioned above

$$\widehat{f}_{\rho_i} = X_G^{\rho_i}(f) = \sum_{t \in G} \rho_i(g) f(g).$$

The Fourier transform may be thought of as taking an arbitrary group matrix X_G into the block diagonal matrix

$$\widehat{X_G}(f) = diag(X_G^{\rho_1}(f), X_G^{\rho_2}(f), \dots, X_G^{\rho_2}(f), \dots, X_G^{\rho_r}(f), \dots, X_G^{\rho_r}(f)), \tag{7.2.5}$$

where $\{\rho_i\}_{i=1}^r = Irr(G)$, the degree of ρ_i is d_i, and where each $X_G^{\rho_i}(f)$ is repeated d_i times. The matrix $X_G(f) = \sum_{g \in G} \pi(g) f(g)$ corresponds to the full group matrix. The matrix P appearing in Theorem 2.4, Chap. 2 produces the similarity transformation

$$P X_G(f) P^{-1} = \widehat{X_G}(f).$$

The inverse Fourier transform takes as input $\widehat{X_G}(f)$ and produces

$$X_G(f) = P^{-1} \widehat{X_G}(f) P.$$

The values $\{f(g)\}$ may be read off from the first column of $X_G(f)$.

The explicit calculation of a single $f(h)$ may be carried out via

$$|G| f(h) = Tr(X_G(f)(\pi(h^{-1}))) = \sum_{g \in G} Tr(\pi(g)\pi(h^{-1})) f(g)$$

$$= \sum_{g \in G} Tr(\pi(h^{-1}g)) f(g) = \sum_{g \in G} \sum_{i=1}^r d_i Tr(\rho_i(h^{-1}g)) f(t)$$

$$= \sum_{i=1}^r d_i Tr(X^{\rho_i}(f)\rho_i(h^{-1})). \tag{7.2.6}$$

The following theorem from [82, p. 49] proves that P above is of the form given in Theorem 2.4, Chap. 2. The reader is reminded of the definition of P.

Let G be a finite group with a given ordering $\{g_i\}_{i=1}^n$ with $g_1 = e$, with $Irr'(G) = \{\rho_1, \ldots, \rho_r\}$, $deg(\rho_i) = d_i$ and ρ_1 the trivial representation. Assume that a choice of a set of representing matrices $\{\rho_i(g)\}_{g \in G}$ has been made for each ρ_i, so that $\rho_i(g)_{jk}$ is the element in the $(j, k)^{th}$ position of $\rho_i(g)$. The vector ψ_i of length d_i^2 is defined as

$$\psi_i(g) = \sqrt{\frac{d_i}{n}} (\rho_i(g)_{11}, \rho_i(g)_{21}, \ldots, \rho_i(g)_{d_i 1}, \rho_i(g)_{12}, \ldots, \rho_i(g)_{d_i d_i}).$$

Then if $\phi(g)$ is the column vector of length n

$$(\psi_1(g), \psi_2(g), \ldots, \psi_r(g))^T,$$

P is constructed as the $n \times n$ matrix whose columns are successively $\phi(g_1), \phi(g_2), \ldots, \phi(g_n)$. Let $X_G = X_G(g_1, g_2, \ldots, g_n)$.

Theorem 7.1 *For P as above $PX_G P^{-1} = \widehat{X}_G$.*

Proof It is a consequence of the Schur relations in Theorem 1.5, Chap. 1 that P is a unitary matrix, and the result of the theorem is that

$$PX_G P^{-1} = PX_G P^* = diag(M_1, M_2, \ldots, M_r)$$

where M_i is the $d_i \times d_i$ matrix

$$diag(X_G^{\rho_i}, \ldots, X_G^{\rho_i}).$$

From Fourier inversion the (i, j)th entry of X_G is

$$\frac{1}{n} \sum_{k=1}^r d_k Tr(X_G^{\rho_k} \rho_k(g_i g_j^{-1})) = \frac{1}{n} \sum_{k=1}^r d_k Tr(X_G^{\rho_k} \rho_k(g_i) \rho_k(g_j^{-1}))$$

$$= \frac{1}{n} \sum_{k=1}^r d_k Tr(\rho_k(g_j^{-1}) X_G^{\rho_k} \rho_k(g_i))$$

and this simplifies to

$$\sum_{k=1}^r \psi_k(g_j)^* M_k \psi_k(g_i).$$

$\square$

7.2.1 The Fourier Transform from Factors of the Group Determinant

If G is a nonabelian group and f is a function on G the Fourier transform $\widehat{f}_\rho$ as defined above depends on a choice of the representing matrices $\{\rho_i(g)\}_{g \in G}$. Consider the following data for $f : G \to \mathbb{C}$. Suppose that for each irreducible representation ρ_i of G the corresponding factor $\phi_{\rho_i} (= \det(\sum_{g \in G} \rho_i(g)x_g))$ is known. This is independent of the choice of $\{\rho_i(g)\}_{g \in G}$. The question arises as to whether there is sufficient information in the ϕ_{ρ_i} to obtain the x_g. This is analogous to the Klein "Normformproblem" (see for example [29]).

On examination of the process of obtaining a function f from its Fourier transform in (7.2.2) above it is necessary to know $\eta^h_{\rho_i}(f) = Tr(\widehat{f}_{\rho_i}\rho_i(h^{-1}))$ for each h in G and each ρ_i. Now

$$\sum X^{\rho_i}_G(f)\rho_i(h^{-1}) = \sum_{g \in G} \rho_i(g)f(g)\rho_i(h^{-1}) = \sum_{g \in G} \rho_i(gh^{-1})f(g) = \sum_{k \in G} \rho_i(k)f(kh)$$

so that if $\eta^h_{\rho_i}$, the trace of $\sum_{g \in G} \rho_i(g)x_{gh}$ is known then $\eta^h_{\rho_i}(f)$ may be obtained by substituting $f(g)$ for x_g in $\eta^h_{\rho_i}$ for all g.

Now $\eta^h_{\rho_i}$ may be calculated by substituting $x_h + \lambda$ for x_h in $\phi^h_{\rho_i}$, taking the coefficient of λ^{d_i-1} in the resulting polynomial and multiplying by the factor $\det(\rho_i(h^{-1}))$. By Theorem 2.2, Chap. 2, $\phi^h_{\rho_i} = \phi_{\rho_i} \det(\rho_i(h))^{-1}$. There is also the formula

$$\eta^h_{\rho_i} = \frac{1}{(d_i - 1)!} \frac{\partial^{d_i-1}}{\partial x_h{}^{d_i-1}} (\phi_i) \det(\rho_i(h^{-1})).$$

Note that $\det(\rho_i(g))$ is the coefficient of $x_g{}^{d_i}$ in ϕ_{ρ_i}. If ρ_i is linear then $\eta^h_{\rho_i} = \phi_{\rho_i}\rho_i(h)^{-1}$.

The following example illustrates.

Example 7.1 Let G be S_3. From Chap. 1, Example 1.11 with an appropriate ordering the factors of Θ_G are $\phi_1 = \sum_{i=1}^6 x_i$, $\phi_2 = \sum_{i=1}^3 x_i - \sum_{i=4}^6 x_i$ and

$$\phi_3 = x_1^2 + x_2^2 + x_3^2 - x_1x_2 - x_2x_3 - x_3x_1 - x_4^2 - x_5^2 - x_6^2 + x_4x_5 + x_5x_6 + x_6x_4.$$

Thus $\eta^{g_j}_{\rho_1} = \phi_1$, $i = 1, \ldots, 6$, and $\eta^{g_j}_{\rho_2} = \phi_2$, $j = 1, \ldots, 3$, $\eta^{g_j}_{\rho_2} = j = -\phi_2$, $j = 4, \ldots, 6$.

Writing $\eta^{g_j}_{\rho_3}$ as η_j,

$$\eta_1 = 2x_1 - x_2 - x_3, \quad \eta_2 = 2x_2 - x_1 - x_3, \quad \eta_3 = 2x_3 - x_1 - x_2$$

$$\eta_4 = 2x_4 - x_5 - x_6), \quad \eta_5 = 2x_5 - x_4 - x_6, \quad \eta_6 = 2x_6 - x_4 - x_5.$$

Suppose that $f(g_i) = i$, $i = 1, \ldots, 6$. The values of $\phi_1(f)$ and $\phi_2(f)$ are respectively 21 and -9. The vector of values of the $\eta_i(f)$ is $[-3, 0, 3, -3, 0, 3]$. It may be seen that for $i = 1, \ldots, 6$, $f(g_i) = \frac{1}{6}(\eta_{\rho_1}^{g_i}(f) + \eta_{\rho_2}^{g_j}(f) + 2\eta_i(f))$.

Note the linear relationships between the η_i:

$$\eta_1 + \eta_2 + \eta_3 = 0, \eta_4 + \eta_5 + \eta_6 = 0.$$

Remark 7.1 It has already been indicated in Sect. 7.3 that if the irreducible factors of the group determinant are given without a specification of the variable corresponding to e, for each h in G a group $\widetilde{G}$ isotopic to G can be constructed for which h is the identity element (of $\widetilde{G}$). Then $\eta_h^{\rho_i}(G)$ is up to the factor $\rho_i(h)^{-1}$ the ordinary trace of $X_{\rho_i}(\widetilde{G})$.

For random walks the inverse Fourier transform of $(\widehat{X}_G(p))^k$ is needed where p is a probability. Since Fourier inversion can be carried out using the $\eta_{\rho_i}^{g_j}$ it would be useful to find the corresponding functions $\eta_{\rho_i}^{g_j}[(\widehat{X}_G(p))^k]$. In the above example, $\eta_i((X_G^\rho)^k)$, $k = 2, 3$ may be calculated from the formulae below, in which ρ_3 is abbreviated to ρ and θ denotes $\det(X_G^\rho)$.

$$\eta_1((X_G^\rho)^2) = (\eta_1(X_G^\rho))^2 - 2\theta = (\frac{\partial\theta}{\partial x_1})^2 - 2\theta.$$

$$\eta_i((X_G^\rho)^2) = \eta_1(X_G^\rho)\eta_i(X_G^\rho) + \theta = \frac{\partial\theta}{\partial x_1}\frac{\partial\theta}{\partial x_i} + \theta, i = 2, 3.$$

$$\eta_i((X_G^\rho)^2) = \eta_1(X_G^\rho)\eta_i(X_G^\rho) = \frac{\partial\theta}{\partial x_1}\frac{\partial\theta}{\partial x_i}, i = 4, 5, 6.$$

$$\eta_1((X_G^\rho)^3) = \eta_1(X_G^\rho)(\eta_1((X_G^\rho)^2) - \theta) = \eta_1(X_G^\rho)((\eta_1(X_G^\rho))^2 - 3\theta).$$

$$\eta_i((X_G^\rho)^3) = \eta_i(X_G^\rho)(\eta_1((X_G^\rho)^2) + (\eta_1(X_G^\rho) - \eta_i(X_G^\rho))\theta, i = 2, 3.$$

$$\eta_i((X_G^\rho)^3) = \eta_i(X_G^\rho)(\eta_1((X_G^\rho)^2) + \theta) = \eta_i(X_G^\rho)((\eta_1(X_G^\rho))^2) - \theta), i = 4, 5, 6$$

These formulae suggest that there may be corresponding formulae for $\eta_i((X_G^{\rho_i})^k)$ for arbitrary k and arbitrary ρ_i which would be useful.

7.3 Induced Class Functions and Conditional Expectations

This follows closely [260].

Let Ω be a set. A σ-**field** on Ω is a set $\mathcal{F}$ of subsets of Ω which contains Ω and is closed under complementation and countable unions. A **probability measure** on the σ-field $\mathcal{F}$ is a function $\mu : \mathcal{F} \to [0, \infty]$ such that $\mu(\Omega) = 1$ and

$$\mu(\cup_{i=0}^{\infty} A_i) = \sum_{i=0}^{\infty} \mu(A_i)$$

for any family $\{A_i\}_{i \in \mathbb{N}}$ of mutually disjoint elements of $\mathcal{F}$. A **probability space** is a triple $(\Omega, \mathcal{F}, \mu)$ where $\mathcal{F}$ is a σ-field on Ω and μ is a probability measure on $\mathcal{F}$. A **complex random variable** X on $(\Omega, \mathcal{F}, \mu)$ is a function $X : \Omega \to \mathbb{C}$ such that for any real number r the sets

$$\{\omega \in \Omega | \operatorname{Re} X(\omega) \le r\} \text{ and } \{\omega \in \Omega | \operatorname{Im} X(\omega) \le r\}$$

lie in $\mathcal{F}$. If these sets all lie in a σ-field $\mathcal{G}$ contained in $\mathcal{F}$ then X is said to be $\mathcal{G}$-**measurable**. X is said to be **integrable** if the Lebesgue integrals

$$\int_{\Omega} \operatorname{Re} X d\mu \text{ and } \int_{\Omega} \operatorname{Im} X d\mu$$

exist, which implies that

$$\int_{A} X d\mu = \int_{A} \operatorname{Re} X d\mu + i \int_{A} \operatorname{Im} X d\mu$$

for any A in $\mathcal{F}$.

Now let X is an integrable random variable on the probability space $(\Omega, \mathcal{F}, \mu)$. A **conditional expectation** $E(X|\mathcal{G})$ of X with respect to a σ-field $\mathcal{G}$ contained in $\mathcal{F}$ is a random variable $E(X|\mathcal{G})$ satisfying the conditions

(i) $E(X|\mathcal{G})$ is integrable

(ii) $E(X|\mathcal{G})$ is $\mathcal{G}$-measurable, and

$$(iii) \forall A \in \mathcal{G}, \quad \int_{A} E(X|\mathcal{G}) d\mu = \int_{A} X d\mu. \qquad (7.3.1)$$

Conditional expectations satisfying the above conditions exist, but they are not necessarily unique in that they can differ on a set with measure 0.

The following proposition lays out a necessary and sufficient condition that a random variable be a conditional expectation.

Proposition 7.1 Let $(M, \cap)$ be a subsemilattice of the meet semilattice reduct $(\mathcal{G}, \cap)$ of the σ-field $\mathcal{G}$. Suppose that $\mathcal{G}$ is the smallest σ-subfield of $\mathcal{F}$ containing M. Then an integrable random variable Y is a conditional expectation $E(X|\mathcal{G})$ if and only if it is $\mathcal{G}$-measurable and

$$\forall A \in M, \quad \int_{A} Y d\mu = \int_{A} X d\mu. \qquad (7.3.2)$$

7.3.1 Probability on Association Schemes

Suppose that $\mathcal{X} = (Q, \{\mathcal{R}_i\}_{0 \le i \le d})$ is an association scheme. Let $\Omega = Q \times Q$. The power set $\mathbf{F}$ of Ω is a σ-field. Define the normalized counting measure

$$\mu : \mathbf{F} \to [0, 1]; \ A \longmapsto |A|/|Q \times Q|.$$

Then $(\Omega, \mathbf{F}, \mu)$ is a probability space. Since the subsets $\mathcal{R}_i$ are pairwise disjoint, $\mathbf{M} = \{\mathcal{R}_i\} \cup \{\emptyset\}$ forms a subsemilattice $(\mathbf{M}, \cap)$ of $(\mathbf{F}, \cap)$. Let $\mathbf{G}$ be the smallest σ-subfield of $\mathbf{F}$ containing $\mathbf{M}$,

$$\mathbf{G} = \{\mathcal{R}_{i_1} \cup \ldots \cup \mathcal{R}_{i_s}\}.$$

Any complex-valued function $X : \Omega \to \mathbb{C}$ is an integrable random variable, with

$$\int_{\mathcal{R}_i} X d\mu = |Q \times Q|^{-1} \sum_{z \in \mathcal{R}_i} X(\mathcal{R}_i)$$

for any $\mathcal{R}_i$. Moreover, X is $\mathbf{G}$-measurable if and only if its restriction to each $\mathcal{R}_i$ is constant. In accordance with the definition in Chap. 4 for association schemes arising from quasigroups, a **class function** is defined to be any function $f : \Omega \to \mathbb{C}$ which is constant on the classes of the scheme, or in other words for each i, $f(a_1, b_1) = f(a_2, b_2)$ whenever the pair (a_1, b_1) and (a_2, b_2) lie in $\mathcal{R}_i$. This implies that the $\mathbf{G}$-measurable random variables on $(\Omega, \mathbf{F}, \mu)$ are precisely the class functions on $\mathcal{X}$.

Now let $\mathcal{Y} = (P, \{\mathcal{S}_i\})$ be a refinement subscheme of $\mathcal{X}$. For $f : P \times P \to \mathbb{C}$, define the random variable $X_f : \Omega \to \mathbb{C}$ by

$$X_f(x, y) = \frac{|Q \times Q|}{|P \times P|} f(x, y) \text{ if } (x, y) \in P \times P, 0 \text{ otherwise.}$$

The result of Smith follows [260].

Theorem 7.2 *For a class function f in $\mathbb{C}\{\mathcal{S}_i\}$, the induced class function f^Q in $\mathbb{C}\{\mathcal{R}_i\}$ is the conditional expectation $E(X_f|\mathbf{G})$.*

Proof In this setting, conditional expectations are unique, since for $A \in \mathcal{F}$, $\mu(A) = 0$ if and only if $A = \emptyset$. The class function f^Q is integrable and $\mathbf{G}$-measurable, and

$$\int_\emptyset f^Q d\mu = \int_\emptyset X_f d\mu = 0.$$

Now consider an arbitrary $\mathcal{R}_i$. Let $\mathcal{B}_i = \{\cup \mathcal{S}_j \mid \mathcal{S}_j \subseteq \mathcal{R}_i\}$. Then

$$\int_{\mathcal{R}_i} f^Q d\mu = |Q \times Q|^{-1} \sum_{(x,y) \in \mathcal{R}_i} f^Q(x, y) = |P \times P|^{-1} \sum_{(z,t) \in \mathcal{B}_i} f(z, t)$$

$$= |Q \times Q|^{-1} \sum_{(z,t) \in \mathcal{B}_i} X_f(z, t) = |Q \times Q|^{-1} \sum_{(x,y) \in \mathcal{R}_i} X_f(z, t)$$

$$= \int_{\mathcal{R}_i} X_f d\mu.$$

Thus, by Proposition 7.1, f^Q is the conditional expectation $E(X_f | \mathcal{G})$. $\square$

The properties of induction of class functions are related to properties of conditional expectations. The fact that $\mathbb{C}\{\mathcal{S}_i\} \uparrow_P^Q$ is an ideal of $\mathbb{C}\{\mathcal{R}_i\}$ is a special case of the relation

$$E(XY | \mathcal{G}) = X E(Y | \mathcal{G})$$

for any random variable X and integrable random variables Y and XY. If $\mathcal{Z} = (U, \{\mathcal{T}_i\})$ is a refinement subscheme of $\mathcal{Y} = (P, \{\mathcal{S}_i\})$ which is in turn a subscheme of $\mathcal{X} = (Q, \{\mathcal{R}_i\})$, the transitivity property

$$\uparrow_U^Q = \uparrow_U^P \uparrow_P^Q : \mathbb{C}\{\mathcal{T}_i\} \rightarrow \mathbb{C}\{\mathcal{R}_i\}$$

is a consequence of the relation

$$E(E(X | \mathbf{G}_2) | \mathbf{G}_1) = E(X | \mathbf{G}_1).$$

7.4 Finite Fourier Transforms

7.4.1 The Discrete Fourier Transform on $\mathbb{Z}_n$

The discrete Fourier transform (DFT) on $\mathbb{Z}_n$ arises as follows. Experimental information on a function (or **signal**) $f(t)$ is obtained from observations (or **samples**) on an interval $[a, b]$ of real numbers. Examples include the readings of weather observation stations, the intensity of the radiation on a certain frequency coming from the sun or seismograph readings over time. Many more are given in the literature. This information is used to approximate the Fourier transform

$$\mathcal{F}(s) = \int_a^b f(t) e^{-2\pi i s t} dt. \tag{7.4.1}$$

Usually the samples are equally spaced over $[a, b]$ and then without loss of generality it can be assumed that $a = 0$ and $b = 1$.

The basic problem is: for a function $f(t)$, given values $f(t_j)$, $t_j = \frac{1}{n}, \frac{2}{n}, \dots, \frac{n}{n}$ approximate $\mathcal{F}(s)$.

A natural approximation is

$$\mathcal{F}(s) \cong \frac{1}{n} \sum_{j=1}^{n} f(j) e^{-2\pi i s t_j} = \frac{1}{n} \sum_{j=1}^{n} f(j) e^{-\frac{2\pi i s j}{n}} \tag{7.4.2}$$

where the approximation increases in accuracy with the size of n. This is the DFT on $\mathbb{Z}_n$.

According to Heideman et al. [140], the first application of the DFT was by Clairaut in 1754 when he used it in astronomical calculations long before Fourier's work. It is better known that Gauss used it in his work on astronomy.

Consider $G = C_n$ written additively as $\mathbb{Z}_n$. The formula (7.4.2) may be interpreted as follows. Let $\{\rho_j\}_{j=1}^{n}$ be the irreducible representations of $\mathbb{Z}_n$ given by $\rho_j(s) = e^{\frac{2\pi i j s}{n}}$. Then for $f \in L^2(\mathbb{Z}_n)$

$$\widehat{f}(\rho_i) = \widehat{f_i} = \sum_{j \in \mathbb{Z}/n\mathbb{Z}} f(j) \rho_i(-j).$$

This definition is analogous to that used for infinite groups. For finite groups the slightly different version given in Sect. 7.2 is used by Diaconis

$$\widehat{f}(\rho_i) = \widehat{f_i} = \sum_{j \in \mathbb{Z}_n} f(j) \rho_i(j). \tag{7.4.3}$$

In the finite context Fourier analysis can proceed with either definition, but since the second definition is closer to the group matrix approach that in (7.4.3) will be used here.

The following properties of the DFT on $G = \mathbb{Z}_n$ follow from those given in Sect. 7.2 for arbitrary groups. However there is a difference in the cyclic case in that if $\mathcal{F}(f)(y) = \widehat{f}_{\rho_y} = \widehat{f}(y) \in \mathbb{C}$, $\widehat{f}$ may be regarded as a function from G to $\mathbb{C}$, i.e. $\widehat{f} \in L^2(G)$ and $\mathcal{F} : L^2(G) \to L^2(G)$ is a linear bijection. Then Fourier inversion becomes

$$f(j) = \sum_{y \in G} \widehat{f}(y) \rho_y(j) = \mathcal{F}\mathcal{F}(f)(-j).$$

The Plancherel theorem may be expressed as follows. For any f_1, f_2 in $L^2(G)$ an equivalent definition of the inner product is

$$\langle f_1, f_2 \rangle = \frac{1}{n} \sum_{t \in G} f_1(t) \overline{f_2(t)}.$$

where $\bar{a}$ denotes complex conjugate of a. Then

$$\langle f, f \rangle = \langle \hat{f}, \hat{f} \rangle.$$

It also follows from Lemma 7.1 that for f_1 and f_2 in $L^2(G)$

$$\mathcal{F}(f_1 * f_1) = \mathcal{F}(f_1).\mathcal{F}(f_2). \qquad (7.4.4)$$

where the product on the right is the pointwise product. It may be seen that (7.4.4) gives a quick method to calculate convolution. The significance of this is reflected in the following quotations which appear in [276].

> ...linear convolution is one of the most frequent computations carried out in digital signal processing (from [279]).

> Convolution and its behaviour under the DFT is probably the most important and powerful tool in modern analysis (from [31, p. 58]).

An equivalent definition of the DFT appears in [72] is as follows. Recall that the Fourier matrix F_n defined in Chap. 1 is given by

$$F^* = F_n^* = \frac{1}{\sqrt{n}} \begin{bmatrix} 1 & 1 & 1 & \cdots & 1 \\ 1 & \sigma & \sigma^2 & \cdots & \sigma^{n-1} \\ 1 & \sigma^2 & \sigma^4 & \cdots & \sigma^{2(n-1)} \\ & & \cdots & & \\ 1 & \sigma^{n-1} & \sigma^{2n-2} & \cdots & \sigma^{(n-1)^2} \end{bmatrix}$$

where σ is a primitive nth root of 1. The vector $Z = (f(0), f(1), \ldots f(n))^T$ has Fourier transform

$$\hat{Z} = \sqrt{n} F^* Z$$

where the vector $\hat{Z} = (\hat{f}(0), \hat{f}(1), \ldots, \hat{f}(n-1))$. Fourier inversion is then given by

$$Z = \frac{1}{\sqrt{n}} F \hat{Z}.$$

This process may be explained in terms of the group matrix as follows. As is set out in Chap. 1 the group matrix X_G of $C_n \simeq \mathbb{Z}_n$ with appropriate ordering is the circulant $C(x_0, x_1, x_2, \ldots, x_{n-1})$. If each x_j is replaced by $f(j)$, the matrix $X_G(f)$ is obtained which is the Fourier transform of f with respect to the regular representation of $\mathbb{Z}_n$. If F is the Fourier matrix, $F^{-1} X_G F$ is the diagonal matrix with entries given in (1.3.3) Chap. 1,

$$(x_1 + x_2 + \ldots + x_n), (x_1 + \sigma x_2 + \ldots + \sigma^{n-1} x_n), (x_1 + \sigma^2 x_2 + \ldots + \sigma^{2(n-1)} x_n),$$

$$\ldots, (x_1 + \sigma^{n-1} x_2 + \ldots + \sigma^{(n-1)^2} x_n).$$

and it is readily seen that

$$\widehat{X_G(f)} = F^{-1} X_G(f) F = diag(\widehat{f_0}, \ldots \widehat{f_{n-1}}),$$

where $\widehat{f_i} = f(1) + \sigma^i f(2) + \ldots + \sigma^{(n-1)i} f(n)$.

In this case, the calculation of the DFT is equivalent to the calculation of the values obtained for each irreducible factor of the group determinant $\Theta(G)$ by replacing each x_t by $f(t)$ for all $t \in G$. The DFT is also equivalent to the factorization of the group determinant $\Theta_G(f)$ in the sense that the information in it can be obtained by calculating the set of irreducible factors of Θ_G and substituting $f(t)$ for x_t for all $t \in G$ in each factor.

7.4.2 The Fast Fourier Transform: An Explanation in Terms of the Group Matrix

Consider the calculation of the Fourier transform of a function f by using the DFT on $\mathbb{Z}_n$. In the matrix version of the DFT and its inverse given above:

$$\widehat{Z} = \sqrt{n} F^* Z, \ Z = (1/\sqrt{n}) F \widehat{Z}$$

the direct computation of $F^* Z$ requires n^2 multiplications (usually the additions are neglected). The **Fast Fourier Transform** (FFT) is an algorithm which carries out this computation in $n \log n$ steps. The ideas appeared first in a paper by Gauss which was written about 1805 but which only appeared in his collected works in 1866. It is called the Cooley-Tukey algorithm after the modern rediscoverers (see [65, 140]). The range of the discussion in the literature associated to this algorithm is indicated in the following quote from [267, p. 37]:

> Presently there is variety of algorithms in the quite voluminous literature on FFT, each of them suitable with respect to some a priori assumed criteria of optimality. These criteria are very different and range from the reduction of the time and memory resources needed for the computation to the use of some particular properties of the functions whose DFT is to be determined, or the use of the properties of spectral coefficients which should be calculated. Note for example, the real or pure imaginary functions, the symmetric functions, the functions with a lot of zero values, and similarly for spectral coefficients. For more details see, for example, [1, 40, 225].

The discussion here will be restricted to the case $n = 2^m$, which is commonly used in practice. Consider the construction of the group matrix X_f of $\mathbb{Z}_n$ with respect to the cosets of $H = 2\mathbb{Z}_n$ as in Lemma 1.2, Chap. 1. On listing the elements of G as $\{0, 2, \ldots, n-2, 1, 3, \ldots, n-1\}$, X_G has the form

$$X_G = \begin{bmatrix} C(x_0, x_2, \ldots, x_{n-2}) & C(x_{n-1}, x_1, x_3, \ldots, x_{n-3}) \\ C(x_1, x_3, \ldots, x_{n-1}) & C(x_0, x_2, \ldots, x_{2(n-2)}) \end{bmatrix}. \tag{7.4.5}$$

As indicated above, the calculation of $\widehat{f}$ for a function f on G may be obtained from a list of the irreducible factors of $\Theta(G)$ by replacing x_i by $f(i)$ for $i = 0, \ldots, 2n-1$. If ψ is a irreducible factor of the group determinant of a cyclic group K, and $f : K \to \mathbb{C}$ is given, denote by $f(\psi)$ the scalar obtained by replacing x_g by $f(g)$ for all $g \in K$.

If $m = |H| = n/2$, the factors $\phi_1, \ldots, \phi_m$ of the determinant of $C(x_0, x_2, \ldots, x_{n-2})$ given in (1.3.3), Chap. 1 are

$$(x_0+x_2+\ldots+x_{n-2}), (x_0+\sigma x_2+\ldots+\sigma^{s-1} x_{n-2}), \ldots, (x_0+\sigma^{m-1} x_2+\ldots+\sigma^{(m-1)^2} x_{n-2})$$

where $\sigma = e^{4\pi i/n}$. Assume inductively that the list of the values $f(\phi_1), \ldots, f(\phi_m)$,

$$(f(0) + f(2) + \ldots + f(n-2)), (f(0) + \sigma f(2) + \ldots + \sigma^{s-1} f(n-2)), \ldots,$$
$$(f(0) + \sigma^{m-1} f(2) + \ldots + \sigma^{(m-1)^2} f(n-2))$$

has been obtained in $m \log m$ steps, and if $\psi_1, \ldots, \psi_m$ are the factors of

$$\det(C(x_1, x_3 \ldots, x_{n-3}, x_{n-1})),$$

the values of $f(\psi_1), \ldots, f(\psi_m)$ been obtained in $m \log m$ steps, where

$$f(\psi_{i+1}) = (f(1) + \sigma^i f(3) + \ldots + \sigma^{i(m-1)} f(n-1)).$$

If the same algorithm were applied to $C(f(n-1), f(1), f(3), \ldots, f(n-3))$, the $(i+1)^{st}$ value is

$$f(n-1) + \sigma^i f(1) + \sigma^{2i} f(3) + \ldots + \sigma^{i(m-1)} f(n-3)) = \sigma^i \psi_{i+1}.$$

Since by (7.4.5) X_G may be written in the form

$$X_G = \begin{bmatrix} A & B \\ C & D \end{bmatrix}$$

where A, B, C and D are circulants and therefore commute (Lemma 1.1, Chap. 1), it follows that

$$\det(X_G(f)) = \det(AD - BC),$$

by (2.4.7) Chap. 2. If it is assumed that A, B, C, D have been diagonalized, the matrix $AD - BC$ is diagonal, the $(i + 1)^{st}$ element on the diagonal being

$$\phi_{i+1}^2 - \sigma^i \psi_{i+1}^2 = (\phi_{i+1} + \sqrt{\sigma^i} \psi_{i+1})(\phi_{i+1} - \sqrt{\sigma^i} \psi_{i+1}).$$

The extra calculation to obtain the factors of $\det(X_G(f))$ is thus the calculation of the (substituted) factors of $X_G(f)$, from $f(\phi_1), \ldots, f(\phi_m)$ and $f(\psi_1), \ldots, f(\psi_m)$ and involves only the m multiplications to obtain $\sqrt{\sigma^i} f(\psi_{i+1}), i = 0, \ldots, m - 1$. Therefore the terms in $\hat{f}$ can be obtained with the use of $2m \log m + m \cong n \log(n)$ multiplications.

There are interesting connections between the FFT and other parts of mathematics given in [7]. It may be interesting to look at these from the point of view of group matrices and group determinants. It may be remarked that the extension of the methods using group determinant factors from finite cyclic groups to arbitrary finite abelian groups are not difficult.

7.4.3   Fast Fourier Transforms for Arbitrary Finite Groups

The basic facts about Fourier analysis on finite groups have been set in Sect. 7.2. The process takes a function f in $L^2(G)$ and calculates the Fourier transform $X_G^{\rho_i}(f)$ for each irreducible representation ρ_i. One application is an efficient calculation of the convolution of f_1 and f_2 by calculating the products $X_G^{\rho_i}(f_1) X_G^{\rho_i}(f_2)$ and then using Fourier inversion to obtain $f_1 * f_2$. For arbitrary groups, there are correspondingly more complications in that there may not be an effective method to calculate the irreducible representations in a reasonable amount of time, and even if they are known, matrix products are involved. The tools to partially diagonalize a group matrix discussed in Chap. 2 become relevant, in that they require little knowledge of the character table or the representations of G.

Example 7.2 Consider Fr_{21}. The construction of the group matrix on the cosets of the normal subgroup of order 7 has already been presented in Example 1.9, Chap. 1. It is of the form of a block matrix

$$X = \begin{bmatrix} C_{11} & C_{12} & C_{13} \\ C_{21} & C_{22} & C_{23} \\ C_{31} & C_{32} & C_{33} \end{bmatrix},$$

where the blocks are 7×7 circulants. If F_7 is the Fourier matrix of order 7 and P is the block diagonal matrix $diag(F_7, F_7, F_7)$ then

$$\tilde{X} = P^{-1} X P = \begin{bmatrix} \Lambda_{11} & \Lambda_{12} & \Lambda_{13} \\ \Lambda_{21} & \Lambda_{22} & \Lambda_{23} \\ \Lambda_{31} & \Lambda_{32} & \Lambda_{33} \end{bmatrix},$$

where each Λ_{ij} is a diagonal matrix. In some sense $\widetilde{X}$ is almost as simple as the Fourier transform $\widehat{X}$ and the convolution of functions f_1 and f_2 may be carried out by calculating the product of the relatively sparse matrices $\widetilde{X}(f_1)$ and $\widetilde{X}(f_2) = \widetilde{X}(f_1 * f_2)$ and then calculating $P\widetilde{X}(f_1 * f_2)P^{-1}$. Refinements of the method of this calculation could be made, for example by using faster algorithms to diagonalize the circulants.

There is a body of work on fast Fourier transforms for arbitrary groups due to Diaconis and Rockmore (for example see [87, 214],[215]). Some of the techniques in [214] seem close to using the partial diagonalizations arising from the group matrix obtained from an ordering via the cosets of a subgroup.

Although there are many interesting mathematical problems associated with using the various ways of regarding group matrices to speed up the calculation of the Fourier transform, which would improve existing methods, this is not developed here since according to Diaconis (personal communication), at present those applying Fourier transforms for non-abelian groups find more basic methods adequate.

7.5 Markov Chains and Random Walks

7.5.1 Historical Remarks

Markov chains seem to have occured first in work on a branching process by the French probabalist Bienaymé in the mid nineteenth century. An early introduction of a group occurs in the book by Poincaré [235], where card shuffling is introduced as a random walk on $\mathcal{S}_{52}$. It is shown that the pack is eventually random provided the corresponding probability does not have its support contained in a coset of a subgroup. It is also interesting that in [235] he is thinking of the techniques described here. The following passage occurs (in the context of card shuffling [234]).

> I refer to several of the works which discuss complex numbers (*i.e. hypercomplex numbers-author's comment*) and their relationship with groups. In the first place I cite the works of M. Frobenius... and then a memoir of M. Cartan "on bilinear groups and systems of complex numbers".... I have myself been occupied by the question, and in particular I have made an effort to relate the results presented by these two eminent savants in quite different forms in the study [234]).

Other early discussions are in [23, 199]. Subsequent work involved the estimation of the number of steps needed to reach approximate stationarity. This led to more precise models and the definitions of distance functions between probability distributions. One example is a model for the riffle shuffle used for card shuffling given by Gilbert and Shannon, used by Aldous [3] to prove that $\frac{3}{2}log_2 n$ riffle shuffles are necessary and sufficient to mix up n cards, as $n \to \infty$. The result of Bayer and Diaconis [13] that seven riffle shuffles are required to mix up a deck of 52 cards attracted attention and surprise among bridge players.

7.5.2   The Basic Definition

For simplicity, the following definition is not stated in its most general form. For more complete versions see for example [205, 248] or [54]).

Definition 7.3 Let Ω be a finite set. A **Markov chain** is defined by a matrix $K(x, y)$ with

$$K(x, y) \geq 0, \quad \sum_{y \in \Omega} K(x, y) = 1,$$

for each x in Ω.

Thus each row is a probability measure so K can direct a random walk as follows: from x, choose y with probability $K(x, y)$; from y choose z with probability $K(y, z)$; etc. The outcomes $X_0 = x$, $X_1 = y$, $X_2 = z, \ldots$ are referred to as a run of the chain starting at x. Then if $P(X_1 = y \mid X_0 = x)$ denotes the probability that $X_1 = y$ subject to $X_0 = x$,

$$P(X_1 = y \mid X_0 = x) = K(x, y), \quad P(X_1 = y; \ X_2 = z \mid X_0 = x) = K(x, y)K(y, z).$$

From this,

$$P(X_2 = z, \ X_0 = x) = \sum_{y \in \Omega} K(x, y)K(y, z)$$

and so on. The nth power of the matrix $K(x, y)$ has (x, y) entry $P(X_n = y, X_0 = x)$.

A Markov chain has a **stationary distribution** if there exists a function $\upsilon(x) : \Omega \to [0, 1]$ with $\sum_{x \in \Omega} \upsilon(x) = 1$, satisfying

$$\sum_{x \in \Omega} \upsilon(x)K(x, y) = \upsilon(y).$$

Markov chains involving groups occur in the following cases:

(a) A random walk on a finite group G corresponding to the probability $p :$ $G \to [0, 1]$, introduced in Sect. 7.1. Here K may be regarded as $(X_G(p))^T$. The convolution $p * p$ is defined as above and p^{*r} is used to denote $p * p * \ldots * p$ (r copies of p).

(b) A random walk on a homogeneous space obtained from a finite group G by the action on the right cosets of a subgroup H. This includes random walks on objects with G acting as symmetry group. Here the transition matrix K is obtained from the group matrix corresponding to the permutation representation of G.

The **uniform probability distribution** U on a group G of order n is defined as $U(g) = \frac{1}{n}$ for all $g \in G$, and the analysis of the walk often involves the closeness of p^{*r} to U. Similarly if the walk is on the set X with $|X| = m$ the uniform distribution is $U(x) = \frac{1}{m}$ for all $x \in X$.

A Markov chain is called an **ergodic** chain if the probability of movement from an arbitrary state to every other state is greater than 0.

The following four examples appear in many presentations.

1. The drunkard's walk. Let $G = C_n = \mathbb{Z}_n$. The elements of G are regarded as the vertices of a regular n-gon and the walk is defined as follows. At a given stage there is an equal probability of moving to either of the nearest vertices. This is defined $p : G \to [0, 1]$ with $p(\pm 1) = \frac{1}{2}$ and $p(j) = 0$ otherwise. Then $p^{*r}(j)$ is the chance that a simple random walk starting at 0 is at j after r steps. It is usually assumed that n is odd to avoid the possibility of periodic walks. The irreducible representations of G are as described above: $\rho_h(j) = e^{2\pi i j h / n}$. The Fourier transform of p is

$$\widehat{p}(h) = \frac{1}{2} e^{2\pi i h / n} - \frac{1}{2} e^{-2\pi i h / n} = \cos(2\pi h / n).$$

The analysis in this case shows that for $r > n^2$, p^{*r} is close to U (see [82, p. 29]).

2. Random transpositions in $\mathcal{S}_n$. The model is a deck of n playing cards face down in an ordered row with card 1 on the left, card n on the right. The cards are mixed as follows. The left hand touches a random card and the right hand touches a random card, and these cards are transposed, and if the cards are the same nothing is done. These are random transpositions in $\mathcal{S}_n$ and the corresponding probability p is defined by $p(e) = \frac{1}{n}$, $p((i, j)) = \frac{2}{n^2}$ for all i, j in $\{1, \ldots, n\}$ with $i \neq j$ and $p(\sigma) = 0$ otherwise. Then $p^{*r}(j)$ becomes close to U after approximately $\frac{1}{2} n \log n$ steps (see [85, p. 126] for further details).

3. The Ehrenfest diffusion model. Two urns are numbered 0, 1. There are n balls numbered $1, 2, \ldots, n$. A **configuration** is a placement of the balls into urns. There are 2^n configurations, each of which can be identified with the subset B of $\{1, 2, \ldots, n\}$ corresponding to the set of balls in urn 0. At the initial stage, all balls are in urn 0. At each step a ball is chosen randomly and moved to the other urn.

4. The Bernoulli-Laplace diffusion model. Consider two urns numbered 0 and 1 and $2n$ balls numbered $1, 2, \ldots, 2n$. A **configuration** is a placement of the balls into the two urns, with n balls in each. There are $\binom{2n}{n}$ configurations, in 1 : 1 correspondence to the n-subset of balls in urn 0. The initial configuration is when the balls $1, 2, \ldots, n$ are in urn 0 and the rest are in urn 1. At each stage, a ball is selected at random in each urn, and this pair of balls is switched.

The first two examples are random walks on the corresponding groups and the last two are random walks on homogeneous spaces. In these latter cases there is a Gelfand pair (G, K). It is not always the case that the pair (G, K) arising from a random walk on a set is a Gelfand pair. For example the Bernoulli-Laplace urn model with 3 urns, the first containing n red, the second n blue and the third n white balls. At each stage a pair of urns is chosen at random, then a randomly picked pair of balls is switched. The associated action is that of S_{3n} on the cosets of $S_n \times S_n \times S_n$ and the corresponding centralizer ring is not commutative for $n \geq 2$.

A general method was developed as follows. A Markov chain is **irreducible** if it is possible to get from a given state to any other state. A state i has **period** k if any return to state i must occur in multiples of k time steps. If $k = 1$ for each state the chain is **aperiodic**. If K is the transition matrix of a Markov chain which is assumed to be irreducible and aperiodic, from the Perron–Frobenius theorem the largest eigenvalue of K is 1 and the **spectral gap** is the gap between this and the second largest eigenvalue. This may be used to show convergence to stationarity for ergodic random walks on groups (or more generally, ergodic Markov chains) at an exponential rate governed by the spectral gap. The limitations of this method are illustrated by the example of the Gilbert–Shannon–Reeds model for riffle shuffles mentioned above where the spectral gap is $1/2$, independent of n, and thus it is not possible to obtain a specific result for a fixed value of n.

7.5.3   Random Walks on Groups

7.5.3.1   The Cut-off Phenomenon

In order to discuss questions such as how quickly card shuffling produces a random deck the **cut-off phenomenon** was introduced by Aldous, Diaconis and Shahshahani and formalized in [3] and [84]. For card shuffling, an informal statement of this is that up to a certain number r of shuffles the cards retain their order, and after $r + c$ shuffles they are close to random, where c is usually small. The following is a direct quote from [248]:

> It is believed that the cut-off phenomenon is widespread although it has been proved only for a rather small number of examples. One of the most interesting problems concerning random walks on finite groups is to prove or disprove the cut-off phenomenon for natural families of groups and walks. Focussing on walks associated with small sets of generators, one wants to understand how group theoretic properties relate to the existence or non-existence of a cut-off and, more generally, to the behavior of random walks.

The cut-off phenomenon may be described more precisely. Let P_n, Q_n be Markov chains on sets Ω_n. Let a_n, b_n be functions with $\lim_{n\to\infty} a_n = \lim_{n\to\infty} b_n = \infty$ and $\lim_{n\to\infty} b_n/a_n = 0$. The chains are said to satisfy an a_n, b_n **cut-off** if for some starting states ω_n and all fixed real θ, with $k_n = \lfloor a_n + \theta b_n \rfloor$, then if $\|P_n^{k_n} - Q_n\|$ denotes a suitable distance between $P_n^{k_n}$ and Q_n then

$$\|P_n^{k_n} - Q_n\| \to c(\theta),$$

with $\lim_{\theta\to\infty} c(\theta) = 0$ and $\lim_{\theta\to-\infty} c(\theta) = 1$.

If Ω_n is the set of arrangements on n cards, P_n is the Gilbert-Shannon-Reeds distribution, Q_n is the uniform distribution on Ω_n, the following is a theorem of Diaconis.

Theorem 7.3 *Let* P *result from the Gilbert-Shannon-Reeds distribution for riffle shuffling* n *cards. Let* $k = (3/2) \log_2 n + \theta$. *Then*

$$||P_n^{k_n} - Q_n|| = 1 - 2\Phi(-2^{-\theta/4}\sqrt{3}) + O(\frac{1}{\sqrt{n}})$$

with $\Phi(z) = \int_{-\infty}^{z} (e^{-t^2/2}/\sqrt{2\pi})dt$.

In this case $a_n = (3/2)\log_2 n$, $b_n = 1$. There is a sharp cut-off at $(3/2)\log n$; the distance tends to 0 exponentially past this point. It tends to 1 doubly exponentially before this point. This is explained in [84].

7.5.4 Random Walks that Become Uniform in a Finite Number of Steps

The following question was introduced by Vishnevetskiy and Zhmud in [285]:

Consider the random walk on a finite group G associated to a probability p. What are the conditions on G and p such that

$$p^{*n} = U \tag{7.5.1}$$

for some finite positive integer n?

Define $\Lambda(G)$ to be the set of probabilities p such that (7.5.1) holds. Their results may be summarized as follows.

(a) The groups for which $\Lambda(G) = U$ are either abelian groups or Hamiltonian 2-groups (i.e. groups of the form $A \times Q_8$ where A is an elementary abelian 2-group).
(b) For any $p \in \Lambda(G)$, $p^{*m} = U$ where m is the maximal degree of an irreducible representation of G.
(c) If $f \in \Lambda(G)$ and f is constant on conjugacy classes then $f = U$.

7.5.4.1 An Approach via Group Matrices

There is the following alternative formulation in terms of group matrices. For all $f \in L^2(G)$ consider the full group matrix

$$X_G(f) = \sum_{g \in G} \rho(g) f(g)$$

where ρ is the regular representation of G. Since

$$X_G(f_1 * f_2) = X_G(f_1)X_G(f_2)$$

it follows that a probability $p \in \Lambda(G)$ if and only if for some k, $(X_G(p))^k = U$. Here $U = X_G(U)$ is the $|G| \times |G|$ matrix with all entries $1/|G|$. Let the set of irreducible representations of G be $\{\rho_i\}_{i=1}^r$ with $\rho_1 = 1$. Let d_i be the degree of ρ_i. The Fourier transform $\widehat{X_G}(f)$ has been defined in (7.2.5).

Lemma 7.4 *The Fourier transform $\widehat{X_G}(U)$ is the diagonal matrix $diag(1, 0, 0, \ldots, 0)$.*

Proof Since U is constant on conjugacy classes, $\widehat{X_G}(U)$ is a reduced group matrix, which by Chap. 1, (1.6.2) is a diagonal matrix with entries $X_G^{\rho_i}(U) = \varphi_i$. Now if $i > 1$

$$\varphi_i = \frac{1}{|G|}\frac{1}{d_i} \sum_{g \in G} \chi_i(g) = \frac{1}{d_i}\langle \chi_i, \chi \rangle = 0,$$

and $\varphi_1 = \sum_{j=1}^{|G|} \frac{1}{|G|} = 1$. $\qquad\qquad\qquad\qquad\qquad\qquad\qquad\qquad\qquad\qquad\quad \square$

Lemma 7.5 *If $(X_G(p))^k = U$ then for each ρ_i, $i > 1$, $(X_G^{\rho_i}(p))^k = 0$.*

Proof Since there is a matrix P such that $P^{-1}X_G(p)P = \widehat{X_G}(p)$ it follows that if $(X_G(p))^k = U$ then

$$(\widehat{X_G}(p))^k = P^{-1}(X_G(p))^k P = P^{-1}UP = \widehat{X_G}(U) = diag(1, 0, 0, \ldots, 0).$$
$$(7.5.2)$$

Since the blocks of $\widehat{X_G}(p)$ are precisely the $X_G^{\rho_i}(p)$ the result follows. $\qquad \square$

Note that $X_G^{\rho_1}(p) = \sum_{g \in G} p(g) = 1$.

Corollary 7.1 *If $(X_G^{\rho_i}(p))^k = 0$ for some positive integer k then $(X_G^{\rho_i}(p))^{d_i} = 0$.*

Proof

(i) If M is any constant matrix of size $m \times m$ such that $M^s = 0$ for some positive integer s then since M is similar to an upper triangular matrix the eigenvalues $\{\epsilon_i\}_{i=1}^m$ of M are all 0.

(ii) Since each coefficient of the Cayley-Hamilton polynomial $f(x)$ for M is a symmetric polynomial in the ϵ_i (except for the coefficient of x^m) it follows that these are all 0 and therefore $f(x) = x^m$ so that by the Cayley-Hamilton theorem $M^m = 0$.

The conclusion follows from the application of (i) and (ii) to $X_G^{\rho_i}(p)$. $\qquad \square$

Corollary 7.2 *If $(X_G(p))^k = U$ then $(X_G(p))^m = U$ where $m = \max(\deg(\rho_i))$.*

This follows immediately from the previous two corollaries and implies (b) above.

Lemma 7.6 *If f is constant on the conjugacy classes of G and $f \in \Lambda(G)$ then $f = U$.*

Proof From the theory of association schemes each $X_G^{\rho_i}(f)$ is a diagonal matrix. Therefore if $i > 1$, $X_G^{\rho_i}(f) = 0$ and since $X_G^{\rho_1}(f) = 1$

$$\widehat{X_G}(f) = \widehat{X_G}(U)$$

which implies that $f = U$. □

There is an immediate extension of Lemma 7.6 to the following.

Lemma 7.7 *Suppose that* $\mathbf{S}$ *is any commutative S-ring on G. If $f \in \Lambda(G)$ is constant on the classes of* $\mathbf{S}$ *then $f = U$.*

Proof If f is constant on the classes of $\mathbf{S}$ then from association scheme theory $X_G(f)$ is diagonalizable and if $f \in \Lambda(G)$ it follows that $f = U$. □

Proposition 7.2 *Suppose that ρ_i is a representation of degree d_i with corresponding determinant factor θ^{ρ_i}. Suppose that the coefficients of λ^{d_i-j}, $j = 1, 2, \ldots, r_i$ in the expansion of $\det(\lambda I - X_G^{\rho_i}(f))$ are*

$$s_1^{\rho_i}(f) = -Tr(X_G^{\rho_i}(f)), \ s_2^{\rho_i}(f), \ldots, (-1)^{d_i} s_{r_i}^{\rho_i}(f) = \theta^{\rho_i}(f).$$

Then $X_G^{\rho_i}(f)^{d_i} = 0$ if and only if $s_j^{\rho_i}(f) = 0$ for $j = 1, 2, \ldots, d_i$.

Proof Suppose that $s_j^{\rho_i}(f) = 0$ for $j = 1, 2, \ldots, d_i$. Thus the coefficients of the characteristic equation for $X_G^{\rho_i}(f)$ are all zero except for that of λ^{d_i}. Therefore from the Cayley-Hamilton theorem $X_G^{\rho_i}(f)^{d_i} = 0$. Conversely, if $X_G^{\rho_i}(f)^{d_i} = 0$ then the eigenvalues of $X_G^{\rho_i}(f)$ are all 0 and since the s_j are polynomials in these eigenvalues it follows that they are all zero. □

In [285] a probability p in $\Lambda(G)$ is associated to the function

$$q : G \to [-1/|G|, 1 - 1/|G|]$$

defuned by $q(g) = p(g) - 1/|G|$. It is shown that

$$p^{*k} = U \text{ if and only if } q^{*k} = 0.$$

An equivalent statement is in the following lemma.

Lemma 7.8 $X_G(p)^k = U$ *if and only if* $X_G(q)^k = 0$.

Proof Since the Fourier transform is linear, $\widehat{X_G}(q) = \widehat{X_G}(p) - \widehat{U}$. Now $\widehat{U} = diag(1, 0, \ldots, 0)$. Thus $(\widehat{X_G}(q))$ has the first entry on the diagonal 0 and the remaining diagonal blocks identical to the blocks $X_G^{\rho_i}(p)$ of $\widehat{X_G}(p)$. Therefore $\widehat{X_G}(p))^k = \widehat{U}$ if and only if $\widehat{X_G}(q))^k = 0$ and the Lemma follows using the inverse Fourier transform.   □

Thus in order to find $\Lambda(G)$ it is sufficient to find all functions

$$q : G \to [-1/|G|, 1 - 1/|G|]$$

with $\sum q(g) = 0$ such that $X_G(q)^k = 0$ for some k. As is explained in [285], this is equivalent to finding the set $Nil(\mathbb{R}G)$ of all nilpotent elements in $\mathbb{R}G$. This is in turn equivalent to finding the q (with the above restrictions) such that $X_G(q)^{d_i} = 0$ for $i = 1 \ldots k$. Define $\Lambda'(G)$ to be the set of all functions q such that the function p defined by $p(g) = q(g) + 1/|G|$ for all g in G lies in $\Lambda(G)$.

Proposition 7.3 *For an arbitrary finite group G if a function q lies in $\Lambda'(G)$ then for each conjugacy class C_i of G*

$$\sum_{g \in C_i} q(g) = 0.$$

In particular $q(z) = 0$ for all $z \in Z(G)$, the center of G.

Proof Consider a function q in $\Lambda'(G)$. For each linear character χ_i of G corresponding to the representation ρ_i it follows using Corollary 7.1 that

$$X_G^{\rho_i}(q) = \sum_{g \in G} \chi_i(g)q(g) = 0. \tag{7.5.3}$$

For a non-linear representation ρ_j with character χ_j it follows from Proposition 7.2 that since

$$(X_G^{\rho_j}(q))^{d_j} = 0$$

all the coefficients of the characteristic polynomial of $X_G^{\rho_j}(q)$ are zero and hence

$$Tr(X_G^{\rho_j}(q)) = \sum \chi_j(g)q(g) = 0.$$

Thus for each character $\chi_i \in Irr(G)$, q satisfies the linear constraint.

$$\sum_{g \in G} \chi_i(g)q(g) = 0. \tag{7.5.4}$$

If f is the class function on G whose value on an element $g \in C_i$ is $\sum_{g \in C_i} q(g)$, the equations (7.5.4) imply that f is orthogonal to all the irreducible characters of G. Thus f is the zero map and the proposition follows.   □

Corollary 7.3 *The function f satisfies the conditions (7.5.4) if and only if for each conjugacy class C_i of G*

$$\sum_{g \in C_i} f(g) = 0. \tag{7.5.5}$$

Proof This follows since the number of independent linear constraints (7.5.4) is equal to the number of conjugacy classes of G and the fact that the constraints (7.5.5) are clearly linearly independent. $\square$

7.5.4.2   The Case Where G has Representations of Degree at Most 2

Suppose now that G is a group which has representations of degree at most 2. It follows from Proposition 7.2 that if ρ_i is a representation of degree 2 then $X_G^{\rho_i}(f)^2 = 0$ if and only if both $\theta^{\rho_i}(f) = 0$ and $Tr(X_G^{\rho_i}(f)) = 0$. To each p in $\Lambda(G)$ there corresponds a function q in $\Lambda'(G)$ whose values satisfies the following set of equations. These are of two types. As above, there are the linear equations (7.5.4), or equivalently (7.5.5). The second type are quadratic equations: to each non-linear representation ρ_i of G there is the equation

$$\Theta_{\rho_i} = 0. \tag{7.5.6}$$

Any solution of the set of equations (7.5.5) and (7.5.6) which satisfies the constraints $-1/|G| \le q(g) \le 1 - 1/|G$, for all $g \in G$ lies in $\Lambda'(G)$ gives rise to a probability in $\Lambda(G)$.

Example 7.3 Let $G = Q_8$ with the ordering $\{1, -1, i, -i, j, -j, k, -k\}$. Then the equations (7.5.5) imply that $q \in \Lambda'(G)$ then

$$q(1) = q(2) = 0, q(3) + q(4) = 0, q(5) + q(6) = 0, q(7) + q(8) = 0, \tag{7.5.7}$$

and the quadratic equation arising from equating the irreducible factor corresponding to the representation of degree 2 to zero is

$$(q(1) - q(2))^2 + (q(3) - q(4))^2 + (q(5) - q(6))^2 + (q(7) - q(8))^2) = 0 \tag{7.5.8}$$

Since the $q(i)$ are all real, from (7.5.8)

$$q(3) = q(4), \ q(5) = q(6), \ q(7) = q(8).$$

On inserting these values into (7.5.7) it follows that $q(i) = 0$ for all i and hence $\Lambda(G) = U$.

It is clear that if G is abelian then $\Lambda(G) = U$ since if $q \in \Lambda'(G)$ Proposition 7.3 implies that $X_G(q) = 0$ and therefore $q = 0$.

Example 7.4 Let $G = S_3$. As in Chap. 1 with the ordering

$$\{e, (1, 2, 3), (1, 3, 2), (1, 2), (1, 3), (2, 3)\},$$

$X_G(q)$ is similar to the block diagonal matrix whose blocks are

$$\sum_{i=1}^{6} q(i), \quad \sum_{i=1}^{3} q(i) - \sum_{i=4}^{6} q(i),$$

$$B_3 = \begin{bmatrix} q(1) + \omega q(2) + \omega^2(q(3) & q(4) + \omega q(5) + \omega^2(q(6) \\ q(4) + \omega^2 q(5) + \omega(q(6) & q(1) + \omega^2 q(2) + \omega(q(3) \end{bmatrix}, B_3$$

where $\omega = e^{\frac{2\pi i}{3}}$. The determinant of B_3 is, as given previously

$$\theta(q) = q(1)^2 + q(2)^2 + q(3)^2 - q(1)q(2) - q(2)q(3) - q(1)q(3)$$
$$-(q(4)^2 + q(5)^2 + q(6)^2 - q(4)q(5) - q(5)q(6) - q(4)q(5)).$$

Therefore using Proposition 7.3, $q \in \Lambda'(G)$ if and only if the following system of equations is satisfied.

$$q(1) = 0, q(2) = -q(3) \text{ and } q(6) = -(q(4) + q(5)), \quad \theta(q) = 0$$

It follows that, after obvious substitutions into $\theta(q) = 0$,

$$q(2) = \pm\sqrt{q(4)^2 + q(5)^2 + q(4)q(5)}.$$

If $|\alpha|, |\beta| \leq 1/6$ the largest possible value for $\alpha^2 + \beta^2 + \alpha\beta$ is $1/12$. Thus for each pair (α, β) inside the ellipse $x^2 + y^2 + xy = 1/12$ such that $|\alpha + \beta| \leq 1/6$ there is the following solution for q

$$[0, \sqrt{\alpha^2 + \beta^2 + \alpha\beta}, -\sqrt{\alpha^2 + \beta^2 + \alpha\beta}, \alpha, \beta, -(\alpha + \beta)]. \tag{7.5.9}$$

The corresponding p is obtained by adding $1/6$ to each member of (7.5.9).

Example 7.5 Let $G = D_4$. Since the character table of D_4 is the same as that of Q_8 if q lies in $\Lambda'(G)$ by Proposition 7.3

$$q(1) = q(2) = 0, q(3) + q(4) = 0, q(5) + q(6) = 0, q(7) + q(8) = 0,$$

The quadratic constraint given from the determinant factor of corresponding to the representation of degree 2 given in Chap. 1 is

$$(q(1) - q(2))^2 + (q(3) - q(4))^2 - (q(5) - q(6))^2 - (q(7) - q(8))^2 = 0,$$

and after eliminating variables the system reduces to

$$q(3)^2 = q(5)^2 + q(7)^2, \text{ or } q(3) = \pm\sqrt{q(5)^2 + q(7)^2}$$

Thus if $|\alpha|, |\beta| \le 1/8$, the largest possible value for $\alpha^2 + \beta^2$ is $1/32$ an hence for any pair (α, β) inside the circle $\alpha^2 + \beta^2 = 1/32$ there is a solution for q:

$$(0, 0, \sqrt{\alpha^2 + \beta^2}, -\sqrt{\alpha^2 + \beta^2}, \alpha, -\alpha, \beta, -\beta).$$

Example 7.6 Consider an arbitrary dihedral group $G = D_n$. The group matrix of G with the ordering of G on the left cosets of the cyclic group of order n is given in Example 1.7, Chap. 1 as

$$\begin{bmatrix} C(x_1, x_2, \ldots, x_n) & C(x_{n+1}, x_{2n}, x_{2n-1}, \ldots, x_{n+2}) \\ C(x_{n+1}, x_{n+2}, \ldots, x_{2n}) & C(x_1, x_n, x_{n-1}, \ldots, x_2) \end{bmatrix}.$$

It is possible to obtain an infinite subset of $\Lambda'(G)$ as follows. If n is odd, take the set of values of the function q as

$$[0, a, -a, a, -a, \ldots, a, -a, 0, a, -a, a, -a, \ldots, a, -a]$$

with $|a| \le \frac{1}{2n}$. Then

$$X_G(q) = \begin{bmatrix} A & -A \\ A & -A \end{bmatrix}, \tag{7.5.10}$$

where the $n \times n$ matrix A is of the form

$$C(0, a, -a, a, -a, \ldots, a, -a),$$

and by direct calculation $X_G(q)^2 = 0$. Therefore $q \in \Lambda'(G)$. If n is even, then if the set of values of q is

$$q(i) = 0, \ i = 1, \frac{n}{2}, n + 1, \frac{3n}{2}, \ q(i) = a, \ i = 2, \ldots, \frac{n}{2} - 1, \frac{n}{2} + 1, \ldots, \frac{3n}{2} - 1,$$

$$q(i) = -a, \ i = \frac{n}{2} + 1, \ldots, n, \ i = \frac{3n}{2} + 1, \ldots, 2n$$

with $|a| \le \frac{1}{2n}$ then $X_G(q)$ is also of the form (7.5.10) where in this case $n \times n$ matrix A is

$$C(0, a, -a, a, -a, \ldots, 0, a, -a, \ldots, a, -a),$$

and thus $X_G(q)^2 = 0, q \in \Lambda'(G)$.

In general it seems difficult to give a complete description of $\Lambda'(G)$, even in the dihedral case. For an arbitrary group G a natural question in the light of the fact that for any suitable norm form on an algebra the s_j may be calculated rationally from the s_1, s_2 and s_3 is whether $s_i^{\rho_j}(q) = 0$ for $i = 1, 2, 3$ and for all ρ_j necessarily implies that $(X_G^{\rho_j}(q))^{r_j} = 0$ for all j. The answer is negative. The case of the Frobenius group of order 20 provides a counterexample. There are functions q such that for the unique representation ρ_5 of degree 4 the s_1, s_2 and s_3 corresponding to $X_G^{\rho_5}(q)$ are 0 but $\det(X_G^{\rho_5}(q)) \neq 0$.

For arbitrary groups a function q lies in $\Lambda'(G)$ if and only if it satisfies the following.

$$\sum_{g \in C_i} q(g) = 0 \text{ for all conjugacy classes } C_i, \qquad (7.5.11)$$

or equivalently

$$s_1^{\rho_i}(q) = 0 \text{ for all irreducible } \rho_i,$$

$$s_2^{\rho_i}(q) = 0 \text{ for all irreducible } \rho_i \text{ with } d_i > 1, \qquad (7.5.12)$$

$$\cdots$$

$$s_k^{\rho_i}(q) = 0. \text{ for all } \rho_i \text{ with } \deg(\rho_i) \geq k \qquad (7.5.13)$$

$$\cdots$$

Define $\Lambda'_j(G)$ to be the set of $q : G \to [-1/|G|, 1 - 1/|G|]$ which satisfy the constraints (7.5.13) for $k = 1, \ldots, j$ so that if $k = \max\{r_i\}$ then $\Lambda'_k(G) = \Lambda'(G)$. Then since $\Lambda'_1(G)$ consists of all q which satisfy (7.5.11) it is the solution of a set of linear equations which is easily described, but as j increases the $\Lambda'_j(G)$ become more complicated geometrically. It may be interesting to examine these invariants of a group.

7.5.4.3 The Connection with Hamiltonian Groups

For $X_G(f)$ let

$$|X_G(f)| = \sum_{g \in G} f(g).$$

Thus $|X_G(f)|$ corresponds to the trivial factor of Θ_G. Then it follows directly that

$$|X_G(f_1) + X_G(f_2)| = |X_G(f_1) + X_G(f_2)|, \ |X_G(f_1) - X_G(f_2)| = |X_G(f_1)| - |X_G(f_2)|$$

and

$$|X_G(f_1)X_G(f_2)| = |X_G(f_1)||X_G(f_2)|,$$

the latter property being a consequence of Proposition 1.4, Chap. 1.

Let $Nil(\mathbb{R}G)$ be the set of nilpotent elements of the algebra $\mathbb{R}G$. Under the isomorphism described above $Nil(\mathbb{R}G)$ is isomorphic to the algebra of group matrices $X_G(f)$ with $(X_G(f))^k = 0$ for some k.

The lemmas below are consequences of previous work

Lemma 7.9 *For each in $X_G(f)$ in $Nil(\mathbb{R}G)$ then $|X_G(f)| = 0$.*

Proof $(X_G(f))^k = 0$ *implies that* $|(X_G(f))^k| = |(X_G(f))|^k = 0$. □

Lemma 7.10 *For p in $\Lambda(G)$ consider the element $X_G(q)$ in $\mathbb{R}G$ where $q = p - U$. Then if $\alpha = \varphi(q)$, $X_G(p)^k = X_G(q)^k + U$ for any k.*

Proof Since $|X_G(q)| = |X_G(p)| - |U| = 0$, it follows that $UX_G(q) = X_G(q)U = 0$. Therefore

$$X_G(p)^k = (X_G(q) + U)^k = X_G(q)^k + U^k = X_G(q)^k + U.$$

□

Corollary 7.4 *The probability p lies in $\Lambda(G)$ if and only if $X_G(q)$ lies in $Nil(\mathbb{R}G)$.*

Proof $p \in \Lambda(G)$ if and only if $X_G(p)^k = U$ for some k. □

Let $M_n(K)$ be the algebra of all $n \times n$ matrices over a skew field K.

Lemma 7.11 $Nil(M_n(K)) = \{0\}$ *if and only if $n = 1$, i.e. $M_n(K) = K$.*

Proof If $n = 1$, then $M_n(K) = K$ is a skew field, so $Nil(M_n(K)) = \{0\}$. If $n > 1$, matrix units E_{ij} are nilpotent if $i \neq j$. □

Define G to be a **U-group** if $\Lambda(G) = U$.

7.5.4.4 Properties of U-Groups

Theorem 7.4 *The following are equivalent*

(a) G is a U-group.
(b) $Nil(\mathbb{R}G) = \{0\}$.
(c) $\mathbb{R}G$ is a direct sum of skew fields

Proof (a) ⇔ (b). $\Lambda(G) = U$ is equivalent to if $q \in \Lambda'(G)$ then $q = 0$.

(b) ⇔ (c) By Wedderburn's theorem $\mathbb{R}G$ decomposes into a direct sum of matrix algebras $M_n(K_i)$ over skew fields K_i. Then $Nil(\mathbb{R}G) = \{0\}$ is equivalent to $Nil(M_n(K_i)) = \{0\}$ for all such $M_n(K_i)$. Then by Lemma 7.11 this is equivalent to $M_n(K_i) = K_i$ for all $M_n(K_i)$.

□

It follows that by Theorem 1.2, Chap. 1 each K_i is isomorphic to $\mathbb{R}$, $\mathbb{C}$ or $\mathbb{H}$.

A consequence of the above results is that a subgroup H of a U-group G is a U-group. For if q lies in $\Lambda'(H)$ then if the extension $\tilde{q}$ of q to G is defined by $\tilde{q}(g) = 0$ for $g \in G\backslash H$, $\tilde{q}(h) = q(h)$ for $h \in H$, then $\tilde{q}$ lies in $\Lambda'(G)$. It also follows from Proposition 7.3 that if $X_G(q) \in Nil(\mathbb{R}G)$ then $q(z) = 0$ for all z in $Z(G)$. A further consequence of Proposition 7.3 is that if G is abelian then G is a U-group.

A nonabelian group is called **Hamiltonian** if all its subgroups are normal. A well-known result of Dedekind is that a finite Hamiltonian group is of the form

$$A \times B \times Q_8, \tag{7.5.14}$$

where A is abelian of odd order, B is an elementary abelian 2-group, and either or both A and B could be trivial.

Let G be a Hamiltonian U-group. The algebra $\mathbb{R}A$ decomposes as an orthogonal direct sum of fields

$$\mathbb{R}A = \Psi_1 \oplus \Psi_2 \oplus \ldots \oplus \Psi_r \tag{7.5.15}$$

Lemma 7.12 *The following hold*

(a) $\Psi_i \cong \mathbb{R}, i = 1, \ldots, r$
(b) G is a 2-group.

Proof If (a) does not hold then $\Psi_i \cong \mathbb{C}$ for some i. The algebra $\mathbb{R}Q_8$ is an orthogonal sum of skew fields. Then since Q_8 is nonabelian one of these skew fields K is isomorphic to the quaternion skew field $\mathbb{H}$. From (7.5.14) the algebra $\mathbb{R}Q_8$ contains the ideal $\mathcal{I} = \Psi_i.K \cong \mathbb{C} \times \mathbb{H}$. Since $\mathbb{C} \times \mathbb{H}$ is isomorphic to the full matrix ring $M_2(\mathbb{C})$ and $Nil(M_2(\mathbb{C})) \neq 0$ then $Nil(\mathcal{I}) \neq 0$ and hence $Nil(\mathbb{R}G) \neq 0$. This contradicts Theorem 7.4 and therefore (a) follows.

For (b) it is sufficient to prove that $A = \{e\}$. By (7.5.15) an arbitrary element $g \in A$ has a decomposition

$$g = u_1 + u_2 + \ldots + u_r$$

where $u_i \in \Psi_i$. Let $|A| = m$. Since the elements u_1 are mutually orthogonal then

$$1 = g^m = u_1^m + u_2^m + \ldots + u_r^m.$$

Since also

$$1 = e_1 + e_2 + \ldots + e_r$$

where e_i is the unit of Ψ_i then $u_i^m = e_1$ for all i. Therefore u_i is an element of finite order in the multiplicative group of the field Ψ_i. Since $\Psi_i \cong \mathbb{R}$ it follows that $u_i = \pm e_i$ for $i = 1, \ldots, r$ and

$$g^2 = u_1^2 + u_2^2 + \ldots + u_r^2 = 1.$$

But since g lies inside a group of odd order it follows that $g = e$ and therefore $A = \{e\}$. □

Theorem 7.5 *The following are equivalent.*

(a) G is a Hamiltonian 2-group.
(b) G is a nonabelian U-group.

Proof If (a) is satisfied then $G \cong B \times Q_8$ with B an elementary abelian 2-group. The center of G is then of the form $B \times Z$ where Z is the center of Q_8. Therefore if $q \in \Lambda'(G)$ then $q(g) = 0$ for all $g \in B$ and q restricted to Q_8 lies in $\Lambda'(Q_8)$. But it has beem shown in Example 7.3 that $\Lambda'(Q_8) = 0$ and hence $q(g) = 0$ for all $g \in G$ and G is a nonabelian U-group.

Conversely suppose that G is a nonabelian U-group. For arbitrary elements g, h in G let

$$y = (g - 1)h(g^{m-1} + g^{m-2} + \ldots + g + 1) \text{ in } \mathbb{R}G.$$

where m is the order of g. Since

$$(g - 1)(g^{m-1} + g^{m-2} + \ldots + g + 1) = g^m - 1 = 0$$

it follows that $y^2 = 0$. But since from Theorem 7.4 $Nil(\mathbb{R}G) = 0$ it follows that $y = 0$. Since

$$y = ghg^{m-1} + ghg^{m-2} + \ldots + gh - hg^{m-1} - \ldots - h$$

the element gh must be equal to hg^k for some k. Then $h^{-1}gh = g^k$ and hence the subgroup generated by g for each g is normal in G. It follows that every subgroup of G is normal and that G is Hamiltonian. □

Thus the class of U-groups consists of groups which are either Abelian or Hamiltonian 2-groups. It will be seen later in Chap. 10 that this class can be characterized in terms of properties of the group matrix.

7.5.5 Arbitrary Random Walks on Groups

7.5.5.1 Distances Between Probablities

Several measures of distance between probabilities are given in [82]. The most useful one seems to be the following.

Definition 7.4 Let p and q be probabilities on the finite group G. Their **distance** is defined to be

$$||p - q|| = \max_{A \subset G} |p(A) - q(A)| = \frac{1}{2} \sum_{g \in G} |p(g) - q(g)|. \qquad (7.5.16)$$

The Fourier transform may be used to obtain results on distances. As has already been remarked, without loss of generality it can be assumed that every representation ρ of a finite group G has a representing set of matrices $\{\rho(g)\}_{s \in G}$ which are unitary, and in this case $\rho(g^{-1}) = \rho(g)^*$, the conjugate transpose. It is assumed below that all such representing matrices are unitary. The following Lemma is known as the Upper Bound Lemma. It is due to Diaconis and Shahshahani [88]. See also [85]. For a matrix M the trace norm $Tr(MM^*) = ||M||$ has already been introduced.

Lemma 7.13 *Let q be a probability on the finite group G. Then*

$$||q^{*k} - U||^2 \leq \frac{1}{4} \sum_{\rho_i \in Irr(G), \rho_i \neq 1} d_{\rho_i} ||\widehat{q}_{\rho_i}||^{2k}. \qquad (7.5.17)$$

Proof From (7.5.16),

$$4||q^{*k} - U||^2 = (\sum_{g \in G} |q^{*k}(g) - U(g)|)^2 \leq |G| \sum_{g \in G} |q^{*k}(g) - U(g)|^2,$$

where the inequality follows from Cauchy-Schwarz. Then by the Plancherel theorem (Lemma 7.3)

$$|G| \sum_{g \in G} |q(g) - U(g)|^2 = |G| \sum_{g \in G} ((q - U)(g))((q - U)^*(g^{-1}))$$

$$= \sum_{\rho_i \in Irr'(G)} d_i Tr((\widehat{q}_{\rho_i} - \widehat{U}_{\rho_i})(\widehat{q}^*_{\rho_i} - \widehat{U}_{\rho_i})). \qquad (7.5.18)$$

If ρ_i is non-trivial then $\widehat{U}_{\rho_i} = 0$. This follows from Lemma 7.4 because

$$\widehat{X_G}(U) = (1, 0, \ldots, 0)$$

has diagonal entries except for the first of the form $\widehat{U}_{\rho_i}$, $i > 1$. Thus the corresponding term in (7.5.18) becomes $d_i||\widehat{q}_{\rho_i}||^{2k}$.   □

In the context of this book, the question arises: if only the irreducible factors of the group determinant are known, is it possible to calculate the bound (7.5.17)?

The work on walks which reach random in a finite number of steps above suggests that the analysis of p^{*k} for a probability $p : G \to [0, 1]$ may be carried out by considering the function q where $q(g) = p(g) - \frac{1}{|G|}$ for $g \in G$. Then since $p^{*k} = U$ if and only if $q^{*k} = 0$ can a useful measure of closeness to uniform be the sums of the sizes of the invariants $s_k^{\rho_i}(q)$?

Note that the work above provides examples of probabilities where the walk goes from distance $1/4$ from U to U in exactly one step (for example if $q = [0, 1/8, -1/8, 0, 1/8, -1/8]$ for S_3), so that the cutoff is sharp.

An alternative method of analyzing the walk using the group matrix is indicated by the following example. In Example 7.2 the partial diagonalization of the group matrix $X_G(\overline{x})$ of Fr_{21} in the form

$$\widetilde{X} = \begin{bmatrix} \Lambda_{11} & \Lambda_{12} & \Lambda_{13} \\ \Lambda_{21} & \Lambda_{22} & \Lambda_{23} \\ \Lambda_{31} & \Lambda_{32} & \Lambda_{33} \end{bmatrix}$$

is described, where each Λ_{ij} is a diagonal matrix. This is similar to a matrix of the form

$$X' = diag(B_1, B_2, \ldots, B_7)$$

where each B_i is of size 3×3 and only three are distinct, each B_i being the group matrix of an induced representation of a linear representation of the normal C_7. If f is a probability then $X'(f)$ is close to the block matrix whose blocks are the Fourier transforms of f at the irreducible representations. Thus $\widetilde{X}(f)$ contains almost the same information as that in the set of Fourier transforms but is more easily calculated, and the powers of $\widetilde{X}(f)$ could provide a measure of closeness of f^{*r} to U.

7.5.6 Fission of Group Classes

Consider the random walk on S_3. By standard results, if p is a probability which is constant on conjugacy classes then $X_G(p)$ is diagonalizable and the analysis of the random walk associated to p follows in a similar way to that on an abelian group. However, it is not necessary for diagonalizability that p is constant on conjugacy classes. If the following new classes $D_1 = \{e\}$, $D_2 = \{(1, 2, 3)\}$, $D_3 = \{(1, 3, 2)\}$, $D_4 = \{(1, 2), (1, 3), (2, 3)\}$ are taken, it remains true that if p is constant on these new classes then $X_G(p)$ remains diagonalizable.

Thus opens up the question: for any finite group G find of the ways of refining the conjugacy class partition of G such that for any probability p which is constant on the new classes $X_G(p)$ is diagonalizable, and find the maximum size of such a partition. This is discussed in [157] where such a partition is described as a **fission**. The problems are stated in terms of S-rings on groups. An upper bound for the number of classes is described in the following theorem which is a translation of a theorem given there. The **size** of a partition is the number of classes.

Theorem 7.6 *Suppose G is a finite group and $\{\chi_i\}_{i=1}^r = Irr(G)$, where χ_i is of degree d_i. Then the maximal size of a partition of the class algebra such that for any probability p which is constant on the classes of the partition $X_G(p)$ is diagonalizable is at most $\tau_G = \sum_{i=1}^r d_i$.*

The **total character** of a group $\upsilon_G = \sum_{\chi \in Irr(G)} \chi$. It is the permutation character of a Gelfand model for G (if one exists). In general it takes on values which are integers (which could be negative, e.g. for the Mathieu group M_{11}). The integer $\tau_G = \upsilon_G(e)$ seems to have remarkable connections with the geometry of the group. For example, in the case of the S_n, it is the number of involutions and in the case of $GL(n, q)$, it is the number of symmetric matrices. It is also the dimension of the commutative subalgebra of the group algebra of S_n generated by the Jucys-Murphy elements (see [227]).

A fission of a group G may be induced by a subgroup H by taking the classes of the fission to be the orbits of H acting by conjugation on G. After [157] appeared it was found that this goes back to Wigner [295] and is discussed in [187]. See also [243]. The partition μ induced by these classes is a refinement of that induced by the conjugacy class partition. For an arbitrary H there is no guarantee that the corresponding S-ring on G is commutative, or, equivalently, if p is a probability constant on the μ-classes $X_G(p)$ need not be diagonalizable. However, experimental evidence on small groups has shown that nearly all such groups have a commutative fission of dimension υ_G induced by a subgroup H. In the case of $G = S_3$ discussed above, H may be chosen as $\{e, (1, 2, 3), (1, 3, 2)\}$. There is another commutative S-ring with $C_1 = \{e\}$, $C_2 = \{(1, 2, 3), (1, 3, 2)\}$, $C_3 = \{(1, 2)\}$, $C_4 = \{(1, 3), (2, 3)\}$. This is induced by $\{e, (1, 2)\}$, and by symmetry there are two other commutative S-rings induced by the other subgroups generated by involutions. These commutative fissions are maximal since $\upsilon_G = 4$.

Example 7.7 Consider the Frobenius group $G = Fr_{21}$. The partial diagonalization of X_G has already been described in Example 7.2:

$$\widetilde{X} = \begin{bmatrix} \Lambda_{11} & \Lambda_{12} & \Lambda_{13} \\ \Lambda_{21} & \Lambda_{22} & \Lambda_{23} \\ \Lambda_{31} & \Lambda_{32} & \Lambda_{33} \end{bmatrix}. \tag{7.5.19}$$

Now (G, K) is a Camina pair where $K = C_7$ is the Frobenius kernel. The character table is

Class order	1	3	3	7	7
χ_1	1	1	1	1	1
χ_2	1	1	1	ω	ω^2
χ_3	1	1	-1	ω^2	ω
χ_4	3	α	$\bar{\alpha}$	0	0
χ_5	3	$\bar{\alpha}$	α	0	0

where $\omega = e^{\frac{2\pi i}{3}}$ and $\alpha = \frac{-1+\sqrt{3}i}{2}$. It may be seen that $\tau_G = 9$. A maximal commutative fission is given by the action of K, the classes being singletons except for the two coset classes. Thus for any probablity p which is constant on each class of order 7, $X_G(p)$ is diagonalizable. It is easy to show this directly using the partial diagonalization (7.5.19), since in this case the Λ_{ij} for $i \neq j$ have only one non-zero entry which is in the $(1, 1)$ position.

7.5.7 Homogeneous Actions of Finite Groups

7.5.7.1 Finite Spherical Functions

Suppose the finite group G acts faithfully on the set $\Omega = \{1, \ldots, n\}$ and H is the stabilizer of 1. Assume that (G, H) is a Gelfand pair. Let $\mathcal{X}$ be the corresponding association scheme arising from the centralizer ring with $\mathcal{P}$-matrix P. Then if the orbits of H are $\Gamma_0 = \{1\}, \Gamma_2, \ldots, \Gamma_d$, each row of P can be considered as a function v_j from the set $\{\Gamma_i\}_{i=0\ldots d}$ to $\mathbb{C}$. The corresponding spherical function is the function ϕ_j from Ω to $\mathbb{C}$ defined by

$$\phi_j(i) = \frac{v_j(\Gamma_{r(i)})}{|\Gamma_{r(i)}|}.$$

Define the inner product of two functions $f, g : \Omega \to \mathbb{C}$ by

$$\langle f, g \rangle = \sum_{i \in \Omega} f(i)\overline{g(i)}.$$

The spherical functions form a set of mutually orthogonal functions which are constant on the Γ_i, and which take on the value 1 on Γ_0.

7.5.8   An Example

Consider the action of D_4 on the square.

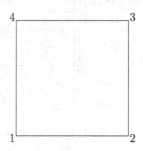

If D_4 is listed as $\{e, a, a^2, a^3, b, ba, ba^2, ba^3\}$ the corresponding permutation action is obtained by

$$a \to (1, 2, 3, 4), b \to (12)(34).$$

The group matrix corresponding to this action is

$$\begin{bmatrix} x_1 + x_8 & x_2 + x_5 & x_3 + x_6 & x_4 + x_7 \\ x_4 + x_5 & x_1 + x_6 & x_2 + x_7 & x_3 + x_8 \\ x_3 + x_6 & x_4 + x_7 & x_1 + x_8 & x_2 + x_5 \\ x_2 + x_7 & x_3 + x_8 & x_4 + x_5 & x_1 + x_6 \end{bmatrix}.$$

This gives rise to an association scheme with $\mathcal{P}$-matrix

$$\begin{bmatrix} 1 & 1 & 2 \\ 1 & 1 & -2 \\ 1 & -1 & 0 \end{bmatrix}.$$

The spherical functions are given by their vector of values on $\{1, 2, 3, 4\}$ as

$$\phi_1 = [1, 1, 1, 1]$$
$$\phi_2 = [1, 1, -1, -1]$$
$$\phi_3 = [1, -1, 0, 0].$$

Suppose the finite group G acts transitively on the set Ω and that $\{v(g)\}_{g \in G}$ is a set of permutations giving this permutation representation of G. Let X_G^v be the corresponding group matrix. (In Chap. 2 a way of constructing this directly from the group matrix appears, using the ordering via the left cosets of H, the point stabilizer.) Then X_G^v "represents" the double coset algebra of H which is of dimension equal to the number of orbits of H on Ω. In the case where (G, H) is a Gelfand pair, a probability p which is constant on each double coset HgH of H has the property that $X_G^\nu(p)$ is diagonalizable. Here the methods of asociation scheme theory can be used.

In the case where the double coset algebra is non-commutative the calculation of $X_G^\nu(p)$ offers a tool for the analysis.

Example 7.8 Consider the action of $G = \mathcal{A}_5$ on the dodecahedron. The stabilizer H of a vertex is a subgroup of order 3 and the permutation representation is of degree 20. The character table of $\mathcal{A}_5$ is

Class representative	e	(12)(34)	(123)	(12345)	(13452)
χ_1	1	1	1	1	1
χ_2	4	0	1	-1	-1
χ_3	5	1	-1	0	-0
χ_4	3	-1	0	α	β
χ_5	3	-1	0	β	α

The character υ of the permutation action of G on the cosets of H decomposes as

$$\chi_1 + 2\chi_2 + \chi_3 + \chi_4 + \chi_5$$

and therefore the centralizer ring is not commutative. If the variables x_g and x_h in X_G^υ are identified to a common variable z_D whenever they lie in the same double coset of D of H in G, then the Fourier transform $\widehat{X_G^\upsilon}(z)$ is a block diagonal matrix with blocks $\{X_G^{\rho_i}(z)\}_{i=1}^5$ where ρ_i is the representation corresponding to χ_i. The blocks $X_G^{\rho_i}(z)$ are of rank 1 for $i \neq 2$ and $X_G^{\rho_2}(z)$ is similar to a block matrix of the form

$$\begin{bmatrix} V & 0 \\ 0 & 0 \end{bmatrix}$$

where V is 2×2. Thus the analysis of the powers of $\widehat{X_G^\upsilon}(z)$ can be carried out in terms of the spherical functions which correspond to the blocks of rank 1, and the powers of V. The rth power of V can be analized by calculating the s_1^r and s_2^r arising from the determinant of V.

A further example is the action of $\mathcal{A}_5$ on the icosahedron.

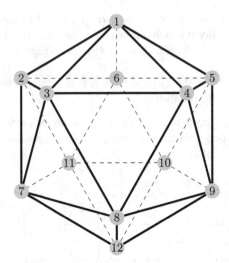

Let the vertices be labelled as in the diagram, then the orbits of the point stabilizer of 1 in $\mathcal{A}_5$ are

$$\{1\}, \{2, 3, 4, 5, 6\}, \{7, 8, 9, 10, 11\}, \{12\}.$$

In this case the centraliser ring is commutative. The $\mathcal{P}$-matrix of the corresponding association scheme is

$$\begin{bmatrix} 1 & 5 & 5 & 1 \\ 1 & -\sqrt{5} & \sqrt{5} & -1 \\ 1 & \sqrt{5} & -\sqrt{5} & -1 \\ 1 & -1 & -1 & 1 \end{bmatrix}$$

and the spherical functions are, ordering the set of vertices as $\{1, 2, \ldots, 12\}$

$\phi_1 = (1, 1, 1, 1, 1, 1, 1, 1, 1, 1, 1, 1)$

$\phi_2 = (1, -1/\sqrt{5}, -1/\sqrt{5}, -1/\sqrt{5}, -1/\sqrt{5}, -1/\sqrt{5}, 1/\sqrt{5}, 1/\sqrt{5}, 1/\sqrt{5}, 1/\sqrt{5}, 1/\sqrt{5}, -1)$

$\phi_3 = (1, 1/\sqrt{5}, 1/\sqrt{5}, 1/\sqrt{5}, 1/\sqrt{5}, 1/\sqrt{5}, -1/\sqrt{5}, -1/\sqrt{5}, -1/\sqrt{5}, -1/\sqrt{5}, -1/\sqrt{5}, -1)$

$\phi_4 = (1, -1/5, -1/5, -1/5, -1/5, -1/5, 1/5, 1/5, 1/5, 1/5, 1/5, -1)$.

Just as in the case of random walks on a group, in the case where a Gelfand pair is present the question arises of whether there is a fission of the double cosets such that $X_G^\gamma(p)$ remains diagonalizable if p is assumed to be constant only on the fissioned classes.

Chapter 8
K-Characters and n-Homomorphisms

Abstract This chapter discusses two situations where the combinatorics behind k-characters appears with no apparent connection to group representation theory. In geometry a Frobenius n-homomorphism is defined essentially in terms of the combinatorics of k-characters. Buchstaber and Rees generalized the result of Gelfand and Kolmogorov which reconstructs a geometric space from the algebra of functions on the space and used Frobenius n-homomorphisms which arise naturally from k-characters. Incidentally they show that given commutative algebras A and B, with certain obvious restrictions on B, a homomorphism from the symmetric product $S^n(A)$ to B arises from a Frobenius n-homomorphism.

The cumulants for multiple random variables may be considered as an alternative method to understand higher connections between distributions. For example, if N billiard balls move randomly on a table, the nth cumulant determines the probability of n balls simultaneously colliding. The FKG inequality can be interpreted as an inequality for the lowest cumulant of the random variables f_1 and f_2 (the case $n = 2$). Richards examined how the inequality could be extended to the higher cumulants. Although there are counterexamples to a direct extension, he stated a result for modified cumulants to which he gave the name "conjugate" cumulants, in the cases $n = 3, 4, 5$. Sahi subsequently explained that Richards' definitions could be incorporated in a more general setting by giving a generating function approach. Richards stated a theorem but Sahi later indicated a gap in the proof, so it remains a conjecture, although Sahi proved a special case.

8.1 Introduction

The combinatorics of k-characters has appeared in two separate areas in the case where the the algebras are commutative. In both cases the full power of the norm form work described in Lemma 1.6 is not needed, but it would seem likely that related results for arbitrary algebras may have applications.

© Springer Nature Switzerland AG 2019

K. W. Johnson, *Group Matrices, Group Determinants and Representation Theory*, Lecture Notes in Mathematics 2233, https://doi.org/10.1007/978-3-030-28300-1_8

Firstly, Buchstaber and Rees generalized the result of Gelfand and Kolmogorov which reconstructs a geometric space from the algebra of functions on the space and used Frobenius n-homomorphisms which arise naturally from k-characters. The ideas came from work on multivalued groups (of which association schemes form an example) and Hopf n-algebras. In order to prove their result, they show that given commutative algebras A and B, with certain obvious restrictions on B, a homomorphism from the symmetric product $S^n(A)$ to B arises from a Frobenius n-homomorphism, which is defined below. In fact for commutative algebras with reasonable assumptions a Frobenius n-homomorphism f reduces to the form $f_1 + f_2 + \ldots + f_n$ where each f_i is an ordinary homomorphism.

The second situation arose in the work of Richards on a generalization of the FKG inequality. This inequality first appeared in [102, 1971]. The inequality now occupies a prominent position because of its simplicity and widespread applicability. Some places where it has appeared are listed in Sect. 8.3.

The cumulants for multiple random variables may be considered as an alternative method to understand higher connections between distributions. For example, if N billiard balls move randomly on a table, the nth cumulant determines the probability of n balls simultaneously colliding. The FKG inequality can be interpreted as an inequality for the lowest cumulant of the random variables f_1 and f_2 (the case $n = 2$). Richards examined how the inequality could be extended to the higher cumulants. Although there are counterexamples to a direct extension, he stated a result for modified cumulants to which he gave the name "conjugate" cumulants, in the cases $n = 3, 4, 5$. Sahi subsequently explained that Richards' definitions could be incorporated in a more general setting by giving a generating function approach. For $n = 3, 4, 5$ Richards' definition coincided with the corresponding objects of Sahi, but they diverge for $n \geq 5$. Richards stated a theorem but Sahi later indicated a gap in the proof, so it remains a conjecture.

Sahi gave a definition for higher cumulants in terms of moments which may be interpreted as equivalent to the expression for the elementary symmetric functions in terms of the power functions, and thus is equivalent on a combinatorial level to that for k-characters. Sahi was able to prove a general result, which is weaker than his extension of Richards' conjectures. Subsequently Blinovsky also published a proof of the extended conjecture of Sahi, but this also has a gap which has yet to be filled. Thus Richards' interesting conjecture remains open. From the point of view of this book, it is intriguing that there is this connection between the combinatorics of k-characters and the variations of cumulants described above.

There remains the possibility that there may be deeper connections to group representation theory in either or both of the above situations.

8.2 Frobenius n-Homomorphisms and Geometry

8.2.1 Frobenius n-Homomorphisms and Related Objects

Let f be a trace-like map from an algebra A to a commutative algebra B. The map $\delta^k : A^{\otimes k} \to B$ (already defined in the case where B is a field) has the inductive definition

$$\delta^{(1)}(f)(a) = f(a),$$

$$\delta^{(k)}(f)(a_1, a_2, \ldots, a_k) = f(a_1)\delta^{(k-1)}(f)(a_2, a_3, \ldots, a_k)$$
$$- \delta^{(k-1)}(f)(a_1 a_2, a_3, \ldots, a_k) - \delta^{(k-1)}(f)(a_2, a_1 a_3, \ldots, a_k)$$
$$- \cdots - \delta^{(k-1)}(f)(a_2, a_3, \ldots, a_1 a_k). \tag{8.2.1}$$

An equivalent definition (consistent with the notation in Chap. 1, Proposition 1.5) is

$$\delta^{(k)} f(a_{i_1}, a_{i_2}, \ldots, a_{i_k}) = \sum_{\tau \in S_k} sgn(\tau) f_\tau(a_{i_1}, a_{i_2}, \ldots, a_{i_k}). \tag{8.2.2}$$

It may be noted that (8.2.1), with appropriate restriction on B is the multilinearization of the formula

$$\delta^{(k)}(f)(a, a, \ldots, a) = f(a)\delta^{(k-1)}(f)(a, a, \ldots, a)$$
$$- \delta^{(k-1)}(f)(a^2, a, \ldots, a) - \delta^{(k-1)}(f)(a, a^2, \ldots, a)$$
$$- \cdots - \delta^{(k-1)}(f)(a, \ldots, a, a^2). \tag{8.2.3}$$

There is a further equivalent definition: $\delta^{(k)}(f)(a, a, \cdots, a)$ is the determinant of the matrix

$$\begin{bmatrix} f(a) & 1 & 0 & 0 & \ldots & 0 \\ f(a^2) & f(a) & 2 & 0 & \ldots & 0 \\ \ldots & f(a^2) & & \ldots & \ldots & 0 \\ \ldots & \ldots & & & & \\ f(a^{k-1}) & f(a^{k-2}) & f(a^{k-3}) & \ldots & f(a) & k-1 \\ f(a^k) & f(a^{k-1}) & f(a^{k-2}) & \ldots & f(a^2) & f(a) \end{bmatrix}.$$

8.2.1.1 Related Work of Khudaverdian and Voronov

Suppose that A and B are associative, commutative and unital algebras over $\mathbb{C}$ or $\mathbb{R}$. Consider a linear map $f : A \to B$ which is not assumed to be an algebra homomorphism. For such a fixed map f, an arbitrary function $f : A \rightarrow B$ is said to be f-**polynomial** if its values $f(a)$, $a \in A$, are given by a polynomial expression in $f(a)$, $f(a^2)$, ... which is universal in the sense that it is independent of a. The ring of f-polynomial functions is naturally graded setting the degree of $f(a)$ to be 1, the degree of $f(a^2)$ to be 2, etc.

Define the **characteristic function** of the linear map of algebras f to be the formal power series with coefficients in B

$$R_f(a, z) := exp(f(\ln(1+az))) = 1+\psi_1(a)z+\psi_2(a)z^2+\psi_3(a)z^3+\cdots. \quad (8.2.4)$$

Regard this as a 'function' of both a and z. The coefficients $\psi_k(a)$ of the series (8.2.4) are f-polynomial functions of a of degree k, so that $\psi_k(\lambda a) = \lambda^k \psi_k(a)$. In fact by differentiating Eq. (8.2.4) with respect to z it may be seen that $\psi_k(a)$ can be obtained by the Newton type recurrence formulae:

$$\psi_k(a) = f(a)$$

and

$$\psi_{k+1}(a) = \frac{1}{k+1}(f(a)\psi_k(a)-f(a^2)\psi_{k-1}(a)+f(a^3)\psi_{k-2}(a)+\ldots+(-1)^{k+1}f(a^{k+1})).$$

Moreover for linear f

$$R_{f+g}(a, z) = R_f(a, z)R_g(a, z).$$

and

$$R_f(a, z)R_f(a', z') = R_f(az + a'z' + aa'zz', 1),$$

or equivalently

$$R_f(a, 1)R_f(b, 1) = R_f(c, 1) \text{ if } 1 + c = (1 + a)(1 + b),$$

this being understood as a formal power series identity.

They show that in the case where A and B are commutative the class of linear maps $f : A \rightarrow B$ whose characteristic functions $R_f(a, z)$ are polynomials of degree n in z coincides with the class of Frobenius n-homomorphisms from A to B. Note that, as mentioned above, for reasonable assumptions on A and B a Frobenius n-homomorphism f may be expressed as

$$f = \rho_1 + \ldots + \rho_n$$

where $\rho_1, \ldots, \rho_n$ are 1-dimensional irreducible representations of A into B.

8.2.2 A Generalization of the Gelfand–Kolmogorov Theorem

The following theorem is well known.

Theorem 8.1 (Gelfand–Kolmogorov) *Let X be a compact Hausdorff space. Then, for an appropriate topology on the space of continuous complex-valued functions $C(X)$ on X, the evaluation map*

$$\mathcal{E} : X \to Hom(C(X), \mathbb{C}), \ \mathcal{E}(x)\varphi = \varphi(x),$$

is a homomorphism onto the set of all ring homomorphisms $C(X) \to \mathbb{C}$.

The original theorem was published in 1939 in [117]. The main result of the paper identifies a compact Hausdorff space X with the space of maximal ideals of the ring of continuous functions on X. It is usually included in monographs and textbooks containing basic functional analysis. Its reformulation above follows from the identification of the set of all ring homomorphisms from an algebra A to $\mathbb{C}$ with the set $m\text{-spec}(A)$ of all maximal ideals of A.

The Gelfand–Kolmogorov theorem may be viewed as a description of the image of the canonical embedding of X into the infinite-dimensional linear space $V = A^*$ where $A = C(X)$ by a system of quadratic equations $f(1) = 1$, $f(a)^2 - f(a^2) = 0$, indexed by elements of A. This aspect was recently emphasized by Buchstaber and Rees (see [36] and the references there), who showed that there is a natural embedding into V not only for X, but also for all its symmetric powers $Sym^n(X)$. In order to do this algebra homomorphisms are replaced by Frobenius n-homomorphisms and the quadratic equations describing the image are replaced by certain algebraic equations of higher degree. The work of Buchstaber and Rees was motivated by their earlier study of Hopf objects for multi-valued groups.

The symmetric power $S^n(X)$ of a space X is the quotient space

$$S^n(X) = X^n/\mathcal{S}_n = \{(x_1, \ldots, x_n) : (x_{\sigma(1)}, \ldots, x_{\sigma(n)}) \sim (x_1, \ldots, x_n), \ \sigma \in \mathcal{S}_n\}.$$

The continuous functions on $S^n(X)$ are the (continuous) symmetric functions on X^n. If A is an arbitrary algebra $S^n(A)$ may be considered as the quotient of $A^{\otimes n}$ under the equivalence

$$(a_{\sigma(1)} \otimes \ldots \otimes a_{\sigma(n)}) \sim (a_1 \otimes \ldots \otimes a_n), \ \sigma \in \mathcal{S}_n, a_i \in A.$$

An element of $S^n(A)$ may then be considered as an n-tuple $(a_1, a_2, \ldots, a_n)$ with the algebra structure induced on $S^n(A)$ from the product

$$(a_1, a_2, \ldots, a_n)(b_1, b_2, \ldots, b_n) = (a_1 b_1, a_2 b_2, \ldots, a_n b_n).$$

This leads to a well-defined multiplication on equivalence classes.

A main result of Buchstaber and Rees is the following.

Theorem 8.2 *Suppose that A, B are commutative associative algebras with unit over $\mathbb{R}$ or $\mathbb{C}$. Then there is a bijection between homomorphisms $f : S^n(A) \to B$ and n-homomorphisms from A to B. The algebra homomorphism $F : S^n(A) \to B$ corresponding to an n-homomorphism $f : A^n \to B$ is given by the formula*

$$F(a_1, \ldots, a_n) = \frac{1}{n!} \delta^n f(a_1, \ldots, a_n),$$

where on the left-hand side a linear map from $S^n(A)$ is written as a symmetric multi-linear function.

Proof (Khudaverdian and Voronov) Let $\Phi_n(A, B)$ be the set of Frobenius n-homomorphisms from an algebra A to an algebra B. Two mutually inverse maps

$$\alpha : \Phi_1(S^n(A), B) \to \Phi_n(A, B)$$

and

$$\beta : \Phi_n(A, B) \to \Phi_1(S^n(A), B)$$

are constructed as follows.

Suppose $F \in \Phi_1(S^n(A), B)$. For $a \in A$ construct the $n \times n$ matrix $\mathcal{L}(a)$ with entries in the algebra $A^{\otimes n} = A \otimes \ldots \otimes A$, where

$$\mathcal{L}(a) = diag(a \otimes 1 \otimes \ldots \otimes 1, 1 \otimes a \otimes \ldots \otimes 1, \ldots, 1 \otimes 1 \otimes \ldots \otimes a).$$

The map $a \longmapsto \mathcal{L}(a)$ is a matrix representation $A \longmapsto Mat(n, A^{\otimes n})$. Consider the equation

$$F(\det(1 + \mathcal{L}(a)z)) = R_f(a, z). \tag{8.2.5}$$

The coefficients of the determinant above lie in $A^{\otimes n}$ but they can be regarded as lying in $S^n(A)$. Suppose (8.2.5) holds identically in z. Then it can be shown that for a given F (8.2.5) uniquely defines f, so that α can be defined by $\alpha(F) = f$, and conversely for a given f (8.2.5) uniquely defines F so that β can be defined by $\beta(f) = F$.

Suppose that $F \in \Phi_1(S^n(A), B)$ is given. Then, comparing the linear terms of (8.2.5), f should be given by the formula

$$f(a) = F(tr(\mathcal{L}(a)) \tag{8.2.6}$$

where

$$tr(\mathcal{L}(a)) = a \otimes 1 \otimes \ldots \otimes 1 + 1 \otimes a \otimes \ldots \otimes 1 + \ldots + 1 \otimes 1 \otimes \ldots \otimes a \in S^n(A).$$

Take (8.2.6) as the definition of f. It is obviously a linear map. The calculation of its characteristic function is

$$R_f(a, z) = e^{f \ln(1+az)} = e^{F(tr\mathcal{L}(\ln(1+az)))} = e^{F(tr(\ln(1+\mathcal{L}(a)z)))}$$

$$= Fe^{tr(\ln(1+\mathcal{L}(a)z))} = F(\det(1 + \mathcal{L}(a)z)$$

noting that since $\mathcal{L}$ and F are algebra homomorphisms they can be switched with the corresponding functions. Thus (8.2.5) is satisfied. In particular the characteristic function of f is a polynomial of degree n and hence f is a Frobenius n-homomorphism from A to B. Thus α has been constructed.

Now suppose that $f \in \Phi_n(A, B)$ is given. In order to define F from (8.2.5), it is necessary to show the existence of a linear map $F : S^n(A) \to B$ which is an algebra homomorphism. Firstly it will be shown that such a map would be unique. On expanding the left-hand side of (8.2.5) it is seen that F is specified on all elements of $S^n(A)$ of the form $tr \wedge^k \mathcal{L}(a) \in S^n(A)$, including $\det(\mathcal{L}(a))$. In particular

$$F(\det(\mathcal{L}(a))) = ber_f(a).$$

where $ber_f(a) = R_f(a-1, 1)$. Take (8.2.2) as the definition of F on such elements. The elements of the form

$$\det(\mathcal{L}(a)) = a \otimes \ldots \otimes a \in S^n(A)$$

linearly span $S^n(A)$ and thus if a linear map F satisfying (8.2.2) exists this formula determines it uniquely. On replacing a by $1 + az$ in (8.2.2) it can be seen that (8.2.5) is automatically satisfied. The existence of such an F follows from $ber_f = \psi_n(a)$ where ψ_n is the restriction to the diagonal of the symmetric multilinear map

$$\frac{1}{n!} \Phi : A \times \ldots \times A \to B$$

which corresponds to a linear map from $S^n(A)$ to B. This is the desired F. It remains to check that F is an algebra homomorphism. This follows from the fact that F is mutiplicative on the elements $a \otimes \ldots \otimes a$ on which it is $\psi_n(a)$. Thus the map β has been produced and the maps α and β are mutually inverse by construction. $\square$

Theorem 8.2 was then used by Buchstaber and Rees to prove the following, see [35]. Let $\mathbb{C}(X)$ be the algebra of continuous functions from X to $\mathbb{C}$ with the supremum norm.

Theorem 8.3 *Let X be a compact Hausdorff space. Define the "evaluation" map* $\mathcal{E} : S^n(X) \to \Phi_n(\mathbb{C}(X))$ *by*

$$[x_1, \ldots, x_n] \to (g \to \sum_{k=1}^{n} g(x_k)).$$

Then $\mathcal{E}$ is a homeomorphism provided that $\Phi_n(\mathbb{C}(X))$ is considered with the weak topology.

Definition 8.1 Let A, B be commutative associative algebras and let $f : A \rightarrow B$ be a ring homomorphism. Then a linear map $\tau : B \rightarrow A$ is an *n-transfer* for f if

   (i) τ is a Frobenius n-homomorphism
  (ii) $\tau (f(a)b) = a\tau(b)$ (i.e. τ is a map of A-modules)
 (iii) $f\tau : B \rightarrow B$ is the sum of the identity and a Frobenus $(n-1)$-homomorphism
        $g : A \rightarrow B$.

Further results relating Frobenius n-homomorphisms appear in [38], in the context of branched coverings and transfer maps in geometry. For example they prove the following theorem

Theorem 8.4 *Let X and Y be compact Hausdorff spaces. Given an n-branched covering $h : X \rightarrow Y$, $t : S^n(Y) \rightarrow X$, the direct image $t_! : \mathbb{C}(X) \rightarrow \mathbb{C}(Y)$ is an n-transfer for the ring homomorphism $h^* : \mathbb{C}(Y) \rightarrow \mathbb{C}(X)$.*

Some further developments involving Frobenius n-homomorphisms have appeared in papers of Gugnin. See [128, 129] and [130].

8.3   The FKG Inequality and Its Generalization

The FKG inequality is a correlation inequality, which has become an important tool in statistical mechanics and probabilistic combinatorics. Some of the applications are listed below.

1. (Statistics): the study of monotonicity properties of power functions of likelihood ratio test statistics in multivariate analysis, association and dependence properties of random variables, and observational studies.
2. Probability theory and mathematical physics: diffusion equations, interacting particle systems, Ising models, reliability theory and percolation.
3. Combinatorial theory, the monotonicity of partial orders, Sperner theory, graph theory and Ramsey theory.

It is due to Fortuin et al. [102]. Informally, it says that in many random systems, increasing events are positively correlated, while an increasing and a decreasing event are negatively correlated. The exposition here follows that in [241].

Let L be a finite distributive lattice with partial ordering $\preceq$, least upper bound $\vee$ and greatest lower bound $\wedge$. A function $f : L \rightarrow \mathbb{R}$ is called **(monotone) increasing** if $f(x) \leq f(y)$ whenever $x \preceq y$. A probability measure is defined in Chap. 7, p. 181. If μ is such a measure on L it is said to be **multivariate totally positive of order** 2 (**MTP$_2$**) if

$$\mu(x \vee y)\mu(x \wedge y) \geq \mu(x)\mu(y)$$

for all $x, y \in L$. In the literature alternative names for MTP$_2$ probability measures
are **FKG measures** or **log-supermodular** measures.

For any probability measure μ on L and any function $f : L \to \mathbb{R}$ denote by

$$\mathbb{E}(f) := \sum_{a \in L} \mu(a) f(a)$$

the **mean** or **average** or **expectation** with respect to μ.

Suppose f_1 and f_2 are both increasing (or both decreasing) real-valued functions
on L, and let μ be an MTP$_2$ probability measure on L. The FKG inequality states
that the covariance

$$\mathrm{Cov}(f_1, f_2) := \mathbb{E}(f_1 f_2) - \mathbb{E}(f_1)\mathbb{E}(f_2) \geq 0. \tag{8.3.1}$$

In mathematical statistics, it is usual to state the FKG inequality for $\mathbb{R}^n$. Consider
vectors $\underline{x} = (x_1, \ldots, x_n)$ and $\underline{y} = (y_1, \ldots, y_n)$ in $\mathbb{R}^n$ with partial order given by
$\underline{x} \preceq \underline{y}$ if $x_i \leq y_i$ for all $i = 1, \ldots, n$. Define

$$\underline{x} \vee \underline{y} = (\max(x_1, y_1), \ldots, \max(x_n, y_n))$$

and

$$\underline{x} \wedge \underline{y} = (\min(x_1, y_1), \ldots, \min(x_n, y_n)).$$

The function $f : \mathbb{R}^n \to \mathbb{R}$ is said to be **increasing** if $f(\underline{x}) \leq f(\underline{y})$ whenever $\underline{x} \preceq \underline{y}$.
A probability density function $K : \mathbb{R}^n \to \mathbb{R}$ is said to be MTP$_2$ if

$$K(x \vee y)K(x \wedge y) \geq K(x)K(y)$$

for all $\underline{x}, \underline{y} \in \mathbb{R}^n$. If the functions $f_1, f_2 : \mathbb{R}^n \to \mathbb{R}$ are both increasing or both
decreasing and the expectations $\mathbb{E}(f_1)$, $\mathbb{E}(f_2)$ and $\mathbb{E}(f_1 f_2)$ with respect to K are
finite then the FKG inequality is stated as

$$\int_{\mathbb{R}^n} f_1(\underline{x}) f_2(\underline{x}) K(\underline{x}) d\underline{x} - \int_{\mathbb{R}^n} f_1(\underline{x}) K(\underline{x}) d\underline{x} \int_{\mathbb{R}^n} f_2(\underline{x}) K(\underline{x}) d\underline{x} \geq 0. \tag{8.3.2}$$

It is well known that (8.3.2) may be deduced by an approximation argument from
the discrete case (8.3.1). Also numerous generalizations of (8.3.1) have appeared in
the literature (see [241] for details).

8.3.1   Extensions to Higher Cumulants

In [241] the extension of the inequality to higher correlations or cumulants is
discussed. The original FKG inequality can be interpreted as an inequality for the

simplest cumulant of the random variables f_1 and f_2. The definition of the higher cumulants is as follows (see [246, 265]). The **moments** γ_r^X of a random variable X (about zero) are $\mathbb{E}(X^r)$, $r = 1, 2 \ldots$. The exponential generating function $M_X(t)$ is known as the **moment generating function (m.g.f)** of X:

$$M_X(t) = \mathbb{E}\{exp(Xt)\}$$

$$= \mathbb{E}(1 + Xt + \frac{1}{2!}X^2t^2 + \ldots)$$

$$= 1 + \gamma_1 t + \frac{1}{2!}\gamma_2 t^2 + \ldots.$$

The **cumulants** $\kappa_r = \kappa_r^X$ of X are defined by their generating function, **the cumulant generating function (c.g.f)**

$$K_X(t) = \log(M_X(t))$$

$$= \kappa_1 t + \frac{1}{2!}\kappa_2 t^2 + \ldots.$$

On expansion the following is obtained

$$\kappa_1 = \gamma_1$$

$$\kappa_2 = \gamma_2 - \gamma_1^2$$

$$\kappa_3 = \gamma_3 - 3\gamma_1\gamma_2 + 2\gamma_1^3, \ldots.$$

there being the inverse relations

$$\gamma_1 = \kappa_1$$

$$\gamma_2 = \kappa_2 + \kappa_1^2$$

$$\gamma_3 = \kappa_3 + 3\kappa_1\kappa_2 + \kappa_1^3, \ldots.$$

Now suppose that $\mathbf{X}$ is a random vector, $\mathbf{X} = (X(1), X(2), \ldots)$ where the number of components is not specified and that $\mathbf{t} = (t(1), t(2), \ldots)$ is an array of indeterminates. Let

$$\mathbf{X[t]} = X(1)t(1) + X(2)t(2) + \ldots$$

The **joint** or **mixed moments** are defined as

$$\gamma(m_1, m_2, \ldots) = \mathbb{E}\{X(1)^{m_1} X(2)^{m_2} \ldots\}$$

where $\gamma(m_1, m_2, \ldots)$ is the coefficient of $t(1)^{m_1} t(2)^{m_2} \ldots$ in the expansion of

$$M_{\mathbf{X}}(\mathbf{t}) = \mathbb{E}\{\exp(\mathbf{X}[\mathbf{t}])\}.$$

The **overall order** of $\gamma(m_1, m_2, \ldots)$ is $m_1 + m_2 + \ldots$ and it is convenient to replace the function γ by a sequence of arrays $\gamma_1, \gamma_2, \ldots$ of moments of order $1, 2, \ldots$ respectively. Namely

$$\gamma_1(l) = \mathbb{E}\{X(l)\}, \quad \gamma_2(l_1, l_2) = \mathbb{E}\{X(l_1)X(l_2)\}, \ldots$$

and where $\gamma_i, \ldots$ are invariant under the action of S_i on the index set of its arguments. Then by a process of multilinearization, using the relationship

$$K_{\mathbf{X}}(\mathbf{t}) = \log(M_{\mathbf{X}}(\mathbf{t}))$$

and abbreviating $\kappa_r(l_1, l_2, \ldots, l_r)$ by $\kappa_{l_1 l_2 \ldots l_r}$ it follows that

$$\kappa_1 = \gamma_1$$
$$\kappa_{12} = \gamma_2 - \gamma_1\gamma_2$$
$$\kappa_{123} = \gamma_{123} - \gamma_{12}\gamma_3 - \gamma_{13}\gamma_2 - \gamma_{23}\gamma_1 + 2\gamma_1\gamma_2\gamma_3, \ldots$$

and

$$\gamma_1 = \kappa_1$$
$$\gamma_{12} = \kappa_{12} + \kappa_1\kappa_2$$
$$\gamma_{123} = \kappa_{123} + \kappa_{12}\kappa_3 + \kappa_{13}\kappa_2 + \kappa_{23}\kappa_1 + \kappa_1\kappa_2\kappa_3, \ldots.$$

The above relationship may be described in terms of the partition lattices $\mathcal{P}(r)$, $r = 1, 2, 3, 4, \ldots$ by associating the terms with elements of the appropriate $\mathcal{P}(r)$, for example $\gamma_{12}\gamma_3$ is associated to the partition $(2 + 1) \vdash 3$ and $\gamma_{123}\gamma_{34}\gamma_5$ to the partition $(3 + 2 + 1) \vdash 6$. Then

$$\kappa_\pi = \sum_{\rho \in \mathcal{P}(r)} \mu(\rho, \pi)\gamma_\rho$$

and

$$\gamma_\pi = \sum_{\rho \in \mathcal{P}(r)} \zeta(\rho, \pi)\kappa_\rho$$

where μ is the Möbius function and ζ is its inverse.

The natural extension of the FKG inequality for three real-valued functions on a lattice L, under the appropriate assumptions would be that

$$\kappa_3(f_1, f_2, f_3) = \mathbb{E}(f_1 f_2 f_3) - \mathbb{E}(f_1 f_2)\mathbb{E}(f_3) - \mathbb{E}(f_1 f_3)\mathbb{E}(f_2) - \mathbb{E}(f_2 f_3)\mathbb{E}(f_1)$$
$$+2\mathbb{E}(f_1)\mathbb{E}(f_2)\mathbb{E}(f_3)$$

is non-negative. However, there are well-known probability distributions for which the FKG inequality holds but for which the higher cumulants are non-positive. Richards defined the third-order "conjugate" cumulant

$$\kappa_3'(f_1, f_2, f_3) := 2\mathbb{E}(f_1 f_2 f_3) - \mathbb{E}(f_1 f_2)\mathbb{E}(f_3) - \mathbb{E}(f_1 f_3)\mathbb{E}(f_2) - \mathbb{E}(f_2 f_3)\mathbb{E}(f_1)$$
$$+\mathbb{E}(f_1)\mathbb{E}(f_2)\mathbb{E}(f_3)$$

and stated the following. As mentioned above, in his original paper it was presented as a theorem, but it remains a conjecture.

Conjecture 8.1 Let L be a finite distributive lattice, let μ be an MTP$_2$ probability measure on L, and let f_1, f_2, f_3 be non-negative increasing functions on L. Then

$$\kappa_3'(f_1, f_2, f_3) \geq 0.$$

He showed that a consequence of Conjecture 8.1 is that

$$\kappa_3(f_1, f_2, f_3) \geq -[\mathbb{E}(f_1 f_2 f_3) - \mathbb{E}(f_1)\mathbb{E}(f_2)\mathbb{E}(f_3)].$$

and showed that it also implies

$$Cov(f_1 f_2, f_3) - \mathbb{E}(f_1)Cov(f_2, f_3) + Cov(f_1 f_3, f_2) \geq 0.$$

He then defined fourth and fifth order conjugate cumulants $\kappa_4'(f_1, f_2, f_3, f_4)$ and $\kappa_5'(f_1, f_2, f_3, f_4, f_5)$, and his conjecture extends to a similar statement that under appropriate assumptions these are non-negative. However for $n = 6$ there are examples where his κ_6' is negative.

The ideas were taken up by Sahi in [249]. There it is first explained that the definition of κ_r may be expressed as follows. Suppose there are given functions $f_1, \ldots, f_n$. If σ is a subset of $\{1, \ldots, n\}$ define E_σ by

$$E_\sigma = E_\sigma(f_1, \ldots, f_n) = \mathbb{E}(\prod_{i \in \sigma} f_i).$$

For any partition $\pi = \pi_1 + \pi_2 + \ldots + \pi_s$ of $\{1, \ldots, n\}$ into disjoint subsets with $|\pi_i| \geq |\pi_{i+1}|$ for all i, define

$$E_\pi = \prod_{i=1}^{s} E_{\pi_i}.$$

Let $\lambda(\pi)$ denote the element $(|\pi_1|, |\pi_2|, \ldots, |\pi_s|)$ of $\mathcal{P}(n)$. Then for any partition $\lambda \vdash n$ let

$$E_\lambda = \sum_{\pi:\lambda(\pi)=\lambda} E_\pi.$$

Then for each partition $\lambda \vdash n$ let c_λ be defined by

$$c_\lambda = (-1)^{\ell(\lambda)-1} 0 \ell(\lambda) - 1)!$$

where $\ell(\lambda)$ is the number of parts of λ, see Chap. 6, Sect. 6.2.2. Then

$$\kappa_n(f_1, \ldots, f_n) = \sum_{\lambda \vdash n} c_\lambda E_\lambda.$$

Now Richards replaced c_λ by

$$c'_\lambda = (-1)^{\ell(\lambda)-1}(\ell(\lambda') - 1)!$$

where λ' is the conjugate partition to λ, and his conjugate cumulants are

$$\kappa'_n(f_1, \ldots, f_n) = \sum_{\lambda \vdash n} c'_\lambda E_\lambda.$$

Sahi suggested that the appropriate formula for the above coefficients should be

$$d_\lambda = (-1)^{l(\lambda)-1} \prod_{i=1}^{l(\lambda)} (\lambda_i - 1)!$$

$$\kappa''_n(f_1, \ldots, f_n) = \sum_{\lambda \vdash n} d_\lambda E_\lambda. \tag{8.3.3}$$

By a coincidence $\kappa'_n(f_1, \ldots, f_n) = \kappa''_n(f_1, \ldots, f_n)$ for $n \leq 5$.

The κ''_n may be defined in terms of power series as follows. Let

$$K(t) = \sum_{r=0}^{\infty} \kappa''_n t^r$$

and

$$P(t) = \sum_{r=0}^{\infty} \gamma_r t^r$$

then

$$P(-t) = K'(T)/K(T) = \frac{d}{dt} \log(K(t)).$$

From now on the "skew cumulant" κ_n'' will be denoted by E_n.

8.3.2   The Conjecture of Sahi

The conjecture of Sahi whose truth would imply that there are generalizations of the FKG inequality for $E_n(f_1, \ldots, f_n)$ for arbitrary n is now presented.

Let X be a finite set and regard the power set 2^X as a partially ordered set L with respect to inclusion. Let $\mathfrak{R} := \mathbb{R}[[t]]$ be the space of formal power series in the variable t with real coefficients. The set

$$\mathfrak{P} = \{a_1 t + a_2 t^2 + \ldots | 0 \le a_i \in \mathbb{R} \text{ for all } i\}$$

is a convex cone in $\mathcal{R}$. Define

$$\mathfrak{R}[X] := \{F | F : 2^X \to \mathfrak{R}\}$$

and

$$\mathfrak{P}[X] := \{F | F : 2^X \to \mathfrak{P}\}.$$

Since $\mathfrak{P}[X]$ is a convex cone in $\mathfrak{R}[X]$, its elements are referred to as **positive** functions on 2^X. The subcone of **increasing** positive functions is defined to be

$$\mathfrak{I}[X] := \{F \in \mathfrak{P}[X] : F(T) - F(S) \in \mathfrak{P} \text{ for all } S \subseteq T \subseteq X\}.$$

A further subcone is introduced. For a function F in $\mathfrak{R}[X]$ define its "cumulation" to be the function F^+ in $\mathfrak{R}[X]$ given by

$$F^+(T) := \sum_{S \subseteq T} F(S) \text{ for } T \subseteq X.$$

The class of cumulations of positive functions is denoted by $\mathfrak{C}(X)$, so that

$$\mathfrak{C}(X) := \{F^+ | F \in \mathcal{P}[X]\}.$$

Then $\mathfrak{C}(X)$ is a subcone of $\mathfrak{I}[X]$.

If μ is a probability measure on 2^X which satisfies

$$\mu(S \cup T)\mu(S \cap T) \geq \mu(S)\mu(T) \text{ for all } S, T \subseteq X \qquad (8.3.4)$$

then μ is an MTP$_2$ measure on L in the above sense.

Sahi's conjecture is

Conjecture 8.2 For any F in $\mathfrak{I}[X]$, and a measure μ satisfying (8.3.4)

$$1 - \prod_{S \subseteq X} (1 - F(S))^{\mu(S)} \geq 0.$$

This implies

Conjecture 8.3 For any n-tuple of functions $f_1, \ldots, f_n$ in $\mathfrak{I}[X]$, and any μ satisfying (8.3.4),

$$E_n(f_1, \ldots, f_n) \geq 0.$$

Thus the truth of Conjecture 8.3 would provide a set of higher "correlation" inequalities.

Although the above conjectures remain open, Sahi proved

Theorem 8.5 *For any F in $\mathfrak{C}(X)$ and any μ satisfying (8.3.4),*

$$1 - \prod_{S \subseteq X} (1 - F(S))^{\mu(S)} \geq 0.$$

A consequence is

Theorem 8.6 *For any n-tuple of functions $f_1, \ldots, f_n$ in $\mathfrak{C}(X)$, and any μ satisfying (8.3.4),*

$$E_n(f_1, \ldots, f_n) \geq 0.$$

8.3.3 The Connection with k-Characters

Sahi's definition (8.3.3) of the $E_k(f_1, \ldots, f_k)$ may be seen to be algebraically very close to the definition in Chap. 7.3, (1.8.2) of $\chi^{(k)}(g_{i_1}, g_{i_2}, \cdots, g_{i_k})$. If in the formula

$$\chi^{(k)}(g_{i_1}, g_{i_2}, \cdots, g_{i_k}) = \sum_{\tau \in \mathcal{S}_k} sgn(\tau)\chi_\tau(g_{i_1}, g_{i_2}, \cdots, g_{i_k}).$$

each term

$$sgn(\tau)\chi_\tau(g_{i_1}, g_{i_2}, \ldots, g_{i_k}).$$

is replaced by

$$d_\lambda E_\lambda(f_1, \ldots, f_k)$$

where lambda is the partition corresponding to τ, then the formula for

$$(-1)^{k-1} E_k(f_1, \ldots, f_k)$$

is produced.

Chapter 9
Fusion and Supercharacters

Abstract In this chapter the idea of fusion of the character table of a group is pursued in more detail. First the question of which groups have the property that their character table is a fusion of that of an abelian group is addressed. It proved difficult to answer this question but many results can be obtained. There is given an explicit description of the finite groups whose character tables fuse from a cyclic group. Then there is given an account of how the idea of fusion was independently discovered and used in the context of upper triangular groups $UT_n(q)$ by Diaconis and Isaacs. Their motive was that whereas the character tables of $UT_n(q)$ are "wild" certain fusions are not and random walks on the groups can be discussed. The interesting result that a fusion of the character table gives rise to a Hopf algebra is presented. There is also given a construction of a fusion of the character table of $UT_4(q)$ by taking the class algebra of a loop constructed by an alternative multiplication on the elements on the elements of the group.

9.1 Introduction

In Chap. 4 the idea of the fusion of a character table of an association scheme is introduced, and it is indicated that it can be used in the calculation of character tables for loops or quasigroups in the cases where their tables fuse from that of a group. For an arbitrary association scheme, as explained in Chap. 4, its character table has a fusion if and only if a "magic rectangle" condition, stated there, is satisfied on the group normalized table. Fusions of the character table of a group are a special case.

Given an arbitrary group H the question has arisen in several contexts of whether it is possible to describe all fusions of its character table. This is equivalent to finding all S-rings which are subrings of the class algebra C. Even in the case of a cyclic group, the classification of such S-rings is quite involved. This case has been addressed in [203, 237] and [204], and a description of all the possible S-rings involved is given. An equivalent account of this also appears in [142]. Applications of this work to circulant graphs appear in [221]. The problem of describing S-rings which arise from the fusion of an arbitrary abelian group has proved difficult and

© Springer Nature Switzerland AG 2019   287
K. W. Johnson, *Group Matrices, Group Determinants*
and Representation Theory, Lecture Notes in Mathematics 2233,
https://doi.org/10.1007/978-3-030-28300-1_9

remains open. It appears to be relatively rare that a fusion of a character table of a group is that of another group, but more often such fusions are character tables of loops.

In [152] a more restricted problem was addressed. If an arbitrary class $\mathcal{C}$ of finite groups is considered, define the class $\mathcal{FC}$ by $G \in \mathcal{FC}$ if and only if there is a group $H \in \mathcal{C}$ such that the character table of H fuses to that of the group G. The case considered in [152] and [153] was that where $\mathcal{C}$ is the class $\mathcal{A}$ of abelian groups. A motivation for this work was that in some sense the groups in $\mathcal{FA}$ should have properties which transfer from abelian groups, especially in that harmonic analysis involving groups in $\mathcal{FA}$ should "pull back" to that on the corresponding abelian group. A precise characterization of the class $\mathcal{FA}$ has proved elusive, even for finite p-groups, where an initial conjecture was that conditions such as regularity or powerfullness could imply that a group lies in $\mathcal{FA}$, but counterexamples exist to show that neither of the above properties ensures that a p-group lies in $\mathcal{FA}$. It is possible to prove that many simple groups cannot lie in $\mathcal{FA}$ and it is conjectured that no non-abelian simple group lies in $\mathcal{FA}$. However, the groups in $\mathcal{FC}$ for $\mathcal{C}$ the class of cyclic groups have been classified, using the classification of S-rings over cyclic groups, see [156].

Whereas it might be expected that the techniques of character theory would be important in addressing the question of fusion between character tables of groups, the techniques used so far were mainly connected with S-ring theory.

The idea of fusion appeared in work on "supercharacters" by Diaconis and Isaacs [86]. If G is a group their definition of the supercharacter table of G is that obtained from a fusion in the sense of [178]. Their discussion is limited to fusion of character tables of groups, and the characters of a fused table are called supercharacters. The results in [178] and [177], valid for arbitrary fusions of association scheme tables, were rediscovered by them in the restricted setting. However, there are very interesting connections arising from their results, motivatated by the examination of random walks on the family of upper triangular groups. The problem of calculating the full character tables of these groups is wild, and this led Diaconis, Isaacs and others to examine certain fusions of these character tables which can be handled asymptotically. A conjecture arose that a Hopf algebra arising from a fusion of the character table of $UT_n(q)$, the group of upper triangular matrices with entries in the finite field $\mathbb{F}_q$, is isomorphic to the Hopf algebra of non-commutative symmetric functions, and this was proved at a workshop at the American Institute of Mathematics [2].

In Sect. 9.3 the results on the class of groups whose character tables fuse from those of abelian groups are described. In the case of p-groups which do not fuse from an abelian group, in all the cases examined they can be embedded in a nontrivial way into a p-group which does fuse.

A description of those groups which fuse from a cyclic group is set out in Sect. 9.3, and in Sect. 9.4 an outline of the results relating supercharacters of upper triangular groups to Hopf algebras is given. There is also a brief account of how a nonassociative loop may be constructed on the set of an upper triangular group such that the class algebra of the loop is a fusion of the class algebra of the group, which is between the class algebra and the fusion mentioned above.

9.2 Fusion from Abelian Groups

9.2.1 The Magic Rectangle Condition

For the purposes of this chapter, the magic rectangle condition is stated in the particular case of a group H. Let $Irr(H) = \{\chi_1, \ldots, \chi_r\}$ with $d_i = \chi_i(e)$, $i = 1, \ldots, r$, and let $\{C_1, \ldots, C_r\}$ be the set of conjugacy classes of H with $|C_i| = k_i$. Then there is a partition $\{B_i\}_{i=1}^f$ of the set $\{C_1, \ldots, C_r\}$ and a partition $\{\psi_i\}_{i=1}^f$ of the set $\{\chi_1, \ldots, \chi_r\}$. The character table of H can then be ordered so that it consists of rectangles, a typical rectangle consisting of the columns corresponding to the elements of $B_j = \{C_{t_1}, \ldots, C_{t_{r_j}}\}$ and rows corresponding to $\psi_i = \{\chi_{w_1}, \ldots, \chi_{w_{v_i}}\}$. If C_{t_i} is abbreviated by C_i and χ_{w_i} by χ_i, the magic rectangle condition states that for a given such rectangle, for each i the value

$$\tau_{ij} = \frac{\sum_{m=1}^{r_j} k_m \chi_i(m)}{[\sum_{m=1}^{r_j} k_m] d_i}$$

is constant and equal to the common value for each j of

$$\frac{\sum_{m=1}^{u_i} d_m \chi_m(j)}{[\sum_{m=1}^{u_i} d_m^2]}.$$

The fused table is an $f \times f$ table which has rows corresponding to the ψ_i and classes corresponding to B_j such that the value of ψ_i on B_j (with slight abuse of notation) is $\eta_i \tau_{ij}$ where

$$\eta_i = \sqrt{\sum_{m=1}^{u_i} d_m^2}.$$

In this case, if $M = Mlt(H)$ is the multiplication group of H, the permutation action of M on H gives rise to the S-ring $\mathcal{Z} = Z(\mathbb{Z}(H))$ (the center of $\mathbb{Z}(H)$) arising from the conjugacy classes, and all fusions of the character table of H arise from S-rings which are subrings of $\mathcal{Z}$. Conversely, any sub-S-ring of $\mathcal{Z}$ gives rise to a fusion. The statement "H fuses from G" will be taken to be equivalent to "the character table of a group H fuses from that of a group G".

The following examples illustrate the magic rectangle condition, and the construction of the corresponding character table.

1. Let G be the cyclic group of order 6 generated by x. Its character table is

	e	x^2	x^4	x^3	x	x^5
χ_0	1	1	1	1	1	1
χ_3	1	1	1	-1	-1	-1
χ_2	1	ω	ω^2	1	ω^2	ω
χ_4	1	ω^2	ω	1	ω	ω^2
χ_1	1	ω	ω^2	-1	ρ	ρ^{-1}
χ_5	1	ω^2	ω	-1	ρ^{-1}	ρ

where $\omega = e^{2\pi i/3}$ and $\rho = e^{2\pi i/6}$. There is a fusion with the partitions indicated, and it can be checked directly that the magic rectangle condition is satisfied in all rectangles. The fused character table is

	B_1	B_2	B_3
ψ_0	1	1	1
ψ_1	1	1	-1
ψ_2	2	-1	0

which is the character table for $\mathcal{S}_3$. Thus $\mathcal{S}_3$ fuses from $\mathcal{C}_6$.

There is another fusion, given by the partitions indicated below

	e	x^2	x^4	x^3	x	x^5
χ_0	1	1	1	1	1	1
χ_3	1	1	1	-1	-1	-1
χ_2	1	ω	ω^2	1	ω^2	ω
χ_4	1	ω^2	ω	1	ω	ω^2
χ_1	1	ω	ω^2	-1	ρ	ρ^{-1}
χ_5	1	ω^2	ω	-1	ρ^{-1}	ρ

The corresponding character table is

	B_1	B_2	B_3	B_4
ψ_0	1	1	1	1
ψ_1	1	1	-1	-1
ψ_2	$\sqrt{2}$	$-1/\sqrt{2}$	$\sqrt{2}$	$-1/\sqrt{2}$
ψ_3	$\sqrt{2}$	$-1/\sqrt{2}$	$-\sqrt{2}$	$1/\sqrt{2}$

It is clear that this cannot be the character table of a group. If this were the character table of a loop, then it would necessarily have a normal subloop of order 2 (the kernel of the character ψ_2). The factor loop would be the group C_3, and its linear characters must pull back to two more non-trivial linear characters of the loop, which is impossible since they do not appear in the table. Note that there is a loop of order 6 with a normal subloop of order 2 with the character table

	B_1	B_2	B_3	B_4
ψ_1	1	1	1	1
ψ_2	1	1	ω	ω^2
ψ_3	1	1	ω^2	ω
ψ_4	$\sqrt{3}$	$-\sqrt{3}$	0	0

the class sizes being $1, 1, 2, 2$ in the order indicated. This fuses from C_6, the corresponding S-ring being $\{e\}, \{x^3\}, \{x, x^4\}, \{x^2, x^5\}$.

2. Consider C_{12} with y as a generator. Let the characters $\{\chi_i\}_{i=0}^{11}$ be indexed by $\chi_i(y) = \upsilon^i$ where $\upsilon = e^{2\pi i/12}$. The following partitions lead to a fusion:

$$\{\psi_i\} = \{\chi_0\}, \{\chi_4\}, \{\chi_8\}, \{\chi_1, \chi_2, \chi_3, \chi_5, \chi_6, \chi_7, \chi_9, \chi_{10}, \chi_{11}\}$$

and

$$\{B_i\} = \{e\}, \{x^3, x^6, x^9\}, \{x, x^4, x^7, x^{10}\}, \{x^2, x^5, x^8, x^{11}\}.$$

The fused table is

	B_1	B_2	B_3	B_4
ψ_1	1	1	1	1
ψ_2	1	1	ω	ω^2
ψ_3	1	1	ω^2	ω
ψ_4	3	-1	0	0

where $\omega = e^{2\pi i/3}$. This table is the character table of A_4.

9.2.2 Basic Results on Fusion

Let G, H be groups and let $\mathcal{Z}$ be an S-ring on G. As mentioned above, the statement "H fuses to $\mathcal{Z}$" will be taken to be equivalent to "there is a subalgebra of $Z(\mathbb{Z}H)$ which is isomorphic to $\mathcal{Z}$". The statement "H fuses to G" will be taken to be equivalent to "H fuses to the class algebra of G".

The results listed below are immediate.

For $g \in G$ let $C(g)$ denote the conjugacy class of g and let $S(g) = g^{-1}C(g)$.

1. For $g \in$, $S(g) \subseteq G'$.
2. If A and B are abelian groups and if A fuses to B, then A and B are isomorphic.
3. If N is a normal subgroup of a group G, then the classes of G contained in N form an S-ring on N. Let $\mathbf{C}_G(N)$ denote the S-ring on these classes; if H fuses to N, H fuses to $\mathbf{C}_G(N)$.

The following lemma indicates how fusion behaves relative to extensions.

Lemma 9.1

(i) *Suppose that H fuses to G and let N be a normal subgroup of G. Then there is a normal subgroup M of H such that M fuses to $\mathbf{C}_G N$ and H/M fuses to G/N.*

(ii) *Suppose that H fuses to G where the corresponding S-ring in H is $R(H)$ and suppose that N is a normal subgroup of H such that $\overline{N} \in R(H)$. Then there is a normal subgroup M of G such that H/N fuses to G/M.*

Proof

(i) Since N is normal in G it is a union of conjugacy classes $\{C_1, \ldots, C_t\}$ of G. The fusion of H to G implies that there is a collection of subsets $\{H_1, \ldots, H_t\}$ of H corresponding to the classes $\{C_1, \ldots, C_t\}$. Since the structure constants for the algebra generated by the $\overline{C_i}$ are the same as those for that generated by the $\overline{H_i}$, $M = \cup_{i=1}^t H_i$ is closed under multiplication and so determines a subgroup of H. But each H_i is a union of conjugacy classes in H and so M is a normal subgroup of H.

It remains to show that H/M fuses to G/N. The ideas in the proof are parallel to those in Lemma 1.2 in [203]. Without loss of generality, one can identify M and N. Let $\{D_i\}_{i \leq s}$ denote the conjugacy classes in G/N. Let $\pi : G \to G/N$ be the quotient map. Now each $\pi^{-1}(D_i)$ is a union of cosets of N: suppose that $G/N = \{w_1, \ldots, w_u\}$ and let $v_i \in G$ where $v_i = w_i$. Denote $\pi^{-1}(D_i)$ by E_i. If

$$D_i = \{w_{i_1}, w_{i_2}, \ldots, w_{i_{k_i}}\},$$

then

$$E_i = \{v_{i_1}N, v_{i_2}N \ldots, v_{i_{k_i}}N\}.$$

Now if $w_i w_j = w_k$, then $(v_i N)(v_j N) = v_k N$ and hence $\overline{v_i N}\,\overline{v_j N} = \overline{v_k N}$. It follows that if $\overline{D_i}\,\overline{D_j} = \sum_k \lambda_{ijk}\overline{D_k}$, then $\overline{E_i}\,\overline{E_j} = |N|\sum_k \lambda_{ijk}\overline{E_k}$. Thus the difference in the structure constants for the D_i and the E_i is the factor $|N|$.

Similarly, let $\pi' : H \to H/N$ be the quotient map and let B_i denote the classes in H/N. Let $F_i = (\pi')^{-1}(B_i)$. Then as in the above the difference in the structure constants for the B_i and the F_i is the factor $|N|$. But in the correspondence between the classes of H and G given by the fusion the F_i correspond to the E_i. Thus the structure constants for the B_i and the D_i are the same. This completes the proof of (i).

(ii) Let the classes in G be $\{C_1, \ldots, C_r\}$ and let the corresponding subsets of H be $\{H_1, \ldots, H_r\}$. Since $\overline{N} \in R(H)$ it has the form $\overline{N} = \sum_k \mu_k \overline{H_k}$, $k \in \{0, 1\}$. Let $\overline{M} = \sum_k \mu_k \overline{C_k}$. Then it easily follows that M is closed under multiplication and since the C_i are invariant under conjugation it follows that M is normal. This is the situation described in (i) and so the rest of (ii) follows from (i). □

Camina pairs and Camina triples have been defined in Chap. 5, Definitions 5.6 and 5.7. If G fuses from an abelian group and N is a normal subgroup of G, then, by Lemma 9.1, so does $\mathbf{C}_G(N)$. Further, if N fuses from an abelian group, then so does $\mathbf{C}_G(N)$. The Camina pair and Camina triple conditions are relevant to the description of properties **P** which may be imposed on a group G which ensure that if a normal subgroup N fuses from an abelian group and G has property **P**, then G fuses from an abelian group. For example, if $N = G'$ the property of (G, G') being a Camina pair is one such property (Proposition 9.1 below). Other examples appear in Propositions 9.2, 9.3, and Lemma 9.2.

Proposition 9.1 *If (G, N) forms a Camina pair and if $\mathbf{C}_G(N)$ and G/N fuse from groups H and K (respectively) then G fuses from $H \times K$. In particular, if (G, G') is a Camina pair and G' is abelian, then G fuses from an abelian group.*

Proof The second statement of the proposition follows from the first, since in this case both G and G/G' are abelian. Thus it is suffient to prove the first statement. Let the classes in $\mathbf{C}_G(N)$ be $\{C_1, \ldots, C_r\}$ and let the corresponding subsets of H be $\{B'_1, \ldots, B'_r\}$; let $B_i = (B'_1, 1) \subset H \times K$. Let

$$G/N = \{N, a_2 N, \ldots, a_u N\},$$

and let $C_{r+1}, \ldots, C_s$ be the classes outside N. Since (G, N) is a Camina pair, it follows that

$$C_i = a_{i,1} N \cup a_{i,2} N \cup \cdots \cup a_{i,i_t} N,$$

where $\{a_2 N, \ldots, a_u N\}$ is partitioned as follows

$$\{a_{i,1} N \cup a_{i,2} N \cup \cdots \cup a_{i,i_t} N\}, i = r+1, \ldots, s.$$

Since K fuses to G/N there is a corresponding partition of

$$K = \{1\} \cup K_{r+1} \cup \ldots \cup K_s$$

where $|C_i| = |K_i|$ for all $i > r$ and $\overline{K_i}\,\overline{K_j} = \sum_k \lambda_{ijk}\overline{K_k}$ if and only if $\overline{C_i}\,\overline{C_j} = \sum_k \lambda_{ijk}\overline{C_k}$ for $i, j > r$. Let $B_i = H \times K_i$ for $i > r$. The B_i have now been defined for $i \leq s$, and it remains to show that the $\{B_i\}$ and the $\{C_i\}$ form isomorphic S-rings, i.e. $\overline{B_i}\,\overline{B_j} = \sum_k \lambda_{ijk}\overline{B_k}$ if and only if $\overline{C_i}\,\overline{C_j} = \sum_k \lambda_{ijk}\overline{C_k}$ for $i, j > s$. This is done by cases.

If $i, j \leq r$, then this case follows from the fact that H fuses to $\mathbf{C}_G(N)$.

For the case where $i \leq r$, $j > r$, it follows that $C_i \subset N$, and since C_j is a union of cosets of N it may be seen that $\overline{C_i}\,\overline{C_j} = |C_i|\overline{C_j}$. But also $B_i = (B_i', 1)$, where $B_i' \subset H$ and since B_j is a union of cosets of H in $H \times K$ it follows that $\overline{B_i}\,\overline{B_j} = |B_i|\overline{B_j} = |C_i|\overline{B_j}$. This does this case and the case $i > r$, $j \leq r$ follows since $\overline{C_i}\,\overline{C_j} = \overline{C_j}\,\overline{C_i}$.

Lastly, assume that $i, j > r$. Let $\pi : G \to G/N$ be the quotient map. The Camina pair hypothesis shows that if $D \neq \{1\}$ is a class in G/N, and $C = \pi^{-1}(D)$ is a class in G. Thus the classes in G/N are

$$\{1\}, \pi(C_{r+1}), \ldots, \pi(C_s).$$

Let $D_i = C_i$ for $i > r$. Then π may be extended to a homomorphism $\pi : \mathbb{Q}G \to \mathbb{Q}G/N$. Let

$$\mathbf{Y}_1 = \langle |N|, \overline{C}_{r+1}, \ldots, \overline{C}_s \rangle$$

be the S-ring on G and let

$$\mathbf{Y}_2 = \langle 1, \overline{D}_{r+1}, \ldots, \overline{D}_s \rangle$$

be the S-ring on G/N. Let $n = |N| = |H|$. Then the map $C_i \to nD_i$ can be extended linearly to a ring isomorphism $\mathbf{Y}_1 \to \mathbf{Y}_2$. Since K fuses to G/N, the S-ring $\mathbf{Y}_2$ is isomorphic to the S-ring

$$\mathbf{Y}_3 = \langle 1, \overline{K}_{r+1}, \ldots, \overline{K}_s \rangle.$$

Now define the S-ring

$$\mathbf{Y}_4 = \langle (\overline{H}, 1), (\overline{H}, \overline{K}_{r+1}), \ldots, (\overline{H}, \overline{K}_s) \rangle.$$

Then there is an isomorphism $\mathbf{Y}_4 \to \mathbf{Y}_3$ determined by $(\overline{H}, \overline{K}_i) \to n\overline{K}_i$ and, composing these three isomorphisms, an isomorphism $\mathbf{Y}_1 \to \mathbf{Y}_4$ is obtained where $C_i \to (\overline{H}, \overline{K}_i)$, as required. This completes the proof of the last case of the Proposition. □

Proposition 9.2 *If a group G has a normal abelian subgroup A of index n, such that any class outside A is a coset of A and G/A is cyclic, then G is in $\mathcal{FA}$.*

Proof Let $g \in G \backslash A$ be such that $\{1, g, g^2, \ldots g^{n-1}\}$ is a set of coset representatives for A in G. Since G/A is abelian it follows that $G' \subseteq A$. Thus $g^n = a \in A$ and it is possible to construct an abelian group B of order $|G|$ from the short exact sequence of abelian groups

$$0 \to A \to B \to \{b\} \to 0$$

where $b^n = a$. Since $G' \subseteq A$ there is a corresponding subgroup D of B.

If C is a class in G which is a subset of A, then let C' denote the corresponding set in B. If C is a class in G which is not in A and $c \in C$, then $c = a_1 g^i$ with $a_1 \in A$ and $i < n$. Define C' to be $a_1 b^i D$. It will be shown that the structure constants obtained from products of the $\overline{C'}$ are the same as those corresponding to the products of the $\overline{C}$.

This is clearly the case if C_1 and C_2 are classes in A. Now if C_1 and C_2 both lie outside A, then $\overline{C}_1 = g^i \overline{G'}$ and $\overline{C}_2 = g^j \overline{G'}$ for some $i, j > 0$, and $\overline{C'}_1 = b^i \overline{D}$ and $\overline{C'}_2 = b^j \overline{D}$. It follows that

$$\overline{C}_1 \overline{C}_2 = |G'| g^{i+j} \overline{G'} \text{ and } \overline{C'}_1 \overline{C'}_2 = |G'| b^{i+j} \overline{D}.$$

Now if $i + j < n$, then these correspond under the correspondence defined above. If $i + j > n$ then write $i + j = n + m$, $m \geq 0$, and then $\overline{C}_1 \overline{C}_2 = |G'| g^m a \overline{G'}$ and $\overline{C'}_1 \overline{C'}_2 = |G'| b^m a \overline{D}$. Thus the structure constants coincide.

Now consider the case where $C_1 = g^i G$ is a class outside A, but C_2 is a class in A. Here if $c \in C_2 \subset A$, then $K = c^{-1} \overline{C}_2 \in \mathbb{Z} G'$. Note also that since $K \subset G'$, the coset cG' is well-defined (i.e. does not depend on the choice of c). Let C_1 and C_2 correspond to the subsets C'_1 and C'_2 of B; let $a_i \in C'_i$. Now since G' is normal in G it follows that $a_i^{-1} C'_i \subseteq D$. This case now follows since

$$\overline{C}_1 \overline{C}_2 = g^i \overline{G'} \times cK = |K| g^i c \overline{G'}$$

while

$$\overline{C'}_1 \overline{C'}_2 = |C'_2| b^i c \overline{D} = |K| g^i c \overline{D}. \qquad \square$$

Proposition 9.3 *If for a prime p a group G has a normal abelian subgroup A of index p, then G lies in $\mathcal{F}A$.*

Proof It is sufficient to show that any conjugacy class outside A must be a coset of A, for then the result follows from Proposition 9.2. Let $g \in G \backslash A$; then $\{1, g, g^2, \ldots, g^{p-1}\}$ is a set of coset representatives for A in G. It will be shown first that any element of G can be written as a product of elements of the form $[g, a]$ for some $a \in A$. Note that since G/A is abelian $G' \subseteq A$. Let $h_1, h_2 \in G$. Then h_1

can be written as $h_1 = g^k a_1 = a_1' g^k$, $h_2 = a_2 g^m$, $a_i \in A$. Set $a_3 = [h_1, a_2] \in A$. Now using the Hall–Witt identities [210, p. 290],

$$[h_1, h_2] = [h_1, a_2]g^m = [h_1, g^m][h_1, a_2][[h_1, a_2], g^m]$$
$$= [a_1, g^m][a_1' g^k, a_2][a_3, g^m] = [g^m, a_1]^{-1}[g^k, a_2][g^m, a_3]^{-1}.$$

Since every element $u \in G'$ is a product of powers of commutators it follows that u is a product of powers of commutators of the form $[g^k, a]$, $a \in A$. Using

$$[g^{k+1}, a] = [g.g^k, a] = [g, a][[g, a], g^k][g^k, a],$$

by induction any $u \in G'$ is a product of powers of commutators of the form $[g, a]$, $a \in A$.

Now any $h \in G'$ can be written as $h = [g^i, a]$, $a \in A$ so that

$$h^{-1}gh = a^{-1}ga = gg^{-1}a^{-1}ga = g[g, a].$$

Thus the class sum for the class containing g is a rational multiple of $\sum_{a \in A} g[g, a]$.

The next step is to show that if $u \in G'$, then there is a in A such that $g^a = gu$. This will show that $\overline{C}(g) = \sum_{u \in G'} gu$, as required. To do this it may be supposed by the above that $u = (ga_1)(ga_2) \ldots (ga_r)$. Then using (9.2.2) and the fact that A is abelian, it follows that

$$g^{a_1 a_2 \ldots a_r} = (g^{a_1})^{a_2 \ldots a_r}$$
$$= g[g, a_1]^{a_2 \ldots a_r}$$
$$= [g^{a_2}[g, a_1]^{a_2}]^{a_3 \ldots a_r}$$
$$= [g[g, a_2][g, a_1]]^{a_3 \ldots a_r} = \ldots$$
$$= g[g, a_r] \ldots [g, a_2][g, a_1] = gu,$$

as required. This completes the proof of the proposition.   □

The following simple result shows that the class of groups which fuse from an abelian group is closed under taking direct products.

Lemma 9.2 *Suppose that $G = H \times K$, where A fuses to H and B fuses to K. Then $A \times B$ fuses to G.*

Proof Let the classes in H be $\{H_i\}$, and the classes in K be $\{K_j\}$. Let the corresponding subsets of A, B be $\{A_i\}$, $\{B_j\}$. Then the classes of G are $\{H_i \times K_j\}$ and define the subsets $A_i \times B_j$ which correspond to these. It follows directly that this determines an isomorphism of S-rings.   □

A useful method to show that certain groups do not lie in $\mathcal{FA}$ is provided by the proposition:

Proposition 9.4 *Suppose that G lies in $\mathcal{FA}$. Let C be a class in G of size k. Then for any prime p at most k of the coefficients of $\overline{C}^p$ are nonzero mod p.*

Proof Let the abelian group H fuse to G via the map ϕ, and let $A = \phi^{-1}(C) \subset H$. Suppose that $k = |C| = |A|$ and that $A = \{a_1, \ldots, a_k\}$. Since H fuses to G the coefficients of each term of $\overline{A}^p$ are the same as the coefficients of the corresponding term of $\overline{C}^p$. However,

$$\overline{A}^p = (a_1 + \ldots + a_k)^p = (a_1^p + \ldots + a_k^p) \mod p.$$

Thus at most k of the coefficients of $\overline{C}^p$ are nonzero mod p. □

The following example illustrates the use of Proposition 9.4.

Example 9.1 Let $G = SL(2,3)$. The class C containing

$$\begin{bmatrix} 0 & 1 \\ 2 & 2 \end{bmatrix}$$

is

$$\left\{ \begin{bmatrix} 0 & 1 \\ 2 & 2 \end{bmatrix}, \begin{bmatrix} 1 & 0 \\ 2 & 1 \end{bmatrix}, \begin{bmatrix} 2 & 1 \\ 2 & 0 \end{bmatrix}, \begin{bmatrix} 1 & 1 \\ 0 & 1 \end{bmatrix} \right\}.$$

The coefficients of the terms in $\overline{C}^2$ are $1, 1, 1, 1, 3, 3, 3, 3$. Hence on taking $p = 2$ it follows that $SL(2,3)$ does not lie in $\mathcal{FA}$.

For a finite group G, let $\mathbb{Q}G$ denote the extension of $\mathbb{Q}$ obtained by adjoining the entries of the character table of G.

Lemma 9.3 *Suppose that the group H fuses to the group G. Then $\mathbb{Q}G$ is a subfield of $\mathbb{Q}H$.*

Proof This follows immediately from the description of fusion in terms of characters given in Sect. 9.3 since this description shows that the entries in the character table of G are rational linear combinations of the entries of the character table of H. □

The next result shows that "fusion is transitive".

Lemma 9.4 *Suppose that the group H fuses to the group G and that G fuses to K. Then H fuses to K.*

Proof This follows immediately from the definition since $Z(\mathbb{C}K)$ is isomorphic to a subalgebra of $Z(\mathbb{C}G)$, and $Z(\mathbb{C}G)$ is isomorphic to a subalgebra of $Z(\mathbb{C}H)$. □

9.2.3   Results on Groups Which Lie in $\mathcal{FA}$

A case-by-case analysis was used to show the following.

Theorem 9.1 *All groups of order less than 48 lie in $\mathcal{FA}$ except for S_4 and $SL(2, 3)$.*

For p-groups the following results were obtained.

Theorem 9.2

(i)   *For an odd prime p any group of order p^n, $n \leq 4$ lies in $\mathcal{FA}$.*
(ii) *All groups of order 2^5 fuse from an abelian group.*
(iii) *For arbitrary p any extra-special p-group fuses from an abelian group.*

The free groups in various varieties were considered. For $m, k, r \in \mathbb{N}$ let $F(m, k, r)$ be the free group in the variety of groups of exponent m of nilpotency class r on k generators, and let $B(m, k)$ be the Burnside group of exponent m on k generators.

Theorem 9.3

(i)   *The group $F(p, 2, 2)$ lies in $\mathcal{FA}$ for all primes p.*
(ii) *For any prime p the group $F(p, 3, 2)$ lies in $\mathcal{FA}$.*
(iii) *The group $B(3, 3)$ lies in $\mathcal{FA}$.*

Some results which hold for solvable groups whose orders are a product of a small number of primes are included in the following theorem.

Theorem 9.4

(i)   *Any group of order pq, where p, q are primes, lies in $\mathcal{FA}$.*
(ii) *Any group of order p^2q, where p, q are primes, lies in $\mathcal{FA}$.*
(iii) *Any group of order pqr, where p, q, r are primes, lies in $\mathcal{FA}$.*
(iv) *Any group of order 2^3q, where $q > 3$ is prime, lies in $\mathcal{FA}$.*

To discuss p-groups further, the following well-known classes are introduced. A p-group is **regular** [134, p. 183] if for all $a, b \in G$ and all $n = p^\alpha$,

$$(ab)^n = a^n b^n s_1^n s_2^n \ldots s_k^n$$

for some $s_1, \ldots, s_k$ in the derived group of $\langle a, b \rangle$. A group which is not regular is said to be **irregular**. A finite p-group G is **powerful** [201, p. 114] if either p is odd and $G' \leq G^p$ or $p = 2$ and $G' \leq G^4$.

A result of [134] states that every regular p-group has a uniqueness basis, i.e. a set $\{a_1, a_2, \ldots, a_\omega\}$ such that if p^{d_i} is the order of a_i, then every element of G can be written uniquely as $a_1^{\varepsilon_1} a_2^{\varepsilon_2}, \ldots, a_\omega^{\varepsilon_\omega}$ where $0 \leq i < d_i$ for all $i \leq \omega$. The list $(d_1, d_2, \ldots, d_\omega\}$ (in increasing order) is the set of e-**invariants** of the regular p-group, this being unique to the group.

The following examples indicate that none of the standard properties of p-groups can serve to describe $\mathcal{FA}$.

Example 9.2 In [158, p. 323] an example is given of a group G of order 3^5 which is regular, but $G \times G$ is not regular. This group lies in $\mathcal{FA}$.

Let $\mathrm{Syl}_p(G)$ denote the Sylow-p-subgroup of G.

Example 9.3 For any prime p the group $\mathrm{Syl}_p(\mathcal{S}_{p^2})$ lies in $\mathcal{FA}$, but is not regular. The group $\mathrm{Syl}_p(\mathcal{S}_{p^2})$ is not powerful if $p > 2$.

Example 9.4 The Blackburn groups [158, p. 334] of order p^{p+1} and of exponent p^2 are all irregular; however, they all lie in $\mathcal{FA}$.

Example 9.5 Any regular p-group with e-invariants $(e, 1, 1)$, $e \geq 1$ lies in $\mathcal{FA}$.

9.2.4 Results on Groups Which Are Not in $\mathcal{FA}$

9.2.4.1 Fusion Invariants

A strategy to show that a given group G is not in $\mathcal{FA}$ is to show that G violates a necessary condition for fusion from an abelian group. The following theorem sets out such a condition.

Theorem 9.5 *Let* $f : A \to G$ *be a fusion of an abelian group A to a group G, where the classes of G are* $C_1, \ldots, C_r$.

(i) *Suppose that there is a class C_i which generates G and a prime p dividing $|G|$ such that* $\overline{C}_i^p = \sum_j \lambda_j \overline{C}_j$; *then* $\lambda_j \equiv 0 \bmod p$.

(ii) *Suppose that there is a prime p dividing G such that G has no normal subgroup of index divisible by p. Suppose further that* $\overline{C}_i^p = \sum_j \lambda_j \overline{C}_j$; *then* $\lambda_j \equiv 0 \bmod p$.

(iii) *Suppose that $C_{i_1}, \ldots, C_{i_r}$ are distinct classes which are in the same orbit under the action of $\mathrm{Aut}(G)$ and let* $\overline{C} = \sum_{k=1}^r \overline{C}_{i_k}$. *Suppose that* $\langle C_{i_1}, \ldots, C_{i_r} \rangle = G$ *and that* $\overline{C}^p = \sum_j \lambda_j \overline{C}_j$. *Then* $\lambda_{i_k} \equiv 0 \bmod p$ *for all* $k = 1, \ldots, r$.

Proof

(i) Since A fuses to G there is a partition $A_1, \ldots, A_r$ of A corresponding to the classes $C_1, \ldots, C_r$. Suppose that C_i generates G, $G = \langle C_i \rangle$, for some $i \leq r$. Then every element of G occurs with nonzero coefficient in some power of C_i. Since $\langle \overline{C}_1, \ldots, \overline{C}_r \rangle \cong \langle \overline{A}_1, \ldots, \overline{A}_r \rangle$, it follows that every element of A occurs as a summand of some power of A_i. Thus $A = \langle A_i \rangle$.

Let $x \in A_i$; then $\overline{A}_i^p = \sum_j \lambda_{ij} \overline{A}_j$ where each λ_{ij} is a nonnegative integer, and so there are λ_{ii} p-tuples $(x_1, \ldots, x_p)$ of elements of A_i such that $x_1 x_2 \ldots x_p = x$. If $\lambda_{ii} = 0$ there is nothing to prove, so assume that $\lambda_{ii} > 0$. Let $\mathfrak{X} = \mathfrak{X}(x) \neq \emptyset$ denote the set of all such p-tuples $(x_1, \ldots, x_p)$. Since A is abelian there is an action of the symmetric group $\mathcal{S}_p$ on $\mathfrak{X}$, where $\sigma \in \mathcal{S}_p$ acts by permuting the subscripts of $(x_1, \ldots, x_p) \in \mathfrak{X}$. This partitions $\mathfrak{X}$ into orbits $O_1, \ldots, O_t$, where each O_j has size $[S_p : H_j]$ for some subgroup H_j of S_p, which is the stabilizer of some $y \in O_j$. Since p is a prime it follows that O_j is divisible by p unless p divides $|H_j|$ in which case $H_j \subset S_p$ contains an element of order p, which is a p-cycle, and it necessarily follows that $O_j = \{(x_1, \ldots, x_1)\}$. Hence $H_j = S_p$ and $x = x_1^p$. Thus in this case A_i consists of pth powers and since p divides $|A| = |G|$, A_i cannot generate A. A contradiction is produced, since C_i cannot generate G. This shows that the case $H_j = S_p$, or equivalently $|O_j| = 1$ can never happen. Thus $|O_j| \equiv 0$ mod p for all j and so $\lambda_{ii} \equiv 0$ mod p and the result (i) follows.

(ii) Following the proof of (i); if $|O_j| = 1$ then also $\langle A_i \rangle$ has index a multiple of p, showing that C_i generates a normal subgroup of index a multiple of p, contradicting the hypothesis. Thus the case $|O_j| = 1$ cannot happen and hence $\lambda_{ii} \equiv 0$ mod p.

(iii) The proof of this again follows the proof of (i) with $\overline{C}$ replaced by $\overline{C} = \sum_{k=1}^r \overline{C}_{i_k}$. The hypothesis implies that each element of $\cup_k C_{i_k}$ has the same order and so the possibility that $|O_j| = 1$ is again ruled out by the assumption that $\langle C_{i_1}, \ldots, C_{i_r} \rangle = G$. This proves (iii) and concludes the proof of the theorem.   □

Example 9.6 Let $G = S_4$ and let $C = C_{(1,2,3,4)}$ denote the class of $(1, 2, 3, 4)$. Then C generates G and $\overline{C}^3 = 20\overline{C} + 16\overline{C}_{(1,2)}$. Applying Theorem 9.5(i), since $20 \equiv 1 \neq 0$ mod 3 it follows that S_4 does not fuse from an abelian group, c.f. Theorem 9.1.

For simple groups there are results as follows.

Theorem 9.6

(a) *For $n \geq 5$ the groups A_n and S_n do not lie in $\mathcal{FA}$.*
(b) *For a prime $p > 3$ the simple group $PSL(2, p)$ does not fuse from an abelian group.*
(c) *None of the sporadic simple groups fuse from an abelian group.*

Cases of free groups in a variety are given in the following theorem.

Theorem 9.7

(i) *The group $F(4, 3, 2)$ does not lie in $\mathcal{FA}$.*
(ii) *The group $F(9, 2, 3)$ does not lie in $\mathcal{FA}$.*

There follows a set of examples of p-groups which are not in $\mathcal{FA}$.

Example 9.7 Let p be a prime and define the p-group G_p by

$$G_p = \langle a, b, c, d, e | a^p, b^p, c^p, d^p, [a, b] = c, [a, c] = d, [a, d], [a, e], [b, c] = e,$$

$$[b, d], [b, e], [c, d], [c, e], [d, e] \; \rangle \; .$$

Then G_3 is of order 3^5 and is not in $\mathcal{FA}$.

Example 9.8 Let

$$G = \langle a, b, c, d, e, f \,|\, a^2, b^2, c^2, d^2, e^2, f^2, [a, b] = d, [a, c] = e,$$

$[b, c] = f$ and d, e, f are central$\rangle$.

Then $|G| = 2^6$ and G does not lie in $\mathcal{FA}$.

The proofs of the above are technical and are given in [152]. The group G in Example 9.8 is interesting in that the conjugacy classes outside G' have order 4 and if for each g in G the set $D(g) = g^{-1}C(g)$ is constructed, then the subsets $D(g)$ may be associated with lines in the Fano plane, and have similar incidences.

Further techniques were developed in [153] to address questions arising in [152]. In the following the notation $D(g) = g^{-1}C(g)$ will be used.

Definition 9.1 A group G is of *subgroup class type* if the following conditions are satisfied.

SCT1 $G' \subseteq Z(G)$.
SCT2 for $g \in G$, $D(g)$ is a subgroup of G'
SCT3 $Z(G)$ and $G/Z(G)$ are elementary abelian p-groups;
SCT4 each class has size 1 or $|G'|/p$; it follows that for all $g, h \in G \backslash Z(G)$ that either $D(g) = D(h)$ or $D(g)D(h) = G'$.
SCT5 for all $g, h \in G \backslash Z(G)$, $D(g) = D(h)$ if and only if $\langle gZG \rangle = \langle hZG \rangle$.

The five conditions above are not independent, but are stated as above for convenience in their use.

Now suppose that G is a group such that $G' \subseteq Z(G)$, which implies that G' is abelian, that G fuses from an abelian group H and that for each $g \in G$, $D(g)$ is a subgroup (necessarily a subgroup of G'). Note that a group of subgroup class type satisfies these requirements.

Under these assumptions, it is easy to show that

$$\text{(i) if } gh \in C(g), \text{ then } D(g) = D(h),$$

and

$$\text{(ii) } \overline{D}(g) \in Z(\mathbb{Z}G).$$

Thus $D(g)$ is a normal subgroup of G which depends only on $C(g)$. Then from (2.1), it follows that H has a subgroup N isomorphic to G'. From now on, N will be identified with G'. From Lemma 9.1, it is also seen that H/N fuses to G/G'. Since G/G' is abelian and H/N fuses to G/G', it follows that $G/G' \cong H/N$. Let $g \in G$ so that $D = D(g)$ is a subgroup of $G' \subseteq ZG$. Using the isomorphism of G' with N, each $D(g)$ will also be considered as a subgroup of N. Then for every $z \in D$,

$$z\overline{C}(g) = gz\overline{D} = g\overline{D} = \overline{C}(g).$$

Let $E(g) \subset H$ be the subset of H corresponding to $C(g)$ under the fusion of H to G. Then for all $z \in S \subseteq N$, it follows that $z\overline{E}(g) = \overline{E}(g)$, and hence $E(g) = dD(g)$ for any $d \in E(g)$.

From the above discussion it is seen that, for every $g \in G$, there is a subgroup $D(g) \leq G$ such that $C(g) = gD(g)$ and $Eg = dD(g)$ for any $d \in E(g)$.

Now write

$$G/G' = \mathbb{Z}_{n_1} \times \mathbb{Z}_{n_2} \times \ldots \times \mathbb{Z}_{n_k}$$
$$= \langle g_1 G' \rangle \times \langle g_2 G' \rangle \times \ldots \times \langle g_k G' \rangle$$
$$\cong \langle h_1 N \rangle \times \langle h_2 N \rangle \times \ldots \times \langle h_k N \rangle.$$

where $g_1, \ldots, g_k \in G$, $h_1, \ldots, h_k \in H$, such that $g_1 G'$ and $h_i N$ have order n_i for $1 \leq i \leq k$. Then for any $0 \leq e_i, f_i < n_i$, $1 \leq i \leq k$, let $a = g_1^{e_1}, \ldots, g_k^{e_k}$, $b = g_1^{f_1}, \ldots, g_k^{f_k}$. Then, for some nonnegative integer ε_z,

$$\overline{C}(a)\overline{C}(b) = \sum_{z \in Z(G)} \varepsilon_z \overline{C}(g_1^{e_1+f_1}, \ldots, g_k^{e_k+f_k} z)$$
$$= \sum_{z \in Z(G)} \varepsilon_z z \overline{C}(g_1^{e_1+f_1}, \ldots, g_k^{e_k+f_k}).$$

Define $E(g_i) = D(g_i) \times h_i N \subset N \times H/N$ for $1 \leq i \leq k$. Now $C(g_i)C(g_j)$ contains $C(g_i g_j)$, which shows that $D(g_i)D(g_j) \times h_i h_j N = E(g_i)E(g_j)$ contains $E(g_i g_j)$. This gives $E(g_i g_j) = zS(g_i g_j) \times h_i h_j N$ for some $z \in G'$. This is the first step of an induction which shows that for any $0 \leq e_i, f_i < n_i$, $1 \leq i \leq k$,

$$E(g_1^{e_1}, \ldots, g_k^{e_k}) = zS(g_1^{e_1}, \ldots, g_k^{e_k}) \times h_1^{e_1}, \ldots, h_k^{e_k} \subseteq N \times H/N,$$

for some $z = z(g_1^{e_1}, \ldots, g_k^{e_k}) \in G'$.

Now let

$$G' = \langle u_1 \rangle \times \langle u_2 \rangle \times \ldots \times \langle u_s \rangle \cong \mathbb{Z}_{m_1} \times \mathbb{Z}_{m_2} \times \ldots \times \mathbb{Z}_{m_s},$$

for $u_i \in G' = N$ and where s is minimal. Then for $1 \leq i \leq k$, it follows that

$$h_i^{n_i} = u_1^{e_{i,1}} u_2^{e_{i,2}} \ldots u_s^{e_{i,s}},$$

where $e_{i,j} \leq m_i$ for $i \leq k$, $j \leq s$. The n_i and $e_{i,j}$ determine the abelian group H. Now the multiplication of classes in G puts restrictions on the integers $e_{i,j}$. For example, $\overline{C}(g_i)^{n_i} = |C(g_i)|^{n_i-1} zS(g_i)$ for some $z = (g_i)^{n_i} \in G'$; it follows from this that

$$h_i^{n_i} = u_1^{e_{i,1}} u_2^{e_{i,2}} \ldots u_s^{e_{i,s}} \in zS(g_i) = g_i^{n_i} S(g_i).$$

More precisely, fix $g \in G$, let $\pi : G' \to G'/D(g)$ be the quotient map, and let

$$Z(G)/D(g) = \mathbb{Z}_{r_1} \times \mathbb{Z}_{r_2} \times \ldots \times \mathbb{Z}_{r_t} = \langle v_1 \rangle \times \langle v_2 \rangle \times \ldots \times \langle v_t \rangle$$

where t is minimal. Then, using the above,

$$1 = \pi(z^{-1}h_i^{n_i}) = (v_1^{\alpha(i,1)} v_2^{\alpha(i,2)} \ldots v_t^{\alpha(i,t)}),$$

where each $\alpha(i, j)$ is a linear polynomial in the $e_{r,s}$. Each of the equations $\alpha(i, j) = 0$ has to be satisfied by the $e_{r,s}$ if H fuses to G, and thus the following proposition holds.

Proposition 9.5 *Let G be a group which fuses from an abelian group and such that $G' \subseteq Z(G)$ and every $D(g)$ is a subgroup. Then the linear equations $\alpha(i, j) = 0$ give necessary conditions for a group of this type to fuse from an abelian group.*

If G is a p-group, and G' and G/G' are elementary abelian, then these equations are all defined over $\mathbb{F}p = \mathbb{Z}/p\mathbb{Z}$. □

Now assume that G is of subgroup class type. Thus it is assumed that

$$G' \subseteq Z(G) \cong \mathbb{Z}_p^s \text{ and } G/Z(G) \cong \mathbb{Z}_p^k.$$

Then

$$n_1 = n_2 = \ldots = n_k = m_1 = m_2 = \ldots = m_s = p$$

and the exponents $e_{i,j}$ can be thought of as indeterminates taking values in $\mathbb{F}_p$. Now suppose that $g = g_1^{e_1} \ldots g_k^{e_k}$ and $g \notin Z(G)$. Then from the above discussion, there is some $z = z(g) \in Z(G)$ such that $D(g) = h_1^{e_1} \ldots h_k^{e_k} z S(g)$.

Let $z_g = g^p \in ZG$, and note that

$$\overline{C}(g)^p = (g\overline{D}(g))^p = g^p \overline{D}(g)^p = z_g \overline{D}(g)^p.$$

Thus $\overline{E}(g)^p = z_g \overline{D}(g)^p$, and since $z^p = 1$, it follows that

$$
\begin{aligned}
\overline{E}(g)^p &= (h_1^{e_1} \ldots h_k^{e_k} z \overline{D}(g))^p \\
&= h_1^{pe_1} \ldots h_k^{pe_k} \overline{D}(g)^p \\
&= (u_1^{e_1,1e_1} u_2^{e_1,2e_1} \ldots u_s^{e_1,s e_1})(u_1^{e_2,1e_2} u_2^{e_2,2e_2} \ldots u_s^{e_2,s e_2}) \ldots \\
&\quad (u_1^{e_k,1e_1} u_2^{e_k,2e_1} \ldots u_s^{e_k,1e_s})\overline{S}(g)^p \\
&= (u_1^{\sum_i e_{i,1}e_i} u_2^{\sum_i e_{i,2}e_i} \ldots u_s^{\sum_i e_{i,s}e_i})\overline{D}(g)^p \\
&= z_g \overline{D}(g)^p.
\end{aligned}
$$

Since $\overline{D}(g)^p = |D(g)|^{p-1}\overline{D}(g)$ and z_g can be expressed as $z_g = u_1^{f_1} \ldots u_k^{f_k}$, this determines a relationship

$$u_1^{-f_1 + \sum_i e_{i,1}e_i} u_1^{\sum_i -f_2 + e_{i,2}e_i} \ldots u_s^{\sum -f_s + e_{i,s}e_i} = z_g^{-1} u_1^{\sum_i e_{i,1}e_i} u_1^{\sum_i e_{i,2}e_i} \ldots u_s^{\sum_i e_{i,s}e_i}$$

which lie in $D(g)$. Now by hypothesis (SCT4), $G'/D(g) \cong \mathbb{Z}_p$, and this latter relationship is uniquely determined by a single linear equation (over $\mathbb{F}_p$) in the $e_{i,j}$, denoted by $\alpha(g) = 0$. It follows that the equations $\alpha(g)$ and $\alpha(gz)$ are the same for $z \in Z(G)$; denote this equation by $\alpha(gZ(G))$. Thus there is one equation for every element of $G/Z(G)$. In fact $\alpha(g^e)$ is equal to an $\mathbb{F}_p$-multiple of $\alpha(g)$ for any e with $\gcd(e, p) = 1$. Thus if the k-tuples $(e_1, \ldots, e_k)$ are thought of as exponents of elements in the projective space $\mathbb{PF}_p^k$ then to each element of $\mathbb{PF}_p^k$ there corresponds a linear equation.

The above discussion leads to

Theorem 9.8 *A group G of subgroup class type lies in $\mathcal{FA}$ if and only if there is a simultaneous solution in $\mathbb{F}_p$ to all the equations $\alpha(gZ(G)) = 0$.*

The proof of Theorem 9.8 is technical, and the reader is referred to [153] for the details. There is also the following corollary.

Corollary 9.1 *Let G be a p-group of subgroup class type such that G has exponent p. Then G fuses from an abelian group.*

9.3   Groups Which Fuse from a Cyclic Group

Camina triples are essential for the classification (see Definition 5.7, Chap. 5). The following conditions also appear. Suppose that G fuses from the cyclic group C_n. If N is a normal subgroup of G the classes of G contained in N form an S-ring on N. This will be denoted by $\mathbf{F}_G(N)$.

(FC1)   For every normal subgroup N of G, there is normal subgroup $H(N)$ of C_n and an S-ring $S(N)$ on $H(N)$ which is isomorphic to $\mathbf{F}_G(N)$.

(FC2)   For every normal subgroup N of G, there is normal subgroup $H(N)$ of C_n such that $C_n/H(N)$ fuses to G/N.

If N is a normal subgroup G then it will be said that N satisfies (FC1)(N) if there is normal subgroup $H(N)$ of C_n and an S-ring $S(N)$ on $H(N)$ which is isomorphic to $\mathbf{F}_G(N)$. The condition (FC2)(N) is similarly defined. From (FC2) it follows that G has cyclic abelianization. Let the class algebra S-ring over G be denoted by $\mathbf{S}$ and let the corresponding S-ring over C_n be denoted by $\mathbf{S}^*$. Thus $\mathbf{S}$ and $\mathbf{S}^*$ are isomorphic S-rings. The classification of S-rings over cyclic groups given in [203, Theorem 3.7] shows that any S-ring over C_n is one of four types:

Type 0:   the S-ring $\mathbf{S}^*$ is trivial.

Type 1: there is a subgroup $\Omega \subseteq \text{Aut}(\mathcal{C}_n)$ such that the basic sets of $\mathbf{C}^*$ are the orbits of Ω.

Type 2: there are non-trivial subgroups H, K of $\mathcal{C}_n$ such that $\mathcal{C}_n = H \times K$ and there are S-rings $\mathbf{S}_H, \mathbf{S}_K$ on H and K (respectively) such that $\mathbf{S}^* = \mathbf{S}_H \times \mathbf{S}_K$; where in $\mathbf{S}_H \times \mathbf{S}_K$ the basic sets are (U, V) where U is a basic set of $\mathbf{S}_H$ and V is a basic set of $\mathbf{S}_K$.

The definition of a Type 3 S-ring is implicitly given in [203, p. 253]. Let H be a subgroup of G and let K be a normal subgroup of G; let $\pi : G \to G/K$ be the quotient map. Let $\mathbf{S}_H$ be an S-ring over H and assume that $\overline{K} \in \mathbf{S}_H$. The Lemma below defines a quotient S-ring:

Lemma 9.5 ([203]) *Suppose that* $K \lhd H$ *and let* $\mathbf{S}_H$ *be an S-ring over* H *with* $\overline{K} \in \mathbf{S}_H$. *Then*

$$\pi * (\mathbf{S}_H) = \oplus_{D \in \mathcal{D}(S_H)} \mathbb{Z}\pi \left(\overline{D}\right)$$

is an S-ring over H/K *satisfying* $\mathcal{D}(\pi * (\mathbf{S}_H)) = \{\pi(D) : D \in \mathcal{D}(S_H)\}$.

Lemma 9.6 ([203]) *Let* $\mathbf{S}_{G/K}$ *be an S-ring over* G/K *and suppose that* H/K *is an* $\mathbf{S}_{G/K}$ *subgroup. Suppose also that* K *is an* $\mathbf{S}_H$-*subgroup and that* $\pi^*(\mathbf{S}_H) = (\mathbb{Z}[H/K]) \cap \mathbf{S}_{G/K}$. *Then there is an S-ring* $\mathbf{S}$ *over* G *such that*

$$\mathcal{D}(\mathbf{S}) = \mathcal{D}(\mathbf{S}_H) \cup \{\pi^{-1}(E) : E \in \mathcal{D}(\mathbf{S}_{G/K}); E \not\subseteq H/K\} \qquad (9.3.1)$$

Further, $\mathbf{S} \cap Z[H] = \mathbf{S}_H$ *and* $\pi^*(\mathbf{S}) = \mathbf{S}_{G/K}$.

The S-ring constructed in Lemma 9.6 is called the **wedge product** of $\mathbf{S}_H$ and $\mathbf{S}_{G/K}$ and is denoted by $\mathbf{S}_H \wedge \mathbf{S}_{G/K}$. These S-rings are said to be of type 3.

The following holds.

Lemma 9.7 *If* G *fuses from the cyclic group* $\mathcal{C}_n$ *and the S-ring* $\mathbf{S}$ *over* $\mathcal{C}_n$ *is a wedge product, then* G *is a Camina triple group.*

Proof Let H, K be the subgroups of $\mathcal{C}_n$ that determine the wedge product S-ring (as constructed above), denoted by $\mathbf{S}$. Then $\overline{H}, \overline{K} \in \mathbf{S}$. Let H_1, K_1 respectively be the subsets of G that correspond to H, K. It will be shown that (G, H_1, K_1) is a Camina triple. Since H is a subgroup, $\overline{H}^2 = |H|\overline{H}$ and hence $\overline{H_1}^2 = |H_1|\overline{H_1}$, showing that H_1 is a (normal) subgroup of G. Similarly K_1 is a normal subgroup of G. Let $\pi : \mathcal{C}_n \to \mathcal{C}_n/K$ be the projection. Now from Eq. (9.3.1) it follows that any basic set $E \in D(\mathbf{S})$ which is not in H has the form $\pi^{-1}(D)$ for some $D \in \mathcal{D}(\mathbf{S}_{G/K})$; but any such set is a union of cosets of K. This implies that $\overline{E}\,\overline{K} = |K|\overline{E}$. Let $E_1 \subset G$ correspond to E. Then $\overline{E_1}\,\overline{K_1} = |K|\overline{E_1} = |K_1|\overline{E_1}$. From Lemma 2.6 it follows that E_1 is a union of cosets of K_1. Thus every basic set outside of H_1 is a union of cosets of K_1 and so (G, H_1, K_1) forms a Camina triple. $\qquad \square$

The classification of groups which fuse from cyclic groups is contained in the following theorem:

Theorem 9.9 *Suppose that G is a non-cyclic group. Then G fuses from a cyclic group if and only if there is a Camina triple (G, H, K) such that $(FC1)(H)$ and $(FC2)(K)$ hold.*

Some properties of groups which fuse from cyclic groups are laid out in the following theorem:

Theorem 9.10 *Assume that G fuses from the cyclic group C_n. Then:*

(i) *The lattice of normal subgroups of G is lattice-isomorphic to a sublattice of the lattice of (normal) subgroups of C_n. Further, distinct normal subgroups of G must have different orders.*

(ii) *If G is nilpotent, then $G \cong C_n$. In particular, if G is a p-group, then $G \cong C_n$.*

(iii) *There is at least one conjugacy class C of G such that $\langle C \rangle = G$.*

(iv) *The group G is solvable.*

For the proofs of the above the reader is referred to [156].

9.4 Supercharacters of Upper Triangular Groups

A particular case of the fusion of the characters of a group with interesting connections to Hopf algebras has appeared in the context of random walks on upper triangular groups.

Let $UT_n(q)$ be the group of $n \times n$ upper triangular matrices with entries in the finite field $\mathbb{F}_q$ and ones on the diagonal. This group is a Sylow p-subgroup of $GL_n(q)$. The problem of describing the conjugacy classes or characters of $UT_n(q)$ is "wild", making it difficult to analyse random walks. In a series of papers, a theory of fused classes was developed by André, the fused classes and characters being described respectively as **superclasses** and **supercharacters**. The resulting theory is well-behaved in that there is a rich combinatorics describing induction and restriction along with an elegant formula for the values of supercharacters on superclasses. The combinatorics is described in terms of set partitions (the symmetric group theory involves integer partitions) and the combinatorics seems akin to tableau combinatorics. At the same time, this fused character theory is rich enough to serve as a substitute for ordinary character theory in some problems (see [5]).

The superclasses are defined as follows. The group $UT_n(q)$ acts on both sides of the algebra of strictly upper triangular matrices $\mathfrak{n}_n$ (which can be thought of as $UT_n(q) - I$). The two sided orbits on $\mathfrak{n}_n$ can be mapped back to $UT_n(q)$ to form "superclasses" by adding the identity matrix. The element sums of these orbits form an S-ring which is a subring of the class algebra of $UT_n(q)$. A similar construction on the dual space $\mathfrak{n}_n^*$ gives a collection of class functions on $UT_n(q)$ which are constant on superclasses, i.e. there is a fusion of the corresponding S-ring.

A normalized version of these orbit sums are the supercharacters, see Chap. 4, Sect. 4.6. Let

$$\mathbf{SC} = \oplus_{n \geq 0} \mathbf{SC}_n$$

where $\mathbf{SC}_n$ is the set of functions from $UT_n(q)$ to $\mathbb{C}$ that are constant on superclasses, and $\mathbf{SC}_0 = \mathbb{C}\text{-span}\{1\}$ is by convention the set of class functions of $UT_0(q) = \{\}$.

The superclasses corresponding to $\mathbf{SC}_n$ may be defined combinatorially. They correspond to certain elements of $\mathfrak{n}_n$ with at most one nonzero entry in each row and column. Every superclass contains a unique such matrix, obtained by a set of elementary row and column operations.

When $n = 3$, there are five such patterns; with $* \in \mathbb{F}_q^*$

$$\begin{bmatrix} 0 & 0 & 0 \\ 0 & 0 & 0 \\ 0 & 0 & 0 \end{bmatrix}, \begin{bmatrix} 0 & * & 0 \\ 0 & 0 & 0 \\ 0 & 0 & 0 \end{bmatrix}, \begin{bmatrix} 0 & 0 & 0 \\ 0 & 0 & * \\ 0 & 0 & 0 \end{bmatrix}, \begin{bmatrix} 0 & 0 & * \\ 0 & 0 & 0 \\ 0 & 0 & 0 \end{bmatrix}, \text{and} \begin{bmatrix} 0 & * & 0 \\ 0 & 0 & * \\ 0 & 0 & 0 \end{bmatrix}.$$

Each representative matrix X can be encoded as a pair (D, ϕ), where

$$D = \{(i, j) | X_{ij} \neq 0\}$$

and $\phi : D \to F_q^*$ is given by $\phi(i, j) = X_{ij}$. There is a slight abuse of notation here since the pair (D, ϕ) does not record the size of the matrix X. Let $X_{D,\phi}$ denote the distinguished representative corresponding to the pair (D, ϕ), and let $\kappa_{D,\phi} = \kappa_{X_{D,\phi}}$ be the function that is 1 on the superclass and zero elsewhere.

The superclasses form a Hopf algebra. Combinatorial expressions for the product and coproduct are as follows. The product is given by

$$\kappa_{X_{D,\phi}} \cdot \kappa_{X_{D',\phi'}} = \sum_{X'} \kappa'_{\begin{bmatrix} X_{D,\phi} & X' \\ 0 & X'_{D',\phi'} \end{bmatrix}}, \qquad (9.4.1)$$

where the sum runs over all ways of placing a matrix X' into the upper right-hand block such that the resulting matrix still has only one non-zero entry in each row and column. This is different from the pointwise product of class functions, which is internal to each $\mathbf{SC}_n$ and does not turn $\mathbf{SC}$ into a graded algebra.

Example 9.9 If $(D, \phi) = (\{\}, \phi) \longleftrightarrow \begin{bmatrix} 0 & 0 \\ 0 & 0 \end{bmatrix}$, and

$$(D', \phi') = (\{1, 2\}, \{2, 3\}, \{\phi'(1, 2) = a, \phi'(2, 3) = b\}) \longleftrightarrow \begin{bmatrix} 0 & a & 0 \\ 0 & 0 & b \\ 0 & 0 & 0 \end{bmatrix},$$

where sizes of the matrices are 2 and 3 respectively, then

$$
\kappa_{X_{D,\phi}} \cdot \kappa_{X_{D',\phi'}} = \kappa
\begin{bmatrix}
0 & 0 & 0 & 0 & 0 \\
0 & 0 & 0 & 0 & 0 \\
0 & 0 & 0 & a & 0 \\
0 & 0 & 0 & 0 & b \\
0 & 0 & 0 & 0 & 0
\end{bmatrix}^{+}
$$

$$
\sum_{c \in F_q^*} (\kappa
\begin{bmatrix}
0 & 0 & c & 0 & 0 \\
0 & 0 & 0 & 0 & 0 \\
0 & 0 & 0 & a & 0 \\
0 & 0 & 0 & 0 & b \\
0 & 0 & 0 & 0 & 0
\end{bmatrix}^{+}
\kappa
\begin{bmatrix}
0 & 0 & 0 & 0 & 0 \\
0 & 0 & c & 0 & 0 \\
0 & 0 & 0 & a & 0 \\
0 & 0 & 0 & 0 & b \\
0 & 0 & 0 & 0 & 0
\end{bmatrix}).
$$

The coproduct on $\mathbf{SC}_n$ is defined as

$$
\Delta(\kappa_{X_{D,\phi}}) = \sum_{\substack{[n]=S \cup S^C \\ (i,j) \in D \text{ only if} \\ i,j \in S \text{ or } i,j \in S^C}} \kappa_{(X_{D,\phi})_S} \otimes \kappa_{(X_{D,\phi})_{S^C}}, \tag{9.4.2}
$$

where X_S is the matrix restricted to the rows and columns of S.

Example 9.10 Let $D = \{(1,4),(2,3)\}$, $\phi(2,3) = a$, and $\phi(1,4) = b$. Then

$$
\Delta(\kappa_{X_{D,\phi}}) = \kappa_{X_{D,\phi}} \otimes 1 + \kappa
\begin{bmatrix} 0 & a \\ 0 & 0 \end{bmatrix} \otimes \kappa
\begin{bmatrix} 0 & b \\ 0 & 0 \end{bmatrix} + \kappa
\begin{bmatrix} 0 & b \\ 0 & 0 \end{bmatrix} \otimes \kappa
\begin{bmatrix} 0 & a \\ 0 & 0 \end{bmatrix} + 1 \otimes \kappa_{X_{D,\phi}}.
$$

With the product and coproduct defined in (9.4.1) and (9.4.2) the space $\mathbf{SC}$ becomes a Hopf algebra. It is graded, noncommutative and cocommutative. It has a unit $\kappa_\emptyset$ in $\mathbf{SC}_0$ and a counit $\varepsilon : \mathbf{SC} \to \mathbb{C}$ obtained by taking the coefficient of $\kappa_\emptyset$.

9.4.1 Symmetric Functions in Non-commuting Variables

The notation will be consistent to that in Chap. 6. Let $\lambda \vdash [n] = \{1,2,\ldots,n\}$. A monomial of shape λ is a product of noncommuting variables $a_1 a_2 \ldots a_k$, where variables are equal if and only if the corresponding indices/positions are in the same block/part of λ. For example, if $\lambda = 135|24$ denotes the partition of $[5]$ with parts $\{1,3,5\}$ and $\{2,4\}$ then $xyxyx$ is a monomial of shape λ. Let m_λ be the sum of all monomials of shape λ. Thus, with three variables

$$
m_{135|24} = xyxyx + yxyxy + xzxzx + zxzxz + yzyzy + zyzyz.
$$

It is usual to work with an infinite number of variables. If Π_n is the $\mathbb{C}$-span of $\{m_\lambda \mid \lambda \vdash [n]\}$ the set Π is defined by

$$\Pi = \bigoplus_{n \geq 0} \Pi_n.$$

The elements of Π are called the symmetric functions in non-commuting variables. They are sometimes denoted by NCSym or WSym. They are not to be confused with those in [120]. See [245] for an introduction to them. The algebra Π is a Hopf algebra and is actively studied as part of the theory of combinatorial Hopf algebras there are many references in [2].

The product and coproduct in Π are as follows. Suppose that $\lambda \vdash [k]$ and $\mu \vdash [n-k]$ with $\lambda = \lambda_1 | \lambda_2 | \ldots | \lambda_a$ and $\mu = \mu_1 | \mu_2 | \ldots | \mu_b$. Consider the poset of partitions of the set $[n]$ under refinement. Define the partition $\lambda | \mu$ as the following partition of $[n]$

$$\lambda_1 | \lambda_2 | \ldots \lambda_a | \mu_1 + k | \mu_2 + k | \ldots \mu_2 + k.$$

Thus if $\lambda = 1|2$ and $\mu = 123$, then $\lambda|\mu = 1|2|345$. Then the product

$$m_\lambda m_\mu = \sum_{\substack{\nu \vdash [n] \\ \nu \wedge ([k]|[n-k]) = \lambda|\mu}} m_\nu. \tag{9.4.3}$$

The coproduct is defined by

$$\Delta(m_\lambda) = \sum_{J \subseteq [l(\lambda)]} m_{st(\lambda_J)} \otimes m_{st(\lambda_{J^c})} \tag{9.4.4}$$

where λ has $l(\lambda)$ parts, $\lambda_J = \{\lambda_j \in \lambda \mid j \in J\}$, such that $J \to [|J|]$ is the unique order preserving bijection, and $J^c = [l(\lambda)] \backslash J$. Thus

$$\Delta(m_{14|2|3}) = m_{14|2|3} \otimes 1 + 2m_{13|2} \otimes m_1 + m_{12} \otimes m_{1|2} + m_{1|2} \otimes m_{12} + 2m_1 \otimes m_{13|2}$$

$$+ 1 \otimes m_{14|2|3}.$$

For $q = 2$ there is a simplified version of the basic theorem.

Theorem 9.11 *For $q = 2$, the function*

$$ch : \mathbf{SC} \to \Pi$$

defined by

$$\kappa_\mu \to m_\mu$$

is a Hopf algebra isomorphism.

The version of the theorem for $q > 2$ is more technical and is stated as follows.

Theorem 9.12 *The map*

$$ch : \mathbf{SC} \to \Pi^{(q-1)}$$

defined by

$$\kappa_\mu \to \kappa_{(D_\mu, \varphi_\mu)}$$

is an isomorphism of Hopf algebras. In particular, $\mathbf{SC}$ *is a Hopf algebra for any* q.

Here $\Pi^{(q-1)}$ is a "colored" version of Π. For the details the reader is referred to [2].

9.5   A Fusion Using Nonassociative Algebra

The following is a description of a fusion of the group $UT_4(q)$ arising from a loop construction. The details appear in [185].

The group elements may be written in the form $I + U_4$ where U_4 is the set of upper triangular matrices with entries in the finite field F. Define the following multiplication. Given elements x and y,

$$x = \begin{bmatrix} 1 & x_{12} & x_{13} & x_{14} \\ 0 & 1 & x_{23} & x_{24} \\ 0 & 0 & 1 & x_{34} \\ 0 & 0 & 0 & 1 \end{bmatrix}, \quad y = \begin{bmatrix} 1 & y_{12} & y_{13} & y_{14} \\ 0 & 1 & y_{23} & y_{24} \\ 0 & 0 & 1 & y_{34} \\ 0 & 0 & 0 & 1 \end{bmatrix}$$

with entries x_{ij}, y_{ij} in F, let

$$x.y = \begin{bmatrix} 1 & x_{12} + y_{12} & x_{13} + y_{13} + x_{12}y_{23} & [x.y]_{14} \\ 0 & 1 & x_{23} + y_{23} & x_{24} + y_{24} + x_{23}y_{34} \\ 0 & 0 & 1 & x_{34} + y_{34} \\ 0 & 0 & 0 & 1 \end{bmatrix}, \quad (9.5.1)$$

where

$$[x.y]_{14} = x_{14} + y_{14} + x_{12}y_{24} + x_{13}y_{34} + x_{12}x_{23}y_{34} + x_{12}y_{23}y_{34}. \quad (9.5.2)$$

Note that under ordinary matrix multiplication the elements are the same as in (9.5.1) except that in the $(4, 4)$ position

$$[xy]_{14} = x_{14} + y_{14} + x_{12}y_{24} + x_{13}y_{34}.$$

The summands in (9.5.2) correspond to paths of respective lengths 1; 2; 3 from 1 to 4 in the chain $1 < 2 < 3 < 4$, with labels chosen from x over the former part of the path, and y over the latter part. The other entries in the product have a similar structure. Under the multiplication (9.5.1) $I + U_4$ forms a loop, UL_4 with identity element I_4, the usual identity for multiplication. Denote UL_4 as the **upper triangular loop of degree** 4. The following hold:

1. UL_4 is neither commutative nor associative.

    Recall that the center of a loop is the set of elements which commute and associate with all elements of the loop.

2. The center $Z = Z(UL_4)$ of UL_4 is $\{x \mid x_{ij} = 0$ if $1 \le j - i < 3\}$ (the set of matrices with zero off the diagonal except for the $(4, 4)$ position).

    The definition of nilpotency for a loop Q is as follows. The ascending central series is

$$Z_0(Q) \le Z_1(Q) \le \cdots \le Z_r(Q) \le \ldots$$

where $Z_0(Q) = \{1\}$ and $Z_{r+1}(Q)/Z_r(Q) = Z(Q/Z_r(Q))$. The loop Q is nilpotent (of class at most c) if $Z_c(Q) = Q$.

3. UL_4 is nilpotent of class 3.

4. The conjugacy classes are as summarized in the following table

Type of element	Size of class	Number of classes

0 0 $*$
  0 0
    0
	1	q

0 $*$ F
  0 $*$
    0
	q	$q^2 - 1$

$*$ F F
  $* \neq 0$ F
    $*$
	q^3	$q^2(q - 1)$

$*$ F F
  0 F
    $*$
	q^2	$q(q^2 - 1)$

5. The classes of UL_4 form a fusion of the conjugacy class algebra of $UT_4(q)$. This is finer than the fusion into the superclasses defined above. In fact there is a superclass of $UT_4(q)$ which is of size $|F|^3$.

A similar construction for a non-associative loop UL_n may be given in the case of $UT_n(q)$ which will again give a fusion of the class algebra of $UT_n(q)$ which is finer that the superclass algebra. Some questions which arise are the following.

1. Are the classes of the UL_n wild?

2. If not, is there a connection with Hopf algebras and symmetric functions similar to that for the superclass algebra?

Chapter 10
Other Situations Involving Group Matrices

Abstract This chapter sketches various situations in which the tools described in this book have appeared. They are:

1. The characterization of those class functions on a group which are group characters.
2. The use of group matrices in the theory of group rings.
3. The application of group matrices to the theory of cogrowth of groups.
4. Work of Poincaré on differential equations.
5. The connection between factors of the group determinant and Chern classes.
6. The appearance of the group matrix as a Gram matrix for tight frames with a symmetry group, which have appeared in the theory of wavelets.
7. The theory of 3-manifolds.
8. Control theory and electrical engineering.
9. The algebra defined by considering the k-classes under multiplication.

10.1 Introduction

In this chapter, short descriptions appear of some of the areas where group matrices, group determinants and related objects have arisen which were not covered in previous chapters. It will be seen that in some cases the original authors were not aware of the connection with Frobenius's work.

A brief list of these occurrences follows.

1. The characterization of characters (for groups and algebras).
2. The properties of group matrices and connections to the theory of group rings.
3. The cogrowth of a group presentation in combinatorial group theory.
4. Poincaré's work on differential equations.
5. The Chern classes of a group representation.
6. Wavelets.

© Springer Nature Switzerland AG 2019
K. W. Johnson, *Group Matrices, Group Determinants and Representation Theory*, Lecture Notes in Mathematics 2233, https://doi.org/10.1007/978-3-030-28300-1_10

314 10 Other Situations Involving Group Matrices

7. Three-manifolds and the virtual Haken conjecture.
8. Electrical engineering and control theory.
9. Higher S-rings.

In Sect. 10.2 the result of Helling in [141] is described. It gives a criterion for deciding whether a class function on a group is the character of a representation. The criterion in the paper can be naturally expressed in terms of k-characters but no explicit connection is made in Helling's work. The result also appeared in a more general setting in [275], again with no connection made with group determinants, although subsequent papers arising from [275] did make the connection. The proof in [172] is more directly derived from the group determinant work and is outlined. The applications in [275] are indicated.

In the theory of group rings, the advantages of using group matrices was pointed out by Taussky-Todd, and more recently Hurley has expanded on this, in particular discussing the case when the group is infinite. This work is described in Sect. 10.3.

In Sect. 10.4 the discussion by Humphries on the cogrowth of a group presentation in combinatorial group theory is given. This is related to amenability in a group. As a tool Humphries introduced a group matrix of the form $X_G(\underline{Z})$ where $\underline{Z}$ is a set of matrices.

The work of Poincaré on differential equations was mentioned at the end of Sect. 7.3. The relevant parts are indicated in Sect. 10.5. There is also given an account of some of his ideas in his paper on Kleinian groups, where he proposes invariants of groups which are closely related to k-characters.

In Sect. 10.6 the Chern classes of a representation of a group are discussed. Although a considerable amount of work has been devoted to them they seem to be hard to calculate. A long-standing conjecture of the author that these Chern classes may be calculated via the group determinant is set out.

In the theory of 3-manifolds Cooper and Walsh obtain a polynomial which gives the rank of the part of the homology carried by the solid tori used for Dehn-filling. This is obtained from the symmetrized form of a group determinant. An account of this is given in Sect. 10.7.

The theory of wavelets now has applications in many areas. A branch of the theory uses a redundant set of vectors in a vector space which has a finite symmetry group G. The "Gram matrix" attached to the set is a group matrix for G. Moreover the projective group matrices described in Chap. 2 are also useful in this context. Sect. 10.8 provides a description of this.

In Chap. 7 the Fourier transform of a function on a group has been introduced in the context of probability. It appears in many other applications. In particular, papers in control theory and other areas related to electrical engineering have brought in the group matrix, sometimes with entries in a finite field. A somewhat superficial summary of this work is given in Sect. 10.9 with references to some of the rather extensive literature.

Finally, in Sect. 10.10 work of Humphries and Rode (Turner) is described on the k-class algebra of a group which sheds more light on the questions of Brauer discussed in Chap. 5.

10.2 Characterization of Group Characters

Let G be a finite group and K be a field. As has already been defined, a class function $\psi : G \to K$ is any function satisfying $\psi(gh) = \psi(hg)$ for all pairs g, h in G, or equivalently any trace-like function on KG. A character is usually defined to be the trace of a representation $\rho : G \to GL(m, K)$ for some m, with several equivalent definitions being given in Sect. 7.3. It is a natural question to ask whether there is a way to distinguish those class functions which are characters.

The theorem often described as "Brauer's characterization of characters" [68, p. 269] provides a criterion for a class function to be a generalized character (i.e. a linear combination $\sum r_i \chi_i$ where the χ_i are irreducible characters and the r_i are integers which are not restricted to be non-negative).

Brauer's result may be used to determine whether a class function is an irreducible character since if ψ is a generalized character which satisfies (a) $\psi(e) > 0$ and (b) $\langle \psi, \psi \rangle = 1$, then it is necessarily an irreducible character, but this does not provide an answer to the above question.

It appears to be less well-known that the question has an answer in the case of K algebraically closed of characteristic 0. This has appeared in several papers. It first appeared in the paper quoted above by Helling. It also appeared in work of Wiles and Taylor leading up to the proof of the Fermat theorem [275] where the result is generalized to finite-dimensional representations of infinite groups and monoids. It appeared in the work of Buchstaber and Rees described in Chap. 8. Neither Helling nor Taylor relate the criterion which they obtain to the work of Frobenius, but papers extending Taylor's results do make the connection. The basic theorem is

Theorem 10.1 *A necessary and sufficient condition for a class function f with values in an algebraically closed field K of characteristic 0 to be a character of a finite dimensional representation of a group G is that $\delta^r f = 0$ for some r. The smallest such r is the degree of the corresponding representation.*

The proof in [275] uses a deep result in invariant theory and that in [141] relies on a complicated combinatorial argument.

In [172] the author and Poimenidou gave a more basic proof of the criterion. Given a class function f a formal power series can be produced (essentially using the methods of Frobenius) whose coefficients can be calculated from $\delta^k f$, $k = 1, 2, \ldots$. This series becomes a polynomial if and only if f is a character. The result is given there for $K = \mathbb{C}$, but the proof is essentially the same for any algebraically closed field of characteristic 0.

10.2.1 The Formal Power Series Attached to a Class Function

Consider a class function ψ on a finite group G with values in $\mathbb{C}$. As is well known there is the following expression for ψ

$$\psi = a_1 \chi_1 + a_2 \chi_2 + \cdots + a_t \chi_t$$

with $a_i \in \mathbb{C}$ and $\chi_i \in Irr(G)$. Associate to ψ the product

$$\phi_1(\underline{x})^{a_1}\phi_2(\underline{x})^{a_2}\ldots\phi_t(\underline{x})^{a_2}$$

where $\phi_i(\underline{x})$ is the factor of Θ_G which corresponds to the character χ_i. There is then a well-defined Taylor series associated to $\psi(\underline{x}+\underline{\varepsilon})$ about the point $(0,0,\ldots,0)$.

For example let $G = C_2 = \{e, g\}$ and ψ be the class function given by $\psi(e) = 0$, $\psi(g) = 2$. The irreducible characters are χ_1 and χ_2 whose vectors of values on (e, g) are $(1, 1)$ and $(1, -1)$ respectively. Then $\psi = \chi_1 - \chi_2$. If x_e is denoted by y and x_g is denoted by z, the irreducible factors of Θ_G are $y + z$ and $y - z$, and $\psi(\overline{x}+\overline{\varepsilon}) = (1 + y + z)(1 + y - z)^{-1}$. The Taylor series expansion as far as the terms of degree 4 is

$$1+2yz-2zy+2z^2+2zy^2-4z^2y+2z^3-2zy^3+6z^2y^2-6z^3y+2z^4+\ldots \quad (10.2.1)$$

It is shown in [172] that the coefficients of this power series may be calculated by the same formulae which Frobenius gave for the calculation of those of the factors of the group determinant. For example, on writing ψ^3 for $\delta^3\psi$ and assuming the Frobenius formula may be used, the coefficient of y^2z is given by

$$\frac{1}{3!}(\psi^3(g, e, e) + \psi^3(e, g, e) + \psi^3(e, e, g)) = \frac{1}{2}\psi^3(g, e, e),$$

where the symmetry of the formula for $\delta^3\psi$ has been used. By the usual formula

$$\psi^3(e, e, g) = \psi(e)^2\psi(g) - 3\psi(e)\psi(g) + 2\psi(g) = 4.$$

Thus the coefficient of y^2z is calculated as 2, which is correct as seen in (10.2.1). The following lemma is an important part of the proof:

Lemma 10.1 *For arbitrary $a_1, a_2, \ldots a_t$ in $\mathbb{C}$ the expansion of*

$$\Phi_{\psi,t}(\overline{x}+\overline{\varepsilon}) = \phi_1^{a_1}(\overline{x}+\overline{\varepsilon})\phi_2^{a_2}(\overline{x}+\overline{\varepsilon})\ldots\phi_t^{a_t}(\overline{x}+\overline{\varepsilon})$$

as a formal power series about $(0, 0, \ldots, 0)$ in $x_{g_1}, x_{g_2}, \ldots, x_{g_n}$ is

$$1 + s_{\underline{a},1} + s_{\underline{a},2} + \ldots + s_{\underline{a},r} + \ldots$$

where $\underline{a} = (a_1, a_2, \ldots, a_t)$ and

$$s_{\underline{a},r} = \frac{1}{r!}\sum_{g_{j_1},g_{j_2},\ldots,g_{j_r}\in G}\delta^r(\sum_{k=1}^{t}a_k\chi_k)(g_{j_1}, g_{j_2}, \ldots, g_{j_r})x_{g_{j_1}}x_{g_{j_2}}\ldots x_{g_{j_r}}.$$

10.2.2 Pseudorepresentations

In [275] Taylor defined a "pseudorepresentation" of a group G. His definition is equivalent to that of an n-homomorphism from G to a commutative ring given in Chap. 8. His work extends that of Wiles in [296]. He proves the following theorem (the group G is not necessarily finite).

Theorem 10.2

(1) *Let R be a ring and $\rho : G \to GL_d(R)$ be a representation. Then $Tr(\rho)$ is a pseudorepresentation of dimension d and if R is a field of characteristic 0 then $ker Tr(\rho)$ is the kernel of the semisimplification of ρ.*

(2) *If R is an algebraically closed field of characteristic 0 and $T : G \to R$ is a pseudorepresentation of dimension d then there is a genuine semisimple representation $\rho : G \to GL_d(R)$ with $Tr(\rho) = T$ with ρ unique up to conjugation.*

(3) *If G is a finitely generated group and d a positive integer, then there is a finite subset $S \subset G$ such that for any $\mathbb{Z}[1/d!]$ algebra R a pseudorepresentation $T : G \to R$ of dimension d is determined on its values on S.*

(4) *If G and R are assumed to be topological, then the results of the first three parts remain true with continuous pseudorepresentation (resp. continuous representation, resp topologically finitely generated) replacing pseudorepresentation (resp. representation, resp. finitely generated).*

It may be seen that Theorem 10.2 is a generalization of the result given above on characterization of characters. The proof is given using results in the invariant theory of matrices, and Taylor states that the theorem may be obvious to anyone sufficiently familiar with invariant theory, but his proof seems much less elementary than the proof in [172] (for finite groups) in terms of factors of the group determinant.

Taylor gives the following example to illustrate the application of Theorem 10.2.

Example 10.1 Let K be a number field and $\mathbb{S}$ be a set of places of K which contains all the finite places. Let l be a finite prime and let d and e be positive rational integers and let I be an infinite index set. Further let $\mathcal{O}$ be the integers in a finite extension of $\mathbb{Q}_l$, let λ be its maximal ideal and $\mathbb{T}$ be a ring with distinguished elements $T_v^{(j)}$ for v a place not in $\mathbb{S}$ for j in $\{1, \ldots, e\}$ and a map $\theta : \mathbb{T} \to \mathcal{O}$. For $i \in I$, let $\mathbb{T}_i$ be a finite $\mathbb{Z}$-algebra with a map $\theta_i : \mathbb{T} \to \mathbb{T}_i$. Let $\{\mathbb{S}_i\}$ be a finite set of places containing $\mathbb{S}$ such that if $v \notin \mathbb{S}$ then $v \in \mathbb{S}_i$ for only finitely many i, and let E_i be a finite extension of the field of fractions of $\mathcal{O}$. For each i there is a continuous representation

$$\rho_i : Gal(\overline{K}/K) \to GL_d(\mathbb{T}_i \otimes_{\mathbb{Z}} E_i)$$

where $\overline{K}$ is the algebraic closure of K, which is unramified outside $\mathbb{S}_i$ and which satisfies

$$Tr(\rho_i)(Frob_v)^j = \theta \mathbb{T}_v^{(j)}$$

for $i = 1, \ldots, e$ and $\upsilon \notin \mathbb{S}_i$. Finally, suppose that there an injection $\iota : \mathbb{N} \to I$ such that for each n there is a surjection of $\mathbb{T}$-algebras

$$\mathbb{T}_{\iota(n)} \otimes_{\mathbb{Z}} \mathcal{O} \to \mathcal{O}/\lambda^n.$$

Then there is a finite extension E of the field of fractions of $\mathcal{O}$ and a continuous representation

$$\rho_i : Gal(\overline{K}/K) \to GL_d(\mathbb{E})$$

which is unramified outside $\mathbb{S}$ and which satisfies

$$Tr(\rho_i)(Frob_\upsilon)^j = \theta \mathbb{T}_\upsilon^{(j)}.$$

The proof goes as follows. A pseudorepresentation

$$\mathbb{T}_{\iota(n)} : Gal(\overline{K}/K) \to \mathbb{T}_{\iota(n)} \otimes_{\mathbb{Z}} \mathcal{O}$$

is obtained, and hence a pseudorepresentation

$$\mathbb{T}_{\iota(n)} : Gal(\overline{K}/K) \to \mathbb{T}_{\iota(n)} \otimes_{\mathbb{Z}} \mathcal{O} \to \mathcal{O}_{E_{\iota(n)}}/\lambda^n$$

which is unramified outside $\mathbb{S}_{\iota(n)}$ (in the sense that $K^{\ker(T_{\iota(n)})}/K$ is unramified outside $\mathbb{S}_{\iota(n)}$) and satisfies

$$T_{\iota(n)}((Frob\ \upsilon)^j) = \theta(\mathbb{T}_\upsilon^{(j)})$$

for $j = 1, \ldots, e$ and $\upsilon \notin \mathbb{S}_{\iota(n)}$. Let $n \geq m$; then because Frobenius elements are dense in the maximal quotient of $Gal(\overline{K}/K)$ unramified outside $\mathbb{S}_{\iota(n)} \cup \mathbb{S}_{\iota(m)}$ it may be seen that $T_{\iota(n)}$ and $T_{\iota(m)}$ are in fact valued in $\theta/\lambda^?$ and that $\mathbb{T}_{\iota(n)}$ mod $\lambda^m = \mathbb{T}_{\iota(m)}$. Thus a continuous pseudorepresentation into $GL_d(E')$ has been obtained with the desired properties, where E' is the algebraic closure of the field of fractions of $\mathcal{O}$, and this must be a genuine representation.

There are interesting extensions of this work in [247] and [226].

10.3   Group Matrices and Group Rings

10.3.1   Taussky-Todd on Normal Group Matrices

The following appeared in [272]. As may be seen it gives an alternative characterization of the class of groups which appear in Chap. 7, Sect. 7.5.4. In Chap. 4 a matrix A of complex numbers has been defined to be normal if $AA^* = A^*A$ where A^* is the transposed, complex conjugate of A. If the matrix also contains indeterminates their conjugates may be introduced formally and normality may be defined as before. If the indeterminates are considered identical with their conjugates they are called real.

Theorem 10.3 *Consider a finite group G and assume that the indeterminates x_g in X_G are real. Then X_G is a normal matrix if and only if G is abelian or a Hamiltonian 2-group. If the x_g are complex then X_G is normal if and only if G is abelian.*

Proof By direct calculation X_G is normal if and only if

$$\sum_{g\in G} x_{hg^{-1}}\overline{x}_{kg^{-1}} = \sum_{g\in G} \overline{x}_{gh^{-1}}x_{gk^{-1}} \text{ for all } h,k \text{ in } G.$$

If the x_g are real this is possible only if to every g,h,k there is an s in G such that one of the two pairs of relations holds:

$$hg^{-1} = sk^{-1} \tag{10.3.1}$$
$$kg^{-1} = sh^{-1}$$

$$hg^{-1} = sh^{-1} \tag{10.3.2}$$
$$kg^{-1} = sk^{-1}.$$

On inserting e for h, (10.3.1) implies that $gk = kg$ and (10.3.2) implies that $kg = gk^{-1}$ so that

$$k^{-1}g^{-1}kg = k^{-2}$$

and hence

$$g^{-1}k^{-1}gk = k^2.$$

On switching the roles of g and k it follows that either g and k commute or

$$g^{-1}k^{-1}gk = g^{-2}(= k^2).$$

Since g,k are arbitrary elements G, it follows that any two elements g,k of G either commute or satisfy the relation

$$g^2 = k^{-2}.$$

If $[g,k] \neq e$ then $[g,k^{-1}] \neq e$ and hence

$$g^4 = e = k^4.$$

A similar calculation gives

$$g^2 = (gk)^2.$$

From the original paper of Dedekind [75], it follows that G is either abelian or a Hamiltonian 2-group. If the x_g are complex, then only the possibility (10.3.1) remains and G is abelian.   □

At the end of the paper, Taussky gives the following indication of a proof of the above result which she attributes to Ky Fan:

Consider $X_G(f)$ where f is a real or complex-valued function on G. Then

1. In order that G be abelian or a Hamiltonian 2-group it is necessary and sufficient that for every 2-valued function f on G taking values 0 or 1 that $X_G(f)$ is a normal matrix.
2. In order that a finite group G be abelian it is necessary and sufficient that for every 3-valued function f on G taking values 0, 1 and i, $X_G(f)$ is a normal matrix.

She also indicates that, writing $f^*(h) = \overline{f}(h^{-1})$ for every $h \in G$ it can be shown that the normality of $X_G(f)$ is equivalent to the condition

$$\sum_{g \in G} f(hg^{-1})f^*(g) = \sum_{g \in G} f^*(hg^{-1})f(g)$$

for all h in G.

Since the class of groups described in Theorem 10.3 coincides with the class of groups with $\Lambda(G) = U$ discussed in Chap. 7, it would be interesting to make a direct connection between the two results, especially along the lines suggested by Ky Fan.

10.3.2   Other Early Work on Group Matrices

The possible uses of group matrices in the theory of group rings are discussed by Taussky in [273], which includes a section on group matrices. She considers group matrices with respect to a given group with integer entries, and indicates the 1:1 correspondence between group matrices and elements of the group algebra described in Chap. 2, Lemma 2.1.

The following is a direct quote.

> To use the group matrices instead of units (in group rings) has certain advantages: the concept of symmetric and normal matrices exist and allow the formulation of further problems.

She notes that the set of such group matrices (with respect to a fixed group with a given ordering), is closed under the Schur–Hadamard product, and this also extends

to positive definite group matrices. The matrix M is **unimodular** if $\det(M) = \pm 1$.
She notes that the Schur–Hadamard product of unimodular group matrices is not
necessarily unimodular. She discusses units in group rings and the result of G.
Higman that in the group ring of a finite abelian group no units exist of finite order
except for group elements and negated group elements, and that units of infinite
order exist unless all elements of the group are of order 2, 3, 4 or 6. She refers to a
proof of this result by herself and Newman for cyclic groups using group matrices
which in this case are circulants. A brief discussion of eigenvalues of such group
matrices is included.

More recently Hurley developed these ideas. He extended the discussion to
infinite groups and showed how various classes of matrices (Toeplitz, Hankel)
appear both in the finite and infinite setting. He and other coauthors have developed
applications in several directions. Much of the following work is due to him.

As mentioned above, the fundamental result for group rings is given in Chap. 2,
Lemma 2.1 . It is restated as follows. Suppose that an ordering of the group G is
fixed. Let the set of (full) group matrices of G with respect to this ordering and with
elements in a domain R be denoted by $\mathfrak{X}_G(R)$. This set forms an algebra over R, the
operations being matrix addition, multiplication and scalar multiplication. There is
an isomorphism σ from group algebra RG to $\mathfrak{X}_G(R)$.

10.3.3   The Integral Group Ring Case

Here $R = \mathbb{Z}$.

Lemma 10.2 *A group matrix $X_G(\underline{a})$ corresponds to a unit in $\mathbb{Z}G$ if and only if it is
unimodular.*

Proof Clearly if a matrix M and its inverse M^{-1} have integer entries both $\det(M)$
and $\det(M^{-1})$ are integers, and since $\det(M^{-1}) = 1/\det(M)$ it must follow that
$\det(M) = \pm 1$. Therefore if M corresponds to a unit in $\mathbb{Z}G$ it must be unimodular.
Conversely, if $M \in \mathfrak{X}_G(\mathbb{Z})$ is unimodular, then its adjoint M^* is also in $\mathfrak{X}_G(\mathbb{Z})$
and therefore $M^{-1} = (1/\det(M))M^*$ is in $\mathfrak{X}_G(\mathbb{Z})$. Thus the element in $\mathbb{Z}G$
corresponding to M is a unit.   □

As in previous chapters (o) is used for the operation of Schur–Hadamard
multiplication. It follows directly that $X_G(\underline{a}) \circ X_G(\underline{b}) = X_G(\underline{a} \circ \underline{b})$, where
$\underline{a} \circ \underline{b} = (a_1 b_1, a_2 b_2, \ldots, a_n b_n)$. Thus for any R, $\mathfrak{X}_G(R)$ is closed under Schur–
Hadamard multiplication. However as indicated above the subgroup of unimodular
group matrices in $\mathfrak{X}_G(R)$ is not closed under Schur–Hadamard multiplication. For
example take $G = C_2$. The matrices

$$A = \begin{bmatrix} 1 & \sqrt{2} \\ \sqrt{2} & 1 \end{bmatrix} \text{ and } B = \begin{bmatrix} 2 & \sqrt{3} \\ \sqrt{3} & 2 \end{bmatrix}$$

have respective determinants -1 and 1, but

$$\det(A \circ B) = \det \begin{bmatrix} 2 & \sqrt{6} \\ \sqrt{6} & 2 \end{bmatrix} = -2.$$

However $\mathfrak{X}_G(R)$ is closed under transposition. This is clear since the transpose of a group matrix X_G is obtained by replacing x_g by $x_{g^{-1}}$ for all g in G. (See Sect. 7.3, Proposition 1.3, (v). The result is stated there for a special ordering but the proof does not rely on the ordering).

From Sects. 7.3 and 1.3.7.1 if A, B are in $\mathfrak{X}_G(R)$ and are obtained from the same group ring element of RG but possibly different listings there exists an $n \times n$ permutation matrix P such that $PAP^T = B$. Thus A and B represent the same element of RG relative to possibly different listings if and only if A is permutation equivalent to B. Note that $P^T = P^{-1}$ for any permutation matrix.

10.3.3.1   Units and Zero Divisors

A non-zero element z in a ring W is said to be a **zero-divisor** in W if and only if there exists a non-zero element $r \in W$ with $zr = 0$. When RG has an identity 1_W then u is a **unit** in RG if and only if there exists an element $w \in RG$ with $uw = 1_W$. The group of units of W is denoted by $U(W)$.

10.3.4   Some Rings of Matrices which Occur as Group Rings

Let $R_{n \times n}$ denote the ring of $n \times n$ matrices with coefficients in a domain R.

When G is C_n, it has been shown in Sect. 7.3 that, with an appropriate ordering on G, X_G is a circulant matrix, and hence the matrices in $\mathfrak{X}_G(R)$ with respect to this ordering are circulant $n \times n$ matrices in $R_{n \times n}$. In the case where G is an elementary abelian 2-group of rank n and order 2^n, where the matrix size is $2^n \times 2^n$, the matrices in $\mathfrak{X}_G(R)$ are called **Walsh-Toeplitz matrices** over R—see for example [220]. A **Toeplitz matrix** is a matrix that is constant along any diagonal running from upper left to lower right. Circulant matrices are special types of Toeplitz matrices. It is easy to show that a Toeplitz $n \times n$ matrix can be embedded in a $2n \times 2n$ circulant matrix (see [72]), and this has proved useful in the study of Toeplitz matrices and their applications. A **Hankel** matrix is an $n \times n$ matrix with constant reverse diagonals. Reverse circulants are special cases of Hankel matrices.

The case of the dihedral group D_n under a special ordering has been discussed in Chap. 2. Using a different ordering, it is shown in [159] that the matrices in $\mathfrak{X}_G(R)$ can be expressed in the form

$$\begin{bmatrix} A & B \\ B & A \end{bmatrix}, \tag{10.3.3}$$

where A is a circulant and B is a reverse circulant. It is easy to see that any block matrix of the form given in (10.3.3) is similar to a matrix of the form

$$\begin{bmatrix} A+B & 0 \\ 0 & A-B \end{bmatrix}.$$

10.3.5   Infinite Groups

If G is a countable group whose elements are listed as $\{g_1, g_2, \ldots, g_n, \ldots\}$ and f is a function on G it is usually assumed that a group matrix $X_G(f)$ has only a finite number of non-zero entries in each row or column, or in other words that f has finite support. Then the set of infinite Toeplitz matrices over R is isomorphic to the group ring RG of the infinite cyclic group C_∞. The set of Walsh-Toeplitz infinite matrices over R is isomorphic to RG where G is the direct product of an infinite number of copies of C_2. When G is the infinite dihedral group then RG is isomorphic to the ring of infinite matrices of the form

$$\begin{bmatrix} A & B \\ B & A \end{bmatrix}$$

where A is an infinite Toeplitz matrix and B is an infinite Hankel matrix. Again this is isomorphic to the ring of matrices of the form

$$\begin{bmatrix} A+B & 0 \\ 0 & A-B \end{bmatrix}.$$

Hurley proves the following theorems.

Theorem 10.4 *When R is a field, $w \neq 0$ in RG is either a unit or a zero divisor, depending on whether $det(\sigma(w)) \neq 0$ or $det(\sigma(w)) = 0$.*

Theorem 10.5 *Let R be a field and G a finite group. Then $\mathbb{W} = U(RG)$ satisfies the Tits' alternative, i.e. $\mathbb{W}$ is either soluble-by-finite or it contains a non-cyclic free group.*

Proof From Theorems 10.4 and 10.5, $\mathbb{W}$ is a linear group, a subgroup of $GL(n; R)$, and hence satisfies the Tits' alternative by Tits' original theorem [277].   □

Corollary 10.1 *Let G be a locally finite group and R a field, then any finitely generated torsion group of $U(RG)$ is finite, i.e. the generalized Burnside problem has a positive answer for $U(RG)$.*

Note that if $R = \mathbb{Z}$ then $\det(\sigma(w)) = 1$, 0 or n where $|n| > 1$ so that three situations can occur here, the first corresponding to a unit, the second to a zero-divisor and in the third case w is neither a unit nor a zero-divisor.

Applications of group matrices are given in [161] and [160].

10.4 Group Matrices and Cogrowth of Groups

The work in this section is due to Humphries, see [150].
    Let

$$1 \to N \to F \to G \to 1 \tag{10.4.1}$$

be a presentation for the group G where $F = \langle a_1, \dots, a_n \rangle$ is a free group where ϕ is the map $F \to G$. The **degree** of the presentation is defined to be n. Let

$$A = \{a_1, \dots, a_n, a_1^{-1}, \dots, a_n^{-1}\}.$$

For $g \in G$, $c \in A$ let $W(k, g, c)$ be the set of freely reduced words $w \in F$ having length k, which end (on the right) in c and which represent the element $g \in G$. Let $w(k, g, c) = \|W(k, g, c)\|$. The function

$$\Gamma(k) = \sum_{c \in A} w(k, id_G, c)$$

is called the **cogrowth function** for the presentation. Let

$$\gamma(k) = \sum_{j=0}^{k} \Gamma(j) \text{ and } \gamma = \lim_{k \to \infty} (\gamma(k))^{1/k}.$$

Then $log(\gamma)/log(2n - 1)$ is called the **cogrowth** of the presentation. By a result of Cohen [61] a group G is amenable if and only if $\gamma = 2n - 1$. In [150] there is given a method for calculating cogrowth functions of various types of groups and examining the properties of these functions. The key idea is to introduce special kinds of group matrices.
    It is shown that starting from the presentation (10.4.1) a recurrence relation for the variables $w(k, g, c)$ is obtained with recurrence matrix R, relative to the ordering of the $w(k, g, c)$, and this matrix is finite if G is finite. The matrix R is a generalized group matrix. The following results rely on the properties of R.

10.4.1 Cogrowth of Finite Abelian Groups

Consider the trigonometric diophantine equation

$$\sum_{i=1}^{n} \cos(2\pi x_i) = \pm\sqrt{2n - 1}. \tag{10.4.2}$$

where $\{x_i\}_{i=1}^n \subset \mathbb{Q}$. Granville in [124], Proposition 2, using results of Conway and Jones [63], shows that there are only a finite number of rational solutions x_i for a general trigonometric diophantine equation and it follows from his result that for each n there is a finite set of numbers $N(n)$ such that there are no solutions to (10.4.2) if the denominators of the x_i are not divisible by any of the elements of $N(n)$. For example it follows from results of Parnami et al. [230] (which also uses the results of [63]) that $N(2) = \{12\}$ and that $N(3) = \varnothing$. The following was obtained by Humphries.

Theorem 10.6 *If (10.4.1) is a presentation for a finite abelian group G, having no elements of order k where $k \in N(n)$, then there are complex numbers w_j and λ_j such that for all $k > 0$,*

$$\Gamma(k+1) = \sum_{j=1}^{m} w_j \lambda_j^k.$$

Moreover $1 \le |\lambda_j| \le 2n - 1$ and λ_1 and w_1 can be chosen respectively to be $2n - 1$ and $\frac{2n}{|G|}$. In general the λ_j are associated to irreducible representations ρ_j of G and have the form $c_j \pm \sqrt{[c_j^2 - (2n-1)]}$ where

$$2c_j = \rho_j(\phi(a_1)) + \ldots + \rho_j(\phi(a_n)) + \rho_j(\phi(a_1)^{-1}) + \ldots + \rho_j(\phi(a_n)^{-1}).$$

The proof given in [150] is constructive in the sense that for a given presentation one can find the cogrowth function. The heart of this construction is the determination of certain eigenvectors for the recurrence matrix R. These eigenvectors involve the term $1/\sqrt{[c_j^2 - (2n-1)]}$. The result of Granville can be used to decide when $\sqrt{[c_j^2 - (2n-1)]}$ is non-zero. An extension of Theorem 10.4.1 to infinite abelian groups is given in Theorem 1.2 of [150].

Humphries also has the following theorem.

Theorem 10.7 *The cogrowth function for the presentation*

$$\langle a, b | a^2, b^n, (ab)^2 \rangle$$

of D_n has the form

$$\Gamma(k+1) = \sum_{j=1}^{m} w_j \lambda_j^k.$$

where the w_j and k_j are complex numbers with $1 \le |\lambda_j|$ for all j and $\lambda_1 = 3$.

Again, the method of proof uses the recurrence matrix R. It is possible to show that R determines the presentation (up to a possible permutation of the generators).

10.4.2   The Connection Between Cogrowth and the Group Matrix

Suppose that there is a presentation of a group G as in (10.4.1). Let $wc \in W(k, g, c)$, where w has length $k - 1$. Then $w \in W(k - 1, g\phi(c)^{-1}, c')$ where $c' \in A \backslash c^{-1}$; conversely, if $c' \in A \backslash c^{-1}$ and $w \in W(k - 1, g\phi(c)^{-1}, c')$, then $wc \in W(k, g, c)$. Thus there is a recurrence for the variables $w(k, g, c)$ as follows:

$$w(k + 1, g, c) = \sum_{d \in A \backslash c^{-1}} w(k - 1, g\phi(c)^{-1}, d), \tag{10.4.3}$$

Hence a recurrence matrix R is obtained of size $2n|G| \times 2n|G|$. Specifically, if G is a finite group, order the elements of G as $\{g_1, \ldots, g_h\}$ with $g_1 = id$, $|G| = h$. For fixed k form an ordered 'basis' for the $W(k, g, c)$ as follows:

$$W(k, g_1, a_1) \ W(k, g_1, a_2) \ \ldots \ W(k, g_1, a_{2n})$$
$$W(k, g_2, a_1) \ W(k, g_2, a_2) \ \ldots \ W(k, g_2, a_{2n})$$
$$\ldots$$
$$W(k, g_h, a_1) \ W(k, g_h, a_2) \ \ldots \ W(k, g_h, a_{2n}).$$

Here $a_{n+i} = a_i^{-1}$ for $i = 1, \ldots, n$. Then by (10.4.3), using these choices, the entries of R are 0 except in the $(W(k, g_p, a_q), W(k, g_r, a_s))$ positions with $g_r = g_p \phi(a_q^{-1})$ and $a_q \neq a_s^{-1}$ where the entry is 1. Since for any pair (g_p, a_q) there is always a g_r with $g_r = g_p \phi(a_q^{-1})$ and since there are $2n - 1$ choices of s with $a_q \neq a_s^{-1}$ it may be seen that each row of R has exactly $2n - 1$ entries equal to 1. It follows that $\frac{1}{2n-1} R$ is a stochastic matrix.

Further, thinking of the above coordinates as being partitioned into subsets according to the g_p (of which there are $|G|$) so that there are $2n$ such coordinates in each such subset, it is seen that R has a block form consisting of blocks of size $2n \times 2n$, there being a $|G| \times |G|$ matrix of these blocks. More specifically the above shows that the matrix R can also be represented as a block sum

$$R = \sum_{g \in G} \sigma_g \otimes H(g)$$

where σ_g is the $|G| \times |G|$ permutation matrix with 1 in the (g_p, g_r) position if $g_r = g_p^{-1}$ and 0 otherwise, and $H(g)$ is the matrix

$$H(g) = \begin{bmatrix} H_{11}(g) & H_{12}(g) \\ H_{21}(g) & H_{22}(g) \end{bmatrix}$$

and where the matrices $H_{ij}(g)$ are constructed as follows;

(i) the only non-zero entries (which are all 1's) of $H_{11}(g)$ occur in the row where $\phi(a_i) = g$ (if there is such an a_i with $1 \le i \le n$);
(ii) similarly, the only non-zero entries (which are all 1's) of $H_{22}(g)$ occur in the ith row where $\phi(a_i^{-1}) = g$;
(iii) if the ith row of $H_{11}(g)$ is non-zero, then the ith row of $H_{12}(g)$ has all 1's except in the ith position, where it is 0 (all other entries of $H_{12}(g)$ being 0);
(iv) if the ith row of $H_{22}(g)$ is non-zero, then the ith row of $H_{21}(g)$ has all 1's except in the ith position, where it is 0 (all other entries of $H_{21}(g)$ being 0).

For example, for the presentation of a group G with 2 generators a_1, a_2 such that $\phi(a_1)$ has order 2 and $\phi(a_2)$ has order greater than 2, the non-zero H matrices are

$$H(\phi(a_1)) = \begin{bmatrix} 1 & 1 & 0 & 1 \\ 0 & 0 & 0 & 0 \\ 0 & 1 & 1 & 1 \\ 0 & 0 & 0 & 0 \end{bmatrix}, \quad H(\phi(a_2)) = \begin{bmatrix} 0 & 0 & 0 & 0 \\ 1 & 1 & 1 & 0 \\ 0 & 0 & 0 & 0 \\ 0 & 0 & 0 & 0 \end{bmatrix},$$

$$H(\phi(a_2^{-1})) = \begin{bmatrix} 0 & 0 & 0 & 0 \\ 0 & 0 & 0 & 0 \\ 0 & 0 & 0 & 0 \\ 1 & 0 & 1 & 1 \end{bmatrix}. \tag{10.4.4}$$

For example the block structure of R for the presentation $\langle (1, 2), (1, 2, 3) \rangle$ of the symmetric group $\mathcal{S}_3$ is the group matrix

$$X_G(H(id), H((1, 2, 3)), H((1, 3, 2))H((1, 2)), H((1, 3)), H((2, 3))),$$

where $H((1, 2))$, $H((1, 2, 3))$ and $H((1, 3, 2))$ are given by (10.4.4) with $\phi(a_1) = (1, 2)$, $\phi(a_2) = (1, 2, 3)$ and all the other $H(g)$ are 0.

If G is an infinite group an infinite recurrence matrix is obtained. In this case it is also seen that

$$R = \sum_{g \in G} \sigma_g \otimes H(g)$$

with the $H(g)$ defined as in (i)–(iv). This is an infinite group matrix.

For the many interesting details of how the group matrix and group determinant intervene in this work the reader is referred to [150].

10.5   Poincaré and the Group Determinant

In his work on linear differential equations Poincaré introduces group determinants in the paper [234], published in 1903. This is an expansion of a paper published in 1883. His initial papers introducing Fuchsian groups and Fuchsian functions appeared in 1881–1884 so that group representations were not available to him. The extremely complicated way in which the theory of linear differential equations developed has been explained in the book [125], the various tools being analysis, invariant theory, topology (avant la lettre).... There is also the book of Poincaré's papers on Fuchsian functions by Stillwell [236] which contains extensive comments. An explanation of Poincaré's point of view is also given in the book by Forsyth [101, p. 517].

In the introduction to [234] the following passage appears. (See also the quote in Chap. 7, Sect. 7.5).

> When an algebraic function satisfies a differential equation whose coefficients are rational, the abelian integrals posess certain curious properties and there are some interesting relations between their periods. M. Frobenius, in 1896 and the following years has published a series of memoires on the characters of groups. These results can be usefully applied to the question which occupies us and I hope to recall them quickly, and in addition I avail myself of the opportunity to relate them to other results obtained by M. Cartan and to see how the theories of these two savants can mutually shed light on each other.

Poincaré starts from the differential equation

$$\sum_{i=1}^{n} \frac{d^i y}{dx^i} P_i(x) = 0, \tag{10.5.1}$$

where the coefficients $P_i(x)$ are polynomials. He discusses the case when a rational solution $F(x, y)$ exists. If H is the monodromy group of the equation of order m, he constructs from F a group matrix $X = X_H(b_1, b_2, \ldots, b_m)$ where the b_i are complex numbers, and he states that a necessary condition that F exists is that "the determinant of X and its first $m - n - 1$ minors vanish". This is equivalent to the fact that the rank of X must be n. He states that the condition that a rational solution exists is that "numbers" $b_1, b_2, \ldots, b_m$ exist satisfying $\text{rank}(X_H(b_1, b_2, \ldots, b_m)) = n$, and states that this condition is also sufficient.

In the context of Chap. 7, Sect. 7.5.4, considering b as a map from $H = \{h_i\}_{i=1}^{m}$ to $\mathbb{C}$ with $b(h_i) = b_i$, Poincaré's condition seems to be equivalent to the existence of a non-trivial function b satisfying $s_k^\rho(b) = 0$ for $k > n$.

The question arises whether the result of Vishnevetskiy and Zhmud implies that such functions always exist if H is nonabelian and is not Hamiltonian of even order.

An additional idea comes from the fourth of the big papers in the book [236] which Stillwell describes as the least well understood. This is the paper entitled "Les groupes des équations linéaires". He introduces the "fundamental invariants" of a group element g of a group G which is represented as a unitary $p \times p$ matrix $\rho(g)$. These are the $p - 1$ coefficients of the characteristic equation of $\rho(g)$. He argues as follows. Suppose that G has a generating set of size n. There are p^2 coefficients of each representing matrix of elements in the set, but since $\det(\rho(g)) = \pm 1$ only $p^2 - 1$ are independent. Since Poincaré is interested in determining the linear group up to a similarity transformation, only $(n - 1)(p^2 - 1)$ coefficients determine the generating set and hence G. He then says that the fundamental invariants of $(n - 1)(p - 1)$ elements also determine G.

This seems to be incorrect. For example there are pairs (G, H) of non-isomorphic equidistributed permutation groups, in that there is a bijection $f : G \to H$ such that each pair of elements g and $f(g)$ have the same cycle type (for example the elementary abelian group of order p^3 and the nonabelian group of the same order and exponent p). It may have been that Poincaré had in mind something like the s_k arising from the determinant.

Poincaré poses the problem of obtaining a direct connection between the coefficient functions $P_i(x)$ above and his fundamental invariants. This does not seem to have been explored.

10.6 Chern Classes

Chern classes are topological invariants associated to vector bundles on a smooth manifold. If two such bundles are given in different ways they can be distinguished if their Chern classes are not the same. In topology, differential geometry and algebraic geometry it is often important to count the number of linearly independent sections of a vector bundle. The Chern classes provide a means of approaching this information through the Riemann–Roch theorem and the Atiyah–Singer index theorem.

In particular, there are Chern classes attached to a group representation as follows. Given a discrete group G there is an associated classifying space BG which is unique up to homotopy, with fundamental group isomorphic to G. There is a universal covering space EG. Let ρ be a finite-dimensional irreducible representation of degree n and construct the vector bundle $\underline{E} = G \times_G \mathbb{C}^n = (EG \times \mathbb{C}^n)/G$ where G acts by $(a, x)^g = (ag^{-1}, \rho(g)x)$. The Chern classes attached to ρ are by definition the Chern classes of the vector bundle $\underline{E}$ over BG.

According to Grothendieck [127] there is an algebraic substitute for BG as follows. Consider the "topos" formed from the G-sets. The abelian sheaves (or abelian groups) of this topos are in effect the G-modules, and the functor "sections" is the functor $M \rightsquigarrow H^0(G, M) = M^G$. The derived functors are the cohomology groups $H^i(G, M)$ of G. It may be interesting to describe this topos in terms of the approach to the representation of the Burnside ring by super group matrices given in Chap. 2.

The Chern classes $c_k(\rho)$ lie in $H^{2k}(G, \mathbb{Z})$ and thus there is associated to the algebraic object ρ the algebraic object $c_k(\rho)$, so the question arises of whether there is a purely algebraic construction of the $c_k(\rho)$. This was raised in the Appendix to [6]. In the work cited above, Grothendieck explains that this is not possible for what he describes as trivial reasons. He points out that such a construction must be independent of the topology of $\mathbb{C}$, and would be available if this field were replaced by any other field K, satisfying suitable conditions which are purely algebraic, such as that of being algebraically closed, or even isomorphic to $\mathbb{C}$ (in addition to being of characteristic 0 and of transcendence degree 1 over the prime field of cardinality of the continuum). Then the Chern classes would be unchanged under an isomorphism $K \to K'$ and in particular by the automorphisms of K. They must coincide with the classes obtained by transcendental methods for $K = \mathbb{C}$. Now if $G = C_n$, the first Chern class $c_1(\rho) \in H^2(G, \mathbb{Z})$ for a 1-dimensional representation ρ may be considered to be a homomorphism $G \to K^*$ or further a homomorphism $G \to \mu_\infty(K)$, the group of units in K. On the other hand, G being finite, $H^i(G, \mathbb{Q}) = 0$ for $i > 0$ and

$$H^2(G, \mathbb{Z}) \simeq Hom_G(\mathbb{Q}/\mathbb{Z})$$

so that $c_1(\rho)$ reduces to a homomorphism $G \to \mathbb{Q}/\mathbb{Z}$. This shows the dependence of $c_1(\rho)$ on a homomorphism

$$\phi : \mu_\infty(\mathbb{C}) \to \mathbb{Q}/\mathbb{Z}.$$

Grothendieck points out that such a homomorphism is not algebraic and uses the topology of $\mathbb{C}$ in an essential manner. He quotes the result of Gauss that composing this isomorphism with the automorphisms of $\mu_\infty(\mathbb{C})$ induced by the automorphisms of $\mathbb{C}$ one finds all possible isomorphisms between $\mathbb{Q}/\mathbb{Z}$ and $\mu_\infty(\mathbb{C})$ which form a torsor (i.e. a principal homogeneous set) over the group

$$Aut(\mathbb{Q}/\mathbb{Z}) \simeq \prod \mathbb{Z}_p^*$$

where $\mathbb{Z}_p^*$ is the group of p-adic integers and the product is over all primes p. Thus $c_1(\rho)$ does not have the algebraic invariance needed. Grothendieck then goes on to explain various ways in which an algebraic construction of Chern classes may be made, replacing the coefficients $\mathbb{Z}$ by different groups attached to the base field.

In [278] which is written from the point of view of finite groups Thomas gives the following properties for the Chern classes $c_k(\rho)$:

CH1.   If $f : H \to G$ is a group homomorphism then

$$c_k(f!\rho) = f^*(c_k(\rho))$$

where $f!$ and f^* are the natural pull-back maps in representation theory and cohomology.

CH2.   If the total Chern class $c(\rho)$ is defined by

$$c(\rho) = 1 + c_1(\rho) + \ldots + c_n(\rho)$$

then

$$c(\rho_1 \oplus \rho_2) = c(\rho_1).c(\rho_2).$$

CH3. If ρ is a one-dimensional representation then $c_1(\rho)$ is the image of the cohomology class of ρ in $H^2(G, \mathbb{Z})$ under the boundary map associated with the exact sequence of coefficients

$$\mathbb{Z} \to \mathbb{C} \to \mathbb{C}^*.$$

Thomas also points out that the cohomology ring $H^*(G, \mathbb{Z})$ should in theory play a similar role in finite group theory to that of the complex representation ring $R(G)$ but recognizes that the number of results proved using it is small. He indicates that the structure of $R(G)$ can be used to study part of $H^*(G, \mathbb{Z})$ via Chern classes.

In [6] Atiyah also posed the problem of computing the Chern classes (and Stiefel-Whitney classes) of an induced representation. This problem was posed and solved in a much wider context by Fulton and Macpherson in [111]. There are interesting parallels in their work with results in group representation theory. In their discussion of the Grothendieck–Riemann–Roch theorem there is even a connection with the polynomials s_k from which the factors of a group determinant are calculated by Frobenius. Nontheless, the Chern classes of a finite group representation seem to remain somewhat mysterious and hard to calculate.

For Lie groups there is an alternative definition of Chern classes using concepts from differential geometry, via the "Chern–Weil" homomorphism. Suppose that G is a Lie group. Consider an m-dimensional vector bundle $X \to M$ over a differential manifold M with G acting. For any connection θ on X an $m \times m$ curvature matrix M_θ can be obtained by standard means, the entries being 2-forms, see [126, p. 400]. The **elementary invariant polynomials** in a matrix A are defined by

$$\det(A + tI) = \sum_{k=0}^{\infty} P^{m-k} t^k.$$

For each P^i, $P^i(M_\theta)$ is a well-defined global $2i$-form on M, the products of forms being cup products, and the cohomology class represented by this is independent of the connection θ. Define

$$c_k(M_\theta) = P^k(\frac{i}{2\pi} M_\theta)$$

and the Chern class $c_k(X)$ is the class of $P^k(\frac{i}{2\pi} M_\theta)$ in $H_{DR}^{2i}(M)$, the De Rham cohomology of M. It can be shown that $c_k(X)$ is naturally in $H^{2i}(X, \mathbb{Z})$. If X is constructed from the representation ρ as above, the total Chern class of $c(\rho)$ may be obtained as the class of $\det(I + \frac{i}{2\pi} M_\theta)$ in $H^*(G, \mathbb{Z})$. In [95] Dieudonné describes a series of "miracles", one of which being the fact that the above class is in $H^*(G, \mathbb{Z})$.

Now consider the case of a finite group G of order n. The matrix $\widehat{X_G}$ is a block diagonal matrix with the group matrices $X_G^{\rho_i}$, $\rho_i \in Irr(G)$, as the blocks (see Chap. 7, Eq. (7.2.4)). If ρ_i is of degree d_i, let M_θ^i be the curvature matrix of $GL(d_i, \mathbb{C})$. Let $(\widehat{X_G})_\theta$ be the matrix obtained by replacing each occurrence of $X_G^{\rho_i}$ in $\widehat{X_G}$ by M_θ^i. If the inverse Fourier transformation (i.e $P^{-1}(\widehat{X_G})_\theta\, P$ where P is the matrix in Theorem 15, Chap. 2) is applied to $(\widehat{X_G})_\theta$ a group matrix $X_G(\theta)$ is produced each of whose entries are 2-forms. The first column of $X_G(\theta)$ provides a function θ from G with values in the 2-forms on $GL(n, \mathbb{C})$, specifically $\theta(g)$ is the entry in $X_G(\theta)$ in the place of the variable x_g in X_G. Now let $\phi_i(x_e, x_{g_2}, \ldots, x_{g_n})$ be the irreducible factor of Θ_G corresponding to ρ_i. Consider

$$\phi_i(1 + (i/2\pi)\theta_e, \theta_{g_2}, \ldots, \theta_{g_n}).$$

This may be regarded as an element of the De Rham cohomology ring of $GL(n, \mathbb{C})$, again with products of forms being cup products, and in turn as an element α of the ring $H^*(GL(n, \mathbb{C}), \mathbb{Z})$. The regular representation map $\upsilon : G \to GL(n, \mathbb{C})$ provides a map

$$\upsilon^* : H^*(GL(n, \mathbb{C}), \mathbb{Z}) \to H^*(G, \mathbb{Z}).$$

Conjecture 10.1 The total Chern class $c(\rho_i)$ is $\upsilon^*(\alpha)$.

It may be seen that the properties CH1–CH3 above are satisfied by $\upsilon^*(\alpha)$. This opens the question of whether the objects described in previous chapters, for example the k-characters, may have applications in this area. It may also be interesting to consider whether the theorem that the 1-, 2- and 3-characters of a group determine the group leads to results in this context.

10.7   Tight Frames and Wavelets

This work appears in papers of Waldron and his collaborators on tight frames.

Example 10.2 Consider the three equally spaced vectors u_1, u_2, u_3 in $\mathbb{R}^2$ given in the figure below. They provide a redundant representation for a function $f : \mathbb{R}^2 \to \mathbb{R}$ as follows

$$f = \frac{2}{3} \sum_{j=1}^{3} \langle f, u_j \rangle u_j. \tag{10.7.1}$$

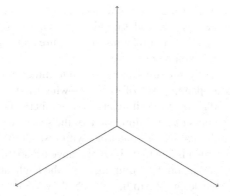

This is the simplest example of a tight frame. It is known as the **Mercedes Benz frame**. Such representations arose in the study of nonharmonic Fourier series in $L^2(\mathbb{R})$ (see [97]) and have recently been used extensively in the theory of wavelets (see [73]).

The philosophy behind the use of frames is that representations such as (10.7.1) are similar to an orthogonal expansion (but with more terms), and that by going to a frame representation the u_j can be chosen to have desirable properties that would be impossible were they to be orthogonal (in the case of wavelets these are certain smoothness and small support properties). In particular, a tight frame may be preferred to an orthogonal basis if there are underlying symmetries which would be desirable for the tight frame, but which cannot be possessed by any orthogonal basis. Of particularly interest are spaces of orthogonal polynomials of several variables with weights with some symmetries, e.g., integration over a polyhedron in $\mathbb{R}^2$. Here the orthonormal bases are replaced with spanning sets which share symmetries of the weight and are tight frames. In particular tight frames can have a group of symmetries, which in the case of the above example is S_3.

10.7.1 Frames and Gramian Matrices

The following is based on the article [53] in the book [52].

The Fourier transform has been a major tool in analysis for over 100 years. However it solely provides frequency information, and hides (in its phases) information concerning the moment of emission and duration of a signal.

In 1946 Gabor in [113] introduced a new approach to signal decomposition, which eventually was recognized as using the idea of a frame. In order to approach problems in nonharmonic series Duffin and Schaeffer in [97] introduced the idea of a Hilbert space frame for which they required a formal structure for working with highly overcomplete families of exponential functions, which fell into the area of time-frequency analysis. The concept was revived in the late 80's by Daubechies, Grossman and Mayer in [74] who showed its importance in data processing.

Traditionally, frames were used in signal and image processing, nonharmonic Fourier series, data compression, and sampling theory. But today, frame theory has ever-increasing applications to problems in both pure and applied mathematics, physics, engineering, and computer science.

The applications mainly require frames in finite-dimensional spaces. In this situation, a frame is a spanning set of vectors—which are generally redundant (overcomplete), requiring control of its condition numbers. Thus a typical frame involves more frame vectors than the dimension of the space, and each vector in the space will have infinitely many representations with respect to the frame. It is this redundancy of frames which is key to their significance for applications. The role of redundancy varies depending on the requirements of the applications at hand. First, redundancy gives greater design flexibility, which allows frames to be constructed to fit a particular problem in a manner not possible by a set of linearly independent vectors.

Two examples are:

(I) In quantum tomography, classes of orthonormal bases with the property that the modulus of the inner products of vectors from different bases are a constant are required.

(II) In speech recognition, when a vector needs to be determined by the absolute value of the frame coefficients (up to a phase factor).

A second major advantage of redundancy is robustness. By spreading the information over a wider range of vectors, resilience against losses (erasures) can be achieved. Erasures are, for instance, a severe problem in wireless sensor networks when transmission losses occur or when sensors are intermittently fading out, or in modeling the brain where memory cells are dying out. A further advantage of spreading information over a wider range of vectors is to mitigate the effects of noise in the signal.

New theoretical insights and novel applications are continually arising due to the fact that the underlying principles of frame theory are basic ideas which are fundamental to a wide canon of areas of research. In this sense, frame theory might be regarded as partly belonging to applied harmonic analysis, functional analysis, operator theory, numerical linear algebra, and matrix theory.

Basically a frame is a set of vectors $(\phi_i)_{i=1}^m$ in $\mathbb{R}^n$ or $\mathbb{C}^n$ withIf $A = B$ then the frame is called an A-tight frame.

The **Gramian matrix** of the frame is the matrix whose (i, j)th entry is $\langle \phi_i, \phi_j \rangle$.

10.7.2   Group Frames

The account here is based on the article by Waldron [286]. A further reference is [240], and there are many other references in [286].

A motivating example of a tight frame is the "Mercedes-Benz frame" described above.

Definition 10.1 Let G be a finite group. A **group frame** or G**-frame** for V is a frame $\Phi = (\phi_g)_{g \in G}$ for which there exists a unitary representation $\rho : G \to U(V)$ with

$$g\phi_h := \rho(g)\phi_h = \phi_{gh}.$$

Early examples of group frames come from the platonic solids. The ϕ_i are the coordinates of the vertices, and the groups are the well-known symmetry groups. Coxeter groups also provide examples of group frames. There can be an infinite number of frames for a given group G if G is nonabelian.

The **Gramian** of a group frame corresponding to a group G is a group matrix of the form $X_G(f)$ for some $f : G \to \mathbb{F}$ for some field $\mathbb{F}$. (Actually Waldron considers the group matrix under left division but this is a relatively minor change). For the Mercedes-Benz frame the group matrix of S_3 is obtained as the Gramian if each vector is repeated twice.

10.7.3 Projective Group Matrices and Finite Frames

The reader is referred to Chap. 2, Sect. 2.7 for projective representations of groups and projective group matrices.

A preprint of Cheng and Han [58] introduces a **twisted (G, α)-frame** where α is a 2-cocycle of G and therefore corresponds to a non-trivial projective representation of G. They assign to such a frame a Gramian matrix which is a left version of the (full) projective group matrix defined by Schur and given in Chap. 2, Sect. 2.7.

This large area of applications should benefit from the techniques given in this book.

10.8 Three-Manifolds

In geometric topology, the virtually Haken conjecture states that every compact, orientable, irreducible three-dimensional manifold with infinite fundamental group is virtually Haken. That is, it has a finite cover (a covering space with a finite-to-one covering map) that is a Haken manifold. It has now been proved.

A Haken manifold is a compact, P^2-irreducible 3-manifold that is sufficiently large, meaning that it contains a properly embedded two-sided incompressible surface. Sometimes one considers only orientable Haken manifolds, in which case a Haken manifold is a compact, orientable, irreducible 3-manifold that contains an orientable, incompressible surface.

Cooper and Walsh in [66] introduce group determinants to prove a result related to the conjecture. Their main result is stated as

Theorem 10.8 *Let Y be a compact, orientable 3-manifold with one torus boundary component. Assume that the interior of Y admits a complete hyperbolic structure of finite volume. Then infinitely many virtually Dehn-fillings of Y are virtually Haken.*

Cooper and Walsh introduced the novel idea of considering a new invariant of the group G of covering translations of a 3-manifold, which they call its "symmetrized group determinant". This is described as follows.

The symmetrized group matrix X_G^{Symm} of a group G, obtained by identifying x_g with $x_{g^{-1}}$ for all g in G, has been mentioned in Chap. 2. It has appeared in Sjogren's work on groups acting on graphs [258]. It is a symmetric matrix. Cooper and Walsh use the group matrix with respect to left division, $X_G^\lambda = \{x_{g_i^{-1}g_j}\}$ and all properties of the usual group matrix translate to properties of this. They also introduce further specialization of the variables. Let H be a subgroup of G of index r and define the map $\sigma : G \to G$ by

$$\sigma(g) = \sigma(g^{-1}), \sigma(gh) = \sigma(g) \text{ for } h \in H.$$

Lemma 10.3 *The map σ is constant on the sets $HgH \cup Hg^{-1}H$.*

Proof For g an element of G and elements h_1, h_2 of H, consider $\sigma(h_1 g h_2)$. Now

$$\sigma(h_1 g h_2) = \sigma(h_1 g) = \sigma((h_1 g)^{-1}) = \sigma(g^{-1} h_1^{-1}) = \sigma(g^{-1}) = \sigma(g),$$

where the equalities follow from the two properties above.   □

They introduce one specialization by identifying x_g with the x_{gh} for all $h \in H$, so that variables correspond to the left cosets of H in G, so that if $G = g_1 H + g_2 H + \ldots + g_r H$ then $x_{g_i H} = x_{g_i}$. They call the matrix obtained from X_G^λ under this specialization $X_{G,H}$. They also introduce a further specialization, where $x_g = x_{\sigma(g)}$ for all $\sigma(g)$, thus there is a single variable corresponding to each subset of G of the form $HgH \cup Hg^{-1}H$. Let the corresponding matrix be $X_{G,H}^{Symm}$. They mention that they do not know the factorization of the symmetrized determinant Θ_G^{Symm}, and it appears to be that there is no general result in this direction because factors which are irreducible as factors of Θ_G may become reducible after the identifications are made to make them factors of Θ_G^{Symm}. For example, in the case where G is the quaternion group the quadratic factor of Θ_G splits into linear factors after symmetrization. They point out that with the ordering of the elements of G by left cosets of H as in Chap. 2, Sect. 2.3.1, then $X_{G,H}^{Symm}$ has the form

$$\begin{bmatrix} B & B & \ldots & B \\ B & B & \ldots & B \\ \multicolumn{4}{c}{\ldots\ldots\ldots\ldots} \\ B & B & \ldots & B \end{bmatrix}$$

where B is the $r \times r$ matrix whose (i, j)th element is the variable $x_{\sigma(g_i^{-1}g_j)}$. If the left group matrix is constructed from the ordering by left cosets of H, it is no longer true that each block is of the form $X_G^\lambda(\underline{u})$, but since the elements in the block indexed by cosets $g_i H$ and $g_j H$ are of the form $h_\alpha^{-1} g_i^{-1} g_j h_\beta$ and $\sigma(h_\alpha^{-1} g_i^{-1} g_j h_\beta) = \sigma(g_i^{-1} g_j)$ the corresponding block of $X_{G,H}^{Symm}$ is $x_{\sigma(g_i^{-1}g_j)}$ multiplied by the matrix J_r. On rearranging rows and columns, a matrix of the above form is obtained where $B_{(i,j)} = x_{\sigma(g_i^{-1}g_j)}$. This illustrates how other symmetric forms of a group matrix may be obtained from those discussed in Chap. 2. Further, by standard row operations $X_{G,H}^{Symm}$ may be transformed by a similarity transformation to the matrix

$$
\begin{bmatrix}
rB & 0 & \cdots & 0 \\
0 & 0 & \cdots & 0 \\
\multicolumn{4}{c}{\cdots\cdots\cdots\cdots} \\
0 & 0 & \cdots & 0
\end{bmatrix}.
$$

In a similar way, matrices B_ρ may be obtained starting from any group matrix X_G^ρ. The following proposition holds.

Proposition 10.1 *Let $\{\rho_i\}_{i=1}^t = Irr(G)$ and let m_i be the rank of B_{ρ_i}. Then*

$$
r = \sum_{\rho_i} \dim(\rho_i) m_i.
$$

With hindsight, since the determinants of the group matrices used factor into linear factors, the results of Cooper and Walsh could be presented in the language of association schemes and Fourier inversion without introducing the group matrix or group determinant, but it should be interesting to pursue further the connection between symmetrized group matrices and 3-manifolds.

10.9   Group Matrices and Electrical Engineering

The group matrix has appeared in what may be loosely described as control theory, the design of digital devices and information theory. The papers have mostly appeared in electrical engineering journals. One group of authors who have produced work on the group matrix is that of Karpovsky and his collaborators. The paper of Trachtenberg [280] has already been cited in Chap. 2. In it appears the following: "Singular value decomposition is a generalization of the eigenvalue-eigenvector representation of a square matrix. It is the only reliable technique for computing ranks of operators". A further quote is "The relatively small class of matrices used as suboptimal approximations was introduced by Frobenius in 1895 (*sic*). Their use as signal processing models is due to the fact that their eigenvalues and their singular values are determined analytically. . . . By choosing various groups of the same order n, we can construct the best models with respect to each of the criteria in ..." Later he states "We give analytical formulas for calculating eigenvalues, singular values, ranks, determinants and pseudo-inverses of Frobenius matrices".

Essentially the techniques appear to rely on the Fourier transform of a finite nonabelian group, and this can be carried out in terms of the group matrix as has been explained in Chap. 7. However, some variations occur, for example the discussion of the group matrix of a representation over a finite field. The discussions quickly turn to practical implementation of the algorithms. The books [189] and [267] have sections on the Fourier transform. Although clearly members of the Karpovsky school were aware of the group matrix, it is difficult to find it in recent papers. Again, there seem to be many opportunities for applications of the techniques presented in this book.

10.10   Higher S-Rings

The following is an account of the work of Humphries and Rode (Turner) in [154] and [155] on the information contained in the k-class algebra of a group. The definition of the k-classes of a group has been given in Chap. 6. In their work they extend the discussion to FC-groups. A group G is an **FC-group** if all its conjugacy classes are finite. If G is an FC-group the k-classes of G are necessarily finite and they form a basis of an S-ring over the group G^k which is called the k-**S-ring** over G and is denoted by $\mathfrak{S}_G^{(k)}$.

Let G be an FC-group. If the set of k-classes is denoted by Γ_i^k, $i = 1 \ldots r$, the positive integer structure constants λ_{ijt} are defined by

$$\overline{\Gamma_i^k}\, \overline{\Gamma_j^k} = \sum \lambda_{ijt} \overline{\Gamma_t^k}.$$

If $k > 1$ the k-S-ring is not necessarily commutative. The groups G_1, G_2 are said to have the **same k-S-ring** if there is a bijection $\phi : G_1 \rightarrow G_2$ which induces an isomorphism between their k-S-rings, or equivalently the corresponding structure constants are equal.

It is well-known that the information in the ordinary characters is identical to that in the (1-) class algebra and this has already appeared in Chap. 5. By contrast for $k = 2$ the information in the 2-S-ring is different from that in the WCT. The following proposition makes this explicit.

Proposition 10.2 *There exist pairs of groups which have the same WCT but not the same 2-S-rings, and pairs of groups which have the same 2-S-rings but not the same WCT.*

Thus the information in the 2-S-ring of a group has a bearing on the questions of Brauer introduced in Chap. 5.

A set of examples which imply Proposition 10.2 have already been discussed in Chap. 5.

(a) The pair (D_4, Q_8) do not have the same WCT. They have the same 2-S-ring.
(b) The pair $(G_1^{p^3}, G_2^{p^3})$ of nonabelian groups of order p^3 where p is a prime, $p > 3$ have the same WCT but nonisomorphic 2-S-rings.

From (a) the 2-S-ring of a group cannot determine G. It follows easily that the inverse map $\mathfrak{I} : g \to g^{-1}$ is determined by the WCT of G, but for arbitrary groups $\mathfrak{I}$ is not necessarily determined by the 2-S-ring, although if $|G|$ is odd it is determined.

In Chap. 5 the result is presented that the WCT does not determine the derived length of a group. The following propositions show in particular that if the information in the 2-S-ring is added the derived length is determined. Groups G_1 and G_2 are said to have the same WCT $\mathfrak{S}^{(2)}$ if there is a weak Cayley isomorphism $\phi : G_1 \to G_2$ which induces an isomorphism between their 2-class algebras.

Proposition 10.3 *Let G be finite. Then the WCT of G and $\mathfrak{S}_G^{(2)}$ together determine the conjugacy classes of G which lie in each term $G^{(i)}$ of the derived series of G. In particular they determine the size of each $G^{(i)}$ and the length of the derived series of G.*

Proposition 10.4 *If G_1 and G_2 have the same WCT $\mathfrak{S}^{(2)}$ determined by $\phi : G_1 \to G_2$ then $\phi(G_1^{(i)}) = G_2^{(i)}$ for all i and in particular the sizes of the derived factors of G_1 and G_2 are the same.*

The information in the WCT and the 2-S-ring is still insufficient to determine a group. There are examples to show that pairs of nonisomorphic groups exist which have the same WCT and 2-S-ring. The smallest example is a Brauer pair of order 2^9.

For any S-ring $\mathbf{S} = \{\Gamma_i\}_{i=1}^r$ on a group G with structure constants λ_{ijk} given by

$$\overline{\Gamma}_i \overline{\Gamma}_j = \sum_t \lambda_{ijt} \overline{\Gamma}_t$$

one may consider the equations

$$y_i y_j = \sum_t \lambda_{ijt} y_t \tag{10.10.1}$$

where $y = \{y_i\}$ is a set of commuting variables. If $\mathbf{S}$ is the class algebra of G, Frobenius showed that there are r independent solutions $\{\underline{y}_t\}_{t=1}^r$ of (10.10.1) where in this case r is the number of conjugacy classes of G and these solutions may be placed into an array which leads to the character table of G.

Humphries and Rode show that if $\mathfrak{S}_G^{(k)}$ is commutative of dimension r then there are precisely r linearly independent solutions to the corresponding equations and that in this case the information in the corresponding $r \times r$ table is the same as that in $\mathfrak{S}_G^{(k)}$. They also show that for any FC-group $\mathfrak{S}_G^{(4)}$ determines G.

In the thesis of Rode [242, 281] the following theorem is proved.

Theorem 10.9 *If G_1 and G_2 are FC-groups with the same 3-S-ring then they are isomorphic.*

Hence the information in the 3-characters and $\mathfrak{S}_G^{(3)}$ coincide. Thus for finite groups the interesting case is where $k = 2$.

Humphries and Rode also note that $\mathfrak{S}_G^{(k)}$ gives complete information about $\mathfrak{S}_G^{(k-1)}$.

The question of Brauer on the information on a group which in addition to the character table determines the group has a natural extension:

Is there a concise way of presenting information which in addition to the WCT of G and $\mathfrak{S}_G^{(2)}$ determines G?

In the paper [155] Humphries and Rhode characterize finite groups G for which $\mathfrak{S}_G^{(3)}$ is commutative. Such groups are either abelian or generalized dihedral groups. They also show that if for a finite group G, $\mathfrak{S}_G^{(4)}$ is commutative then G is necessarily abelian. These results have bearing on the tables of the extended higher characters discussed in Chap. 6.

Acknowledgments The author would like to acknowledge the uses of GAP [115, 118] and the related package LOOPs [222] in calculations for the book.

Appendix A
Spherical Functions on Groups

A.1 Spherical Functions in the Infinite Setting

Spherical functions first appeared on infinite spaces, and have connections to important mathematics, some of which are indicated in [208]. They often occurred before it was realized that groups were present. In Chap. 4 it is explained how spherical functions arise in the case of association schemes, mainly in the case of finite groups, and they appear implicitly in other parts of this book. In some areas of applications, especially in physics, the direct connection to group representations is lost and there is a real possibility that the more robust techniques described in the book can make a significant contribution.

The following account of Gelfand pairs and spherical functions is based on that of Dieudonné in [94]. It will be seen that probabilistic ideas are inherent in the discussion. The definitions below are similar to those in Chap. 4, Sect. 4.3.

Let G be a locally compact topological group. A collection of subsets Σ of G is a σ-**algebra** if it is closed under complementation and countable union, i.e. if $A_1, A_2, A_3, \ldots$ lie in Σ, then

$$\bigcup_{i=1}^{\infty} A_i$$

lies in Σ.

The σ-algebra generated by all the compact subsets of G is called the **Borel algebra** and an element of it is called a **Borel set**. If $g \in G$, $S \subseteq G$, define the **left translate** $gS := \{gs \mid s \in S\}$ and the **right translate** $Sg := \{sg \mid s \in S\}$. Left and right translates map Borel sets into Borel sets.

© Springer Nature Switzerland AG 2019
K. W. Johnson, *Group Matrices, Group Determinants*
and Representation Theory, Lecture Notes in Mathematics 2233,
https://doi.org/10.1007/978-3-030-28300-1

Let Σ be a σ-algebra on G. A function μ from Σ to the extended real number line is called a **measure** if it satisfies the following properties:

1. $\mu(A) \geq 0$ for all $A \in \Sigma$.
2. $\mu(\cup_{i=1}^{\infty} A_i) = \sum_{i=1}^{\infty} \mu(A_i)$.
3. $\mu(\varnothing) = 0$.

A measure μ on the Borel sets of G is called **left-translation-invariant** if and only if for all Borel subsets $S \subseteq G$ and all $g \in G$,

$$\mu(gS) = \mu(S).$$

A result of Haar is that there is a countably additive measure μ on the Borel sets of G, unique up to a constant mulitiple satisfying

1. μ is left-translation invariant.
2. $\mu(K)$ is finite for every compact subset $K \subseteq G$.
3. Every Borel set E is **outer regular**, i.e

$$\mu(E) = \inf\{\mu(U); \; E \subseteq U, U \text{ open}\}.$$

4. Every Borel set E is **inner regular**, i.e

$$\mu(E) = \sup\{\mu(K); \; U \subseteq K, K \text{ compact}\}.$$

Such a measure on G is called a **left Haar measure**. A **right Haar measure** is defined similarly.

A.1.1 Spherical Functions, Special Functions and Gelfand Pairs

The term **special function** is used to describe a collection of functions which have been given more or less established names and notation, due to their importance in mathematical analysis, functional analysis, physics, or other applications. They date back to the eighteenth century. Many of these special functions appeared in the work of Cartan and Weyl on representations of Lie groups and this provided an insight into their properties.

The spherical functions are among the most interesting special functions. Their theory generalizes both the "spherical harmonics" of Laplace and commutative harmonic analysis, and they play an important part in the theory of infinite dimensional linear representations of Lie groups (which Dieudonné indicates may be regarded as noncommutative harmonic analysis). A locally compact group G is called **unimodular** if its left Haar measure is also invariant under right translations. It is then also invariant under the symmetry $x \mapsto x^{-1}$. The family of noncommutative unimodular groups includes compact groups and semi-simple Lie groups.

Let G be a unimodular group with Haar measure m_G. Define $L^1(G) = \{f : G \to \mathbb{C} : \int |f(x)| dx < \infty\}$. The **convolution** of two functions γ_1, γ_2 in $L^1(G, m_G)$ generalizes that given in Sect. 7.3 and is defined by

$$(\gamma_1 * \gamma_2)(x) = \int_G \gamma_1(xt^{-1})\gamma_2(t)dm_G(t).$$

It belongs to $L^1(G)$. Under this operation $L^1(G)$ becomes a Banach algebra under the usual norm, but if G is non-abelian it is the dual of the group algebra and therefore noncommutative. Using a similar construction to that given in Chap. 4 a commutative object may often be obtained as follows. Let K be a compact subgroup of G, and consider the functions in $L^1(G)$ which are invariant under both left and right translations by elements of K,

$$\gamma(tx) = \gamma(xt) = \gamma(x) \text{ for all } t \in K$$

where the equality is understood to be for almost all $x \in G$. The subspace of $L^1(G)$ consisting of these functions is written $L^1(K\backslash G/K)$ and is dual to the double coset algebra of K. It is a closed subalgebra of the Banach algebra $L^1(G)$. As in the finite case $L^1(K\backslash G/K)$ need not be commutative, and in the case where it is, the pair (G, K) is defined to be a **Gelfand pair** (generalizing the definition given in Chap. 4 in the finite case). The following theorem, due to Gelfand, gives a set of conditions ensuring a Gelfand pair.

Theorem A.1 *Let $\tau : G \to G$ be an involutive automorphism ($\tau^2 = id$) of the locally compact unimodular group G, and let K be the closed subgroup of elements of G fixed by τ. Suppose that*

(i) K is compact.
(ii) each $x \in G$ can be written in at least one way as $x = yz$ with $\tau(y) = y$ and $\tau(z) = z^{-1}$.

Then (G, K) is a Gelfand pair.

If G is a commutative locally compact group and the compact subgroup K is $\{e\}$ then (G, K) is automatically a Gelfand pair. Harmonic analysis (interpreted by Dieudonné as the theory of commutative locally compact groups) can be generalized to Gelfand pairs as follows. A set of **characters** (or **spherical functions**) can be defined for an arbitrary commutative Banach algebra, which in the case of the algebra $L^1(K\backslash G/K)$ arising from a Gelfand pair can be written uniquely as

$$\gamma \mapsto \varsigma_\omega(\gamma) = \int_G \gamma(x)\omega(x)dm_G(x)$$

where the complex function ω is a uniformly (left and right) continuous function on G such that $|\omega(x)| < \omega(e) = 1$ and $\omega(tx) = \omega(xt) = \omega(x)$ for $t \in K$ and $x \in G$. The functions ω are called the (zonal) spherical functions for the Gelfand pair (G, K). If ω is such a function, the complex conjugate $\bar{\omega}$ and the function $\tilde{\omega}(x) = \omega(x^{-1})$ are also spherical functions. If G is commutative and $K = \{e\}$, the spherical functions coincide with the usual characters of G, i.e. they are continuous homomorphisms of G into the group of complex numbers of absolute value 1. The following properties generalize those of characters of commutative locally compact groups.

(I) For bounded continuous functions ω on G such that $\omega(tx) = \omega(xt) = \omega(x)$ for $t \in K$ and $x \in G$, the following are equivalent:

   (a) ω is a spherical function.
   (b) If m_K is the Haar measure on K with total mass 1,

$$\int_{t \in K} \omega(xty) dm_K(t) = \omega(x)\omega(y) \text{ for } x, y \in G. \tag{A.1.1}$$

   (c) $\omega(e) = 1$ and $\gamma * \omega = \lambda_\gamma \gamma$ for some scalar $\lambda_\gamma \in \mathbb{C}$ for all $\gamma \in L^1(K \backslash G / K)$.

   If G is commutative and $K = \{e\}$ Eq. (A.1.1) becomes $\omega(xy) = \omega(x)\omega(y)$ and the λ_γ in c) is $\int_G \gamma(t)\overline{\omega(t)} dm_G(t)$.

(II) The set $S(G/K)$ of spherical functions is locally compact for the compact-open topology, which coincides on that set with the weak* topology of $L^\infty(G)$ (when ω is identified with the character ς_ω of $L^1(K \backslash G / K)$). The mapping $(\omega, x) \mapsto \omega(x)$ of $G \times S(G/K)$ into $\mathbb{C}$ is continuous, and every compact subset of $S(G/K)$ is equicontinuous.

   If G is commutative and $K = \{e\}$, $S(G/K)$ is the dual group of G but in general $S(G/K)$ has no group structure.

(III) The **Fourier transform** $\mathcal{F}(\gamma)$ of a function $\gamma \in L^1(G)$ is a function on the space $S(G/K)$ defined by

$$\mathcal{F}(\gamma) : \omega \mapsto \int_G \gamma(x)\omega(x^{-1}) dm_G(x). \tag{A.1.2}$$

$\mathcal{F}(\gamma)$ is continuous and tends to 0 at infinity. Furthermore, if γ, ζ are any two functions in $L^1(K \backslash G / K)$

$$\mathcal{F}(\gamma)(\omega) \le \int_G |\gamma(x)| dm_G(x) = N_1(\gamma) \text{ for all } \omega \in S(G/K) \tag{A.1.3}$$

and

$$\mathcal{F}(\gamma * \zeta) = \mathcal{F}\gamma . \mathcal{F}\zeta. \tag{A.1.4}$$

(IV) An important notion in the theory of representations of locally compact
groups is that of functions of **positive type**. A function of positive type is
a (complex valued) bounded continuous function $x \mapsto f(x)$ on the group G,
such that for any finite subset $\{g_1, g_2, \ldots, g_m\}$ in G,

$$\sum_{i,j} f(g_i g_j^{-1}) \varsigma_i \overline{\varsigma_j} \geq 0 \tag{A.1.5}$$

for all systems $\{\varsigma_1, \varsigma_2, \ldots, \varsigma_n\}$ of complex numbers. Thus for a finite group
a function f is of positive type if and only if the group matrix $X_G(f)$ is
nonnegative definite.

Equivalently, for any function $\gamma \in L^1(G)$,

$$\int_G p(x)(\widetilde{\overline{\gamma}} * \gamma)(x) dm_G(x) \geq 0.$$

It is immediate to verify that for a locally compact commutative group G, the
characters of G are functions of positive type. However for a Gelfand pair
(G, K) it does not follow that spherical functions are necessarily functions of
positive type and this brings into consideration the closed subspace $Z(G/K)$
of $S(G/K)$ consisting of spherical functions of positive type. The most
interesting results are obtained on this space and hence it represents the
closest generalization of commutative harmonic analysis.

(V) Firstly, for a function p of positive type on a non-discrete group G the
sesquilinear form

$$(\varsigma, \gamma) \mapsto \int_G (\widetilde{\overline{\varsigma}} * \gamma)(x) p(x) dm_G(x) \tag{A.1.6}$$

defines a structure of pre-Hilbert space on $L^1(G)$, which by passage to quotient and completion, yields a Hilbert space E_p in which the left translation
by an element $g \in G$ extends to a unitary transformation U_g of E_p. Then
$g \mapsto \underline{U}(g)$ is a unitary representation of G into E_p and there is in E_p a
vector x_0 such that

$$p(g) = (\underline{U}(g).x_0 | x_0) \text{ for all } g \in G. \tag{A.1.7}$$

In particular, if (G, K) is a Gelfand pair, and ω a spherical function of positive
type for (G, K) there is a unitary representation $\underline{U}_\omega$ of G in a Hilbert space
E_ω associated to ω in this manner. A remarkable property is that $\underline{U}_\omega$ is
irreducible and that its restriction to K contains the trivial representation of
K exactly once. Conversely, every irreducible unitary representation of G in
a Hilbert space, whose restriction to K contains the trivial representation of
K at least once, is equivalent to one and only one representation $\underline{U}_\omega$ for a
function $\omega \in Z(G/K)$.

(VI) The concept of function of positive type is a special case of the concept of a **complex measure of positive type** on G. It is a measure μ which satisfies the condition that for any $\gamma \in L^1(G)$,

$$\int_G (\widetilde{\overline{\gamma}} * \gamma)(x)d\mu(x) \geq 0,$$

and therefore the functions of positive type p are those such that $p.m_G$ is a measure of positive type. The same construction as in V), with μ replacing $p.m_G$, again produces a Hilbert space E. Consider the closed subspace $H_\mu \subset E$, the closure of the image of $L^1(K\backslash G/K)$. If $\pi : L^1(K\backslash G/K) \to H_\mu$ is the natural mapping, for ζ, γ in $L^1(K\backslash G/K)$, it follows that $\pi(\zeta * \gamma) = V_\mu(\zeta).\pi(\gamma)$, where V_μ is a continuous homomorphism of the commutative algebra $L^1(K\backslash G/K)$ into the algebra $\mathcal{L}(H_\mu)$ of continuous endomorphisms of H_μ. The Plancherel-Godement theorem may be applied to this homomorphism. It shows that, on the locally compact space $Z(G/K)$ there is a unique positive measure μ^Δ such that, for every function $\gamma \in L^1(K\backslash G/K)$ the cotransform $\overline{\mathcal{F}}(\gamma) = \mathcal{F}(\widetilde{\gamma})$ of γ belongs to $L^2(\mu^\Delta)$ and for any two functions $\gamma, \zeta \in L^1(K\backslash G/K)$

$$\mu^\Delta(\widetilde{\overline{\zeta}} * \gamma) = \int_Z \overline{\gamma}\zeta(\omega)\overline{\overline{\gamma}\gamma}(\omega)d\mu^\Delta(\omega), \tag{A.1.8}$$

and H is naturally isomorphic to $L^2(\mu^\Delta)$. Then μ^Δ is defined to be the **Plancherel transform** of the measure μ of positive type. For example it is easily seen that the Dirac measure ε_e at the neutral element e of G is a measure of positive type, and its Plancherel transform ε_e^Δ is written m_Z and is called the canonical measure on $Z(G/K)$. For $\mu = \varepsilon_e$, (A.1.8) gives

$$\int_G \zeta(x)\gamma(x)dm_G(x) = \int_Z \mathcal{F}\zeta(\omega)\overline{\mathcal{F}\gamma(\omega)}dm_Z(\omega). \tag{A.1.9}$$

If G is commutative and $K = \{e\}$, m_Z is the Haar measure on the dual $\widehat{G}$ associated with m_G and (A.1.9) is the usual Plancherel formula. Equation (A.1.9) shows that the Fourier transform $\gamma \mapsto \mathcal{F}\gamma$ extends to an isomorphism of the Hilbert space $L^2(K\backslash G/K)$ (the closure of $L^1(K\backslash G/K)$ in $L^2(G)$) onto the Hilbert space $L^2(Z(G/K),m_Z)$, which generalizes the well-known isomorphism of $L^2(G)$ onto $L^2(\widehat{G})$ in the commutative case.

For any spherical function $\omega \in Z(G/K)$ of positive type, $\omega.m_G$ is a measure of positive type and its Plancherel transform

$$(\omega.m_G)^\Delta = \varepsilon_\omega$$

is the Dirac measure on $Z(G/K)$ at the point ω.

Finally, it may be shown that any bounded measure on G is a linear combination of measures of positive type and therefore its Plancherel transform μ^Δ is defined. In addition μ^Δ has a density with respect to the canonical measure m_Z which is continuous and bounded. This is written $\mathcal{F}\mu$ and it is given by the formula

$$\mathcal{F}\mu(\omega) = \int_G \omega(x^{-1})d\mu(x). \tag{A.1.10}$$

which extends to bounded measures the definition (A.1.2) of the Fourier transform. If G is commutative and $K = \{e\}$, the relation (A.1.10) is written

$$\mathcal{F}\mu(\hat{x}) = \int_G \overline{\langle x, \hat{x}\rangle}d\mu(x), \tag{A.1.11}$$

and for a bounded measure μ on G the relation in (A.1.8) is written

$$\int_G (\zeta * \gamma)(x)d\mu(x) = \int_{\hat{G}} \overline{\mathcal{F}\zeta(\hat{x})}\overline{\mathcal{F}\gamma(\hat{x})}\mathcal{F}\mu(\hat{x})dm_G(\hat{x}).$$

(the generalized Plancherel theorem).

(VII) When G is a unimodular connected Lie group it may be shown that the spherical functions corresponding to a Gelfand pair (G, K) are of class C^∞ and are eigenvectors of all differential operators which are invariant under left translations by elements of G and right translations by elements of K. This implies that on semisimple Lie groups spherical functions are analytic, because in this case elliptic operators are always among these invariant operators.

A.1.2   Examples of Spherical Functions

The motivating example is the following. Let $G = SO(3)$ and let X be the unit sphere $S^2 \subset \mathbb{R}^3$. Let x_0 be the point $(0, 0, 1)$. Then $K = Stab_G(x_0)$ is the group of rotations about the z-axis and the orbits of K are the horizontal circles of height $\cos\varphi$ where φ denotes the angle between a radial vector and the z-axis. The functions which are constant on the orbits of K are called spherical functions and are of the form $\gamma_n(\varphi) = P_n(\cos\varphi)$ corresponding to the **Legendre polynomials** $P_n(x)$. These are the basic example of a family of spherical functions, and arose before the group theoretic connection was available. For example, they have the generating function

$$(1 - 2xr + r^2)^{-1/2} = \sum_{n=0}^{\infty} P_n(x)r^n,$$

and $P_n(x)$ also satisfies the differential equation

$$(1 - x^2)y'' - 2xy' + n(n + 1) = 0$$

It is the solution of the recurrence relation

$$(2n + 1)P_n(x) = (n + 1)P_{n+1}(x) + nP_{n-1}(x), \quad P_{-1}(x) = 0, \quad P_0(x) = 1,$$

and also is **harmonic**, i.e. $\nabla^2 P_n(\cos\phi) = 0$.

In a more general setting, there are three main types of Gelfand pairs (G, K) where G is nonabelian, each giving rise to spherical functions among which there appear many special functions.

(I) Let G be a linear semisimple compact connected Lie group. If σ is any involutive automorphism of G the conditions of Gelfand's theorem are satisfied, K being as above the set of elements fixed by σ. The set of these automorphisms were determined explicitly by E. Cartan. In this case all spherical functions are of positive type and the space $S(G/K)$ is discrete.

The most interesting example is given by the group $G = SO(n + 1)$ of rotations in $\mathbb{R}^{n+1}$ with $K = SO(n)$. G/K is identified with the sphere S^n and K is the subgroup which fixes the first vector e_0 of the canonical basis of $\mathbb{R}^{n+1}$. Since G is compact, $L^2(G)$ is an algebra (under the convolution product) and hence is the closure $L^2(K \backslash G/K)$ of $L^1(K \backslash G/K)$ in $L^2(G)$ which is commutative by Gelfand's theorem. The functions of $L^2(K \backslash G/K)$ can be identified with the functions in $L^2(S^n)$ which depend only on one variable, namely the angle ϕ between the variable vector $x \in S^n$ with the vector e_0.

For $n \geq 2$ the space $L^2(S^n)$ splits into a Hilbert sum of finite dimensional subspaces $E_m (m = 0, 1 \ldots)$ stable under the action of G. E_m is precisely the space of the restrictions to S^n of the harmonic polynomials which are homogeneous of degree m and the subrepresentation of G into E_m is irreducible. Each E_m contains exactly one spherical function ω_m; let $\omega_m(x) = G_{m,n+1}(\cos\theta)$. The $G_{m,n+1}$ are the **Gegenbauer polynomials** and reduce to the **Legendre polynomials** in the case $n = 2$. They satisfy the differential equation

$$(1 - z^2)y'' - nzy' + m(m + n - 1)y = 0.$$

(II) Let G be a linear non-compact connected semisimple Lie group with finite center, and K be a maximal compact subgroup of G. In this case there also exists an involutive automorphlsm σ of G for which K is the group of fixed points and which satisfies the conditions of Gelfand's theorem. A typical example is given by $G = SL(n, \mathbb{R})$, $K = SO(n)$. The automorphism σ is then the map $X \mapsto (X^{-1})^T$ (the contragredient matrix).

It may then be shown that $G = SK$ where S is a closed solvable subgroup (the so-called Iwasawa decomposition). Suppose a continuous homomorphism

$\alpha : S \mapsto \mathbb{C}^*$ is known.Then α extends to a continuous function on G by taking $\alpha(st) = \alpha(s)$ for $s \in S, t \in K$. The function

$$\omega(x) = \int_K \alpha(tx)dm_K(t) \qquad (A.1.12)$$

satisfies the functional equation

$$\omega(x)\omega(y) = \int_K \alpha(xty)dm_K(t) \qquad (A.1.13)$$

and therefore is a spherical function if it is bounded (solutions of (A.1.13) are called generalized spherical functions). A deep theorem of Harish-Chandra proves that all generalized spherical functions relative to the Gelfand pair (G, K) are given by formula (A.1.12) and in addition all homomorphlsms α can be determined explicitly from a detailed study of the Lie algebra of G.

The simpliest example consists in the pair $G = SL(2, \mathbb{R})$, $K = SO(2, \mathbb{R})$. Here $G = KS = SK$ where S is the solvable group of triangular matrices

$$\begin{bmatrix} a & b \\ 0 & a^{-1} \end{bmatrix}$$

wlth $a > 0$, and the decomposition is unique. It is easily shown that the matrices

$$X = \begin{bmatrix} a & b \\ c & d \end{bmatrix}$$

of a double coset relative to K are those for which

$$Tr(X^T X) = a^2 + b^2 + c^2 + d^2$$

has a given value $2v$ with $v \geq 1$. The functions of $C(K\backslash G/K)$ are therefore the functions $\gamma((a^2+b^2+c^2+d^2)/2)$ where γ is continuous on the half-line $[1, \infty[$. The generalized spherical functions are then

$$P_\rho(v) = \frac{1}{2\pi} \int_0^{2\pi} (v + \sqrt{v^2 - 1}\cos\phi)^\rho d\phi$$

where ρ is any complex number. They are the Legendre functions of index 0 and the functional equation corresponding to (A.1.13) is

$$P_\rho(\cosh(t))P_\rho(\cosh(u)) = \frac{1}{2\pi}\int_0^{2\pi} P_\rho(\cosh(t)\cosh(u)$$
$$+ \sinh(t)\sinh(u)\cos\phi)d\phi$$

where t, u are arbitrary real numbers.

(III) In the third set of examples G is a unimodular group containing a normal
abelian subgroup A with no elements of order 2, and a (non-normal) compact
subgroup K such that the mapping $(t, s) \mapsto ts$ is a diffeomorphism of the
manifold $K \times A$ onto G. Then $\sigma : ts \mapsto ts^{-1}$ is an involution having the
properties required by Gelfand's theorem, so that (G, K) is a Gelfand pair.
Starting from a continuous homomorphism $\alpha : A \to C$ and for $x = ts$ with
$t \in K, s \in A$, define

$$\omega(x) = \int_K \alpha(usu^{-1})dm_K(u).$$

It is easily verified that ω satisfies the functional equation (A.1.13) and is thus a
generalized spherical function and it can be shown further that all generalized
spherical functions are obtained in this manner.

The typical example here is the group G of orientation-preserving isome-
tries of the Euclidean plane $\mathbb{R}^2$. Here G can be identified with the group of
matrices

$$\begin{bmatrix} \cos\theta & \sin\theta & x \\ -\sin\theta & \cos\theta & y \\ 0 & 0 & 1 \end{bmatrix}$$

acting by multiplication on vectors of $\mathbb{R}^2$ represented by column vectors

$$\begin{bmatrix} a \\ b \\ 1 \end{bmatrix}.$$

The normal subgroup A is the group of translations (corresponding to matrices
with $\theta = 0$) and K is the group of rotations (corresponding to $x = y = 0$).
The homogeneous space G/K is here identified with $\mathbb{R}^2$, and the double
cosets KsK are identified with the orbits of K in $\mathbb{R}^2$ which are the circles
$x^2 + y^2 = r^2$. Continuous functions on $K \backslash G/K$ are thus identified with
functions $\psi((x^2 + y^2)^{1/2})$ where ψ is continuous on the interval $[0, +\infty)$.
The continuous homomorphlsms $\alpha : A \to C$ are the exponentials $(x, y) \mapsto$
$\exp(\lambda x + \mu y)$ with λ, μ arbitrary complex numbers.

The formula (A.1.2) therefore identifies the solutions of (A.1.13) with the
continuous functions on $[0, +\infty)$ given by

$$\psi(r) = \frac{1}{2\pi} \int_0^{2\pi} \exp(r(\lambda\cos\phi + \mu\sin\phi))d\phi$$

On taking $\lambda = 0$, $\mu = i$ the Bessel function J_0 is given. Since this is
bounded, a spherical function is obtained.

Appendix B
The Personal Characteristics of Frobenius

Frobenius appears to have been a difficult person as a colleague. The following are quotes from [19].

> Unfortunately we can only obtain a picture of him from his statements on faculty affairs. Accounts of him by contemporaries or in other archives are scarce. He shows himself to us as an extraordinarily argumentative, aggressive man, in his excessive aversion to people and things. . . . .

> The aversion of Frobenius against Klein and Lie knew no bounds, as has already been described. Frobenius was also not capable of unrestricted impartiality towards Hilbert. His "Laudatio" for Hilbert is full of small side blows and malicious statements.

The following is a translation of part of a letter written by Frobenius in reply to a request to comment on candidates for a professorial chair at E.T.H. Zurich. It appears in the original German in [105]. It illustrates some of the above comments. Thanks are due to Maria Peters, the daughter of Max Dehn, who helped with the translation.

Letter on the candidates for Professor at ETH Zurich (June 27 1913).

Among the gentlemen whom you name there are only two geometers, Dehn and Salkowski. Dehn is at present the best in topology, on which he reported (with Heegard) in the Encyclopaedia. The problems on which he works all lie somewhat far away from the mainstream. He is a completely original thinker. Exactly the opposite is true of Salkowski. But I can imagine that your colleagues on the faculty will find his investigations on surfaces and space curves especially appealing.

Perron is more an algebraist and an analyst and works on continued fractions, difference equations and differential equations. He has very original investigations on Dirichlet series, on matrices and irreducibility. For my taste he grasps too much at originality. He would prefer to treat a matter less well than to follow the approach of others.

The other five men named belong to the Göttingen school. From their nominations I see that you have already got to work diligently in Göttingen. Here you can

© Springer Nature Switzerland AG 2019
K. W. Johnson, *Group Matrices, Group Determinants
and Representation Theory*, Lecture Notes in Mathematics 2233,
https://doi.org/10.1007/978-3-030-28300-1

rely well on the opinion of Hurwitz. All these men write on integral equations, on equations and forms with infinitely many variables, on extensions with orthogonal functions and carry out the deepest investigations on set theory. Plus ça varie, plus c'est la même chose. The works of my old friend Geiser, to whom I ask you to give hearty greetings, appear without need of claims. In each of his works there is an original thought which gives his work character and which has not before been expressed in this form. But the works of these men are very general, very deep, so fundamentally deep that a short-sighted person like myself has difficulty in understanding the new thoughts in them.

In the last year many new chairs of mathematics have been filled. In almost all these searches they have sought out my advice, but never followed it (you know well my correspondence with Professor Kleiner). To begin with I do not recommend Professor J. (*sic*) Schur (now in Bonn). He is much too good for Zürich and shall 1 day be my successor in Berlin. Among the younger generation Bieberbach (now in Basel) is the most significant (26 years old).

Among the pure Göttinger, Hellinger (in Marburg) is the best in the theory of integral equations. But he must have somehow fallen out of favor with the Göttingers. His name is not among those put forward in the last searches and also is not on your list.

Of the five men whom you have suggested to me I also would put Weyl in first place. His book on Riemann surfaces which has recently appeared is very famous. Incidentally, he has the prospect of being called to Breslau in place of Carathéodory. The works of the young Riesz are strongly philosophical (and not to my taste). His name is especially known for the theorem which he discovered at the same time as Fischer.

Toeplitz busies himself especially with sharpening and simplifying proofs in the fields named above. He has incidentally studied under Schröter in Breslau and has a good knowledge of analytic and synthetic geometry. He is a superb Dozent, truly the best of the men whom you have named.

Haar has proved a number of beautiful theorems on expansion in terms of orthogonal functions, which go much further than the results of Stekloff and Kneser. Plancherel has worked in the same area, with less success.

To summarize my advice once again: if you call Weyl you will have made an excellent flawless choice. If he cannot be had, then take Dehn or Hellinger. But you would also do well to inquire about Bieberbach who is a brilliant researcher and at least speaks for the future.

Perron and Toeplitz would also be good for the position. The others which you have named are good mathematicians, in particular Riesz, and you should look upon these as being in the second line.

. . . .

Hermann Weyl was called and took up the position.

The letter illustrates the penetration of Frobenius but also his prejudices.

The following further comment on Frobenius appeard in [19].

Compromise-free sharpness and a certain degree of irritability are prominent features of the letters of Frobenius (This is illustrated in the letter above).

There is the statement attributed to Frobenius by Gian-Carlo Rota: "Hilbert is a good mathematician but he will never be as good as Schottky". According to Hawkins, this is apocryphal. However, when Schottky was called to Berlin as "Ordentlicher Professor", the letter which Frobenius wrote in support compared him to Hilbert and seemed to strive to put them on the same level (Schottky was a personal friend of Frobenius).

The harsh portrait of Frobenius in Biermann may be contrasted with the extracts of the letters he wrote to Dedekind during the period described in Sect. 7.3 when group characters and group determinants emerged, quoted in [137]. They contain humorous remarks and seem to be written in a jolly frame of mind.

The reminiscences of C. L. Siegel (Erinnerungen an Frobenius, Frobenius Collected Works Vol. I, p. IV) paint a picture of Frobenius as a deeply inspiring lecturer.

When I matriculated in Fall 1915 from Berlin University a war was in full swing. Perhaps I did not see through the background to the political events, but I took an instinctive aversion to the violent thrusts of men and wished to study in an area such as astronomy.

The Astronomy professor at the University had announced that his course would start 14 days after the start of the semester. At the same days and hours scheduled there was a course by Frobenius on number theory. Since I had no idea what number theory was about I attended 2 weeks of Frobenius' lectures out of pure curiosity, and that decided the lifelong direction of my mathematical interests. I cancelled my enrolment in astronomy and remained in Frobenius' number theory course.

It is hard to explain how this course in number theory made such a great and long lasting impression on me. The material was approximately the classical course of Dirichlet as it has come to us in Dedekind's treatment. Frobenius recommended the "Dirichlet-Dedekind" book for the course, and this was the first scientific book which I bought with the pocket money which I earned from giving private lessons, in the same way that in contemporary times a student spends the first instalment of his stipend on a motor bike. Frobenius lectured completely without notes, and made no errors or miscalculations the entire semester. When he introduced continued fractions at the beginning it obviously gave him great joy to give the various algebraic identities and recursions with great certainty and astonishing rapidity one after the other, meanwhile giving a quick ironical glance at the auditorium where the diligent note-takers could hardly keep up with the quantity of material. He rarely looked at the audience and most of the time faced the blackboard.

By the way, it was unusual then in Berlin that a student and professor had any personal relationship in connection with the course, unless special exercise classes were held, for example with Planck in mathematial physics. Frobenius did not hold any exercise classes, but now and again set exercises on the material in the class. The students could lay their solutions on the desk in the auditorium before the next class. Frobenius then took the work with him and in the next class returned them to the desk signed with a "v". However the correct or best solution was never given or presented by a student.

The exercises were not particularly difficult, so far as I can recall, and always involved special cases, and no generalizations. For example in the theory of continued fractions it was to be shown that the number of divisions in the Euclidean algorithm for two natural numbers is at most five times the the number of digits of the smaller number. Relatively few students gave solutions, but I was very interested and I tried to solve all of them, by which I also learned some algebra and number theory which did not appear in the lectures.

I have already related how I cannot explain properly how Frobenius' lectures created such a strong impression. From my description the nature of the experience was daunting. Without it being clear to me, I was perhaps influenced by the creative personality of the great intellectual which came through from the nature of his lecture. After depressing school years under mediocre or downright bad teachers for me this was a novel and liberating experience...

Siegel unexpectedly received a bursary of 144 marks 50 pfennig in memory of Eisenstein. Schur indicated to him later that Frobenius probably had nominated him as a result of his solutions to the exercises. However he never spoke to Frobenius in person.

Bibliography

1. S. Agaian, J. Astola, K. Egiazarian, *Binary Polynomial Transforms and Nonlinear Digital Filters* (Marcel Dekker, New York, 1995)
2. M. Aguiar, C. André, C. Benedetti et al., Supercharacters, symmetric functions in noncommuting variables, and related Hopf algebras. Adv. Math. **229**, 2310–2337 (2012)
3. D. Aldous, Random walks on finite groups and rapidly mixing Markov chains, in *Séminaire de Probabilités, XVII*. Lecture Notes in Mathematics, vol. 986 (Springer, Berlin 1983), pp. 243–297
4. S.A. Amitsur. Groups with representations of bounded degree. II. Ill. J. Math. **5**, 198–205 (1961)
5. E. Arias-Castro, P. Diaconis, R. Stanley, A super-class walk on upper-triangular matrices. J. Algebra **278**, 739–765 (2004)
6. M.F. Atiyah, Characters and cohomology of finite groups. Publ. Math. I.H.E.S. **9**, 23–64 (1961)
7. L. Auslander, P. Tolmieri, Is computing with finite Fourier transforms pure or applied mathematics? Bull. Am. Math. Soc. (N.S.) **1**, 847–897 (1979)
8. J. Baez, The octonions. Bull. Am. Math Soc. **39**, 145–205 (2002)
9. R.A. Bailey, *Association Schemes: Designed Experiments, Algebra and Combinatorics*. Cambridge Studies in Advanced Mathematics, vol. 84 (Cambridge University Press, Cambridge, 2004)
10. E. Bannai, T. Ito, *Algebraic Combinatorics I: Association Schemes* (Benjamin/Cummings, London, 1984)
11. E. Bannai, S.-Y. Song, The character tables of Paige's simple Moufang loops and their relationship to the character tables of PSL(2,q). Proc. Lond. Math. Soc. **58**, 209–236 (1989)
12. A. Baumgartner, Über Kompositionsalgeben beliebigen Grades. Dissertation Universität Düsseldorf, 1970
13. D. Bayer, P. Diaconis, Trailing the dovetail shuffle to its lair. Ann. Appl. Probab. **2**, 294–313 (1986)
14. F.A. Berezin, *Introduction to Superanalysis* (Reidel, Kufstein, 1987)
15. A. Bergmann, Formen auf Moduln über kommutativen Ringen beliebiger Charakteristik. J. Reine Angew. Math. **219**, 113–156 (1965)
16. A. Bergmann, Hauptnorm und Struktur von Algebren. J. Reine Angew. Math. **222**, 160–194 (1966)
17. A. Bergmann, Reduzierte Normen und Theorie von Algebren, in *Algebra-Tagung Halle 1986, Tagungsband Wiss* (Beiträge Halle (Saale) 1987/33 (M48) 1987), pp. 29–57

© Springer Nature Switzerland AG 2019
K. W. Johnson, *Group Matrices, Group Determinants
and Representation Theory*, Lecture Notes in Mathematics 2233,
https://doi.org/10.1007/978-3-030-28300-1

18. M. Bhargava, Higher composition laws I: a new view on Gauss composition, and quadratic generalizations. Ann. Math. **159**, 217–250 (2004)
19. K.-R. Biermann, *Die Mathematik uhd ihre Dozenten an der Berliner Universität 1810–1933* (Akademie Verlag, Berlin, 1988)
20. W. Blaschke, G. Bol, *Geometrie der Gewebe. Topologische Fragen der Differentialgeometrie* (J.W. Edwards, Ann Arbor, 1944)
21. H.I. Blau, Table algebras. Eur. J. Comb. **30**(6), 1426–1455 (2009)
22. M. Bocher, *Introduction to Higher Algebra* (Dover, New York, 2004)
23. E. Borel, A. Chéron, *Théorie Mathématique du Bridge á la Portée de Tous* (Gauthier-Villars, Paris 1940)
24. R. Bott, R.J. Milnor, On the parallelizability of the spheres. Bull. Am. Math. Soc. **64**, 87–89 (1958)
25. N. Bourbaki, *Algebra I* (Springer, Berlin, 1970)
26. H. Brandt, Der Kompositionsbegriff bei den quaterniären quadratischen Formen. Math. Ann **91**, 300–315 (1924)
27. H. Brandt, Über die Komponierbarkeit der quaterniären quadratischen Formen. Math. Ann. **94**, 179–197 (1925)
28. H. Brandt, Idealtheorie in Quaternionalgebren. Math. Ann. **99**, 1–29 (1928)
29. R. Brauer, Über die Kleinsche Theorie der algebraischen Gleichungen. Math. Ann. **110**, 473–500 (1935)
30. R. Brauer, in *Representations of Finite Groups*, ed. by T.L. Saaty. Lectures in Modern Mathematics, vol. I (Wiley, New York, 1963), pp. 133–175
31. E.O. Brigham, *The Fast Fourier Transform* (Prentice-Hall, Englewood Cliffs, 1974)
32. A.E. Brouwer, A.M. Cohen, A. Neumaier, *Distance-regular Graphs. Ergebnisse der Mathematik und ihrer Grenzgebiete*, vol. 3 (Springer, Berlin, 1989), p. 18
33. R.A. Brualdi, Determinantal identities: Gauss, Schur, Cauchy, Sylvester, Kronecker, Jacobi, Binet, Laplace, Muir and Cayley. Linear Algebra Appl. **52/53**, 769–791 (1983)
34. R.H. Bruck, *A Survey of Binary Systems* (Springer, Berlin, 1958)
35. V.M. Buchstaber, E.G. Rees, A constructive proof of the generalized Gelfand isomorphism. Funct. Anal. Appl. **35**(4), 257–260 (2001)
36. V.M. Buchstaber, E.G. Rees, The Gel'fand map and symmetric products. Sel. Math. (N.S.) **8**, 523–535 (2002)
37. V.M. Buchstaber, E.G. Rees, Rings of continuous functions, symmetric products and Frobenius Algebras. Russ. Math. Surv. **59**, 125–144 (2004)
38. V.M. Buchstaber, E.G. Rees, Frobenius n-homomorphisms, transfers and branched coverings. Math. Proc. Camb. Philos. Soc. **144**, 1–12 (2008)
39. W. Burnside, *The Theory of Groups of Finite Order*, 2nd edn. (Cambridge University Press, Cambridge, 1911)
40. C.S. Burrus, T.W. Parks, *DFT/FFT and Convolution Algorithms: Theory and Implementation* (Wiley, New York, 1985)
41. P.J. Cameron, in *Oligomorphic Permutation Groups*. London Mathematical Society Lecture Notes, vol. 152 (Cambridge University Press, Cambridge 1996)
42. P.J. Cameron, Coherent configurations, association schemes and permutation groups, in *Groups, Combinatorics & Geometry (Durham, 2001)* (World Scientific, River Edge, 2003), pp. 55–71
43. P.J. Cameron, Aspects of infinite permutation groups, in *Groups St. Andrews 2005, I*. London Mathematical Society Lecture Notes vol. 339 (Cambridge University Press, Cambridge, 2007), pp. 1–35
44. P.J. Cameron, K.W. Johnson, An investigation of countable B-groups. Math. Proc. Camb. Philos. Soc. **102**, 223–231 (1987)
45. P.J. Cameron, J.H. van Lint, *Designs, Graphs, Codes and Their Links*. London Mathematical Society Student Texts, vol. 22 (Cambridge University Press, Cambridge, 1991)
46. P.J. Cameron, P.M. Neumann, D.N. Teague, On the degrees of primitive permutation groups. Math. Z. **180**, 141–149 (1982)

47. P.J. Cameron, J.M. Goethals, J.J. Seidel, The Krein condition, spherical designs, Norton algebras and permutation groups. Indag. Math. **40**, 196–206 (1987)
48. A.R. Camina, Some conditions which almost characterize Frobenius groups. Israel J. Math. **31**, 153–160 (1978)
49. P. Cartier, A course on determinants, in *Conformal Invariance and String Theory*, ed. by P. Dita, V. Georgescu (Academic, Boston, 1989), pp. 443–445
50. P. Cartier, Mathemagics (a tribute to L. Euler and R. Feynman), in *Noise, Oscillators and Algebraic Randomness (Chapelle des Bois, 1999)*. Lecture Notes in Physics, vol. 550 (Springer, Berlin, 2000), pp. 6–67
51. P. Cartier, A primer on Hopf algebras. Preprint. IHES, IHES/M/06/40 (2006)
52. P.G. Casazza, G. Kutyniok (eds.), *Finite Frames, Theory and Appications* (Birkhäuser, Basel, 2013)
53. P.G. Casazza, G. Kutyniok, F. Philipp, Introduction to finite frame theory, in *Finite Frames, Theory and Appications*, ed. by P.G. Casazza, G. Kutyniok (Birkhäuser, Basel, 2013), pp. 1–54
54. T. Ceccherini-Silberstein, F. Scarabotti, F. Tolli, *Harmonic Analysis on Finite Groups*. Cambridge Studies in Advanced Mathematics, vol. 108 (Cambridge University Press, Cambridge, 2008)
55. A. Chan, C.D. Godsil, A. Munemasa, Four-weight spin models and Jones pairs. Trans. Am. Math. Soc. **355**, 2305–2325 (2003)
56. G.Y. Chen, A new characterization of finite simple groups. Chin. Sci. Bull. **40**, 446–450 (1995)
57. C. Cheng, A character theory for projective representations of finite groups. Linear Algebra Appl. **469**, 30–242 (2015)
58. C. Cheng, D. Han, On twisted group frames. Linear Algebra Appl. **569**, 285–310 (2019)
59. S.S. Chern, From triangles to manifolds. Am. Math. Mon. **86**, 339–349 (1979)
60. D. Chillag, A. Mann, C.M. Scoppola, Generalized Frobenius groups II. Israel J. Math. **62**, 269–282 (1988)
61. J.M. Cohen, Cogrowth and amenability of discrete groups. J. Funct. Anal. **48**, 301–309 (1982)
62. M.J. Collins, Modular analogues of Brauer's characterisation of characters. J. Algebra **366**, 35–41 (2012)
63. J.H. Conway, A.J. Jones, Trigonometric Diophantine equations (on vanishing sums of roots of unity). Acta Arith. **XXX**, 229–240 (1976)
64. J.H. Conway, D.A. Smith, *On Quaternions and Octonians* (A. K. Peters, Natick, 2003)
65. J.W. Cooley, The re-discovery of the fast Fourier transform algorithm. Mikrochim. Acta **III**, 33–45 (1987)
66. D. Cooper, G.S. Walsh, Three-manifolds, virtual homology, and group determinants. Geom. Topol. **10**, 2247–2269 (2006)
67. R.H. Crowell, The derived group of a permutation representation. Adv. Math. **53**, 99–124 (1984)
68. C.W. Curtis, *Pioneers of Representation Theory* (American Mathematical Society, Providence, 1999)
69. C.W. Curtis, I. Reiner, *Representation Theory of Finite Groups and Associative Algebras*. American Mathematical Society, 1962 (Chelsea Publishing, White River Junction, 2006)
70. E.C. Dade, E, Answer to a question of R. Brauer. J. Algebra **1**, 1–4 (1964)
71. E.C. Dade, M.K. Yadav, Finite groups with many product conjugacy classes. Israel J.Math **154**, 29–49 (2006)
72. P.J. Davis, *Circulant Matrices* (Chelsea, New York, 1994)
73. I. Daubechies, Ten lectures on wavelets, in *CBMS Conference on Series in Applied Mathematics*, vol. 61 (SIAM, Philadelphia 1992)
74. I. Daubechies, A. Grossman, Y. Meyer, Painless nonorthogonal expansions. J. Math. Phys. **27**, 1271–1283 (1985)
75. R. Dedekind, Über Gruppen, deren sämtliche Teiler Normalteilers sind. Math. Ann. **48**, 548–56 (1897). Gesammelte mathematische Werke II, Braunschweig (1931) 87–102

76. P. Delsarte, An algebraic approach to the Association scheme of coding theory. Philips Research Reports: Supplements, No. 10 (1973)

77. P. Deligne, J.W. Morgan, Notes on Supersymmetry (following Joseph Bernstein), in *Quantum Fields and Strings: A Course for Mathematicians*, vol. 1 (American Mathematical Society, Providence, 1999), pp. 41–97

78. J. Dénes, A.D. Keedwell, *Latin Squares and Their Applications* (Academic, New York, 1974)

79. C. Deninger, On the analogue of the formula of Chowla and Selberg for real quadratic fields. J. Reine Angew. Math. **351**, 171–191 (1984)

80. J. Deruyts, Essai d'une théorie générale des formes algébriques, Mém. Soc. R. Sci. Liège **17**, 1–156 (1892)

81. M. Deuring, *Algebren* (Springer, Berlin 1935, 1968)

82. P. Diaconis, *Group Representations in Probability and Statistics* (Institute of Mathematical Statistics, Hayward, 1988)

83. P. Diaconis, Patterned matrices. Matrix theory and applications (Phoenix, AZ, 1989), in *Proceedings of Symposia in Applied Mathematics*, vol. 40 (American Mathematical Society, Providence, RI, 1990), pp. 37–58

84. P. Diaconis, The cutoff phenomenon in finite Markov chains. Proc. Nat. Acad. Sci. **43**, 1659–1664 (1995)

85. P. Diaconis, Random walks on groups: characters and geometry, in *Groups St. Andrews 2001 in Oxford, vol. I*. London Mathematical Society Lecture Note Series, vol. 304 (Cambridge University Press, Cambridge, 2003), pp. 120–142

86. P. Diaconis, M. Isaacs, Supercharacters and superclasses for algebra groups. Trans. Am. Math. Soc. **360**, 2359–2392 (2008)

87. P. Diaconis, D. Rockmore, Efficient computation of the Fourier transform on finite groups. J. Am. Math. Soc. **3**, 297–332 (1990)

88. P. Diaconis, M. Shahshahani, Generating a random permutation with random transpositions. Z. Wahrsch. Verw. Geb. **57**, 159–179 (1981)

89. L.E. Dickson, On the group defined for any given field by the multiplication table of any given finite group. Trans. Am. Math. Soc. **3**, 285–301 (1902)

90. L.E. Dickson, An elementary exposition of Frobenius's theory of group-characters and group-determinants. Ann. Math. **4**(2), 25–49 (1902)

91. L.E. Dickson, Modular theory of group matrices. Trans. Am. Math. Soc. **8**, 389–398 (1907)

92. L.E. Dickson, Modular theory of group characters. Bull. Am. Math. Soc **13**, 477–499 (1907)

93. L.E. Dickson, On quaternions and their generalization and the history of the eight square theorem. Ann. Math. **20**(2) , 155–171 (1919)

94. J. Dieudonné, Gelfand pairs and spherical functions. Int. J. Math. Math.Sci. **2**, 153–162 (1979)

95. J. Dieudonné, Schur functions and group representations. Astérisque **87–88**, 7–19 (1981)

96. P.G.L. Dirichlet, *Lectures on Number Theory, (Supplements by R. Dedekind)*. History of Mathematics, vol. 16 (American Mathematical Society, Providence, 1999)

97. R.J. Duffin, A.C. Schaeffer, A class of nonharmonic Fourier series. Trans. Am. Math. Soc. **72**, 341–366 (1952)

98. S, Evdokimov, I. Ponomarenko, Schurity of S-rings over a cyclic group and the generalized wreath product of permutation groups (Russian). Algebra Anal. **24**(3), 84–127 (2012); translation in St. Petersburg Math. J. **24**(3), 431–460 (2013)

99. W.B. Fite, Certain factors of the group determinant. Am. Math Mon. **13**, 51–53 (1906)

100. E. Formanek, D. Sibley, The group determinant determines the group. Proc. Am. Math. Soc. **112**, 649–656 (1991)

101. A.R. Forsyth, *The Theory of Differential Equations*. Part III, vol. IV (Cambridge University Press, Cambridge, 1902)

102. C.M. Fortuin, P.W. Kasteleyn, J. Ginibre, Correlation inequalities on some partially ordered sets. Commun. Math. Phys. **22**, 89–103 (1971)

103. J.S. Frame, The double cosets of a finite group. Bull. Am. Math. Soc. **47**, 458–467 (1941)

104. J.S. Frame, Double coset matrices and group characters. Bull. Am. Math. Soc. **49**, 81–92 (1943)

105. G. Frei, U. Stammbach, *Hermann Weyl und die Mathematik an der ETH Zürich* (Birkhaüser, Basel, 1992), 1913–1930
106. G. Frobenius, *Über Gruppencharaktere* (Sitzungsber. Preuss. Akad. Wiss, Berlin, 1896), pp. 985–1021; Ges Abh. III, pp. 1–37
107. G. Frobenius, *Über die Primfactoren der Gruppendeterminante* (Sitzungsber Preuss. Akad. Wiss. Berlin, 1896), pp. 1343–1382; Ges Abh. III, pp. 38–77
108. G. Frobenius, *Über die Darstellung der endlichen Gruppen durch lineare Substitutionen I* (Sitzungsber. Preuss. Akad. Wiss. Berlin, 1897), pp. 994–1015; Ges Abh. III, pp. 82–103
109. G. Frobenius, *Über Matrizen aus positiven Elementen* (Sitzungsber Akad. Wiss. Berlin, 1908), pp. 471–476, Ges Abh. III, pp. 404–409
110. G. Frobenius, *Gesammelte Abhandlungen I*, ed. by J-P. Serre (Springer, Berlin, 1968)
111. W. Fulton, R. MacPherson, Characteristic classes of direct image bundles for covering maps. Ann. Math. **125**, 1–92 (1987)
112. W. Fulton, J. Harris, *Representation Theory: A First Course*. Graduate Texts in Mathematics, vol. 129 (Springer, New York, 1991) (Lecture 6)
113. D. Gabor, Theory of communication. J. Inst. Electr. Eng. **93**, 429–457 (1946)
114. P.X. Gallagher, Invariants for finite groups. Adv. Math. **34**, 46–57 (1979)
115. GAP, M. Schönert et al., *Lehrstuhl D für Mathematik*, 5th edn. (Rheinisch Westfälische Technische Hochschule, Aachen, 1995)
116. F.R. Gantmacher, *The Theory of Matrices*, vol. I (Chelsea Publishing, White River Junction, 1959)
117. I.M. Gelfand, A.N. Kolmogorov, On rings of continuous functions on topological spaces. Dokl. Akad. Nauk SSSR **22**, 11–15 (1939); English transl., Selected works of A.N. Kolmogorov, *Mathematics and Mechanics*, vol. I (Kluwer, Dordrecht 1991), pp. 291–297
118. I.M. Gelfand, D.A. Rajkov, Irreducible representations of locally bicompact groups. Mat. Sb. **13**(55), 301–316 (1942); Transl, II Ser, A.M.S. **36**, 1–15 (1964); Collected papers I.M. Gelfand, vol II (Springer, Berlin, 1988), pp 3–17
119. I.M. Gelfand, M. Kapranov, A. Zelevinsky, *Discriminants, Resultants, and Multidimensional Determinants* (Birkhäuser, Boston 1994)
120. I.M. Gelfand, D. Krob, A. Lascoux, B. Leclerc, V.S. Retakh, J.-Y. Thibon, Noncommutative symmetric functions. Adv. Math. **112**, 218–348 (1995)
121. I.M. Gelfand, S. Gelfand, V. Retakh, R.L.Wilson, Quasideterminants. Adv. Math. **193**, 56–141 (2005)
122. M. Giuliani, K.W. Johnson, Right division in Moufang loops. Comment. Math. Univ. Carol. **51**, 209–215 (2010)
123. C.D. Godsil, *Algebraic Combinatorics*. Chapman and Hall Mathematics Series (Chapman & Hall, New York, 1993)
124. A. Granville, Finding integers k for which a given Diophantine equation has no solution in kth powers. Acta Arith. **LX**, 203–212 (1992)
125. J.J. Gray, *Linear Differential Equations and Group Theory from Riemann to Poincaré* (Birkhauser, Boston, 2000)
126. P. Griffiths, J. Harris, *Principles of Algebraic Geometry* (Wiley, New York, 1978)
127. A. Grothendieck, Classes de Chern et representations lineaires des groupes discrets, in *Dix exposés sur la Cohomologie des Schémas*. Advanced Studies in Pure Mathematics, vol. 3 (North-Holland, Amsterdam, 1968), pp. 215–305
128. D.V. Gugnin, Polynomially dependent homomorphisms and Frobenius n-homomorphisms. Proc. Steklov Inst. Math. **266**, 59–90 (2009)
129. D.V. Gugnin, Topological applications of graded Frobenius n-homomorphisms. Tr. Mosk. Mat. Obs. **72**(1), 127–188 (2011); English transl., Trans. Moscow Math. Soc. **72**(1), 97–142 (2011)
130. D.V. Gugnin, Topological applications of graded Frobenius n-homomorphisms II. Trans. Moscow Math. Soc. **73**(1), 167–172 (2012)

131. A.È. Guterman, Transformations of nonnegative integer-valued matrices that preserve the determinant (Russian). Uspekhi Mat. Nauk 58(6), 147–148 (2003); translation in Russian Math. Surveys **58**(6), 1200–1201 (2003)

132. A.È. Guterman, A.V. Mikhalëv, General algebra and linear mappings that preserve matrix invariants (Russian). Fundam. Prikl. Mat. **9**(1), 83–101 (2003); translation in J. Math. Sci. N.Y. **128**(6), 3384–3395 (2005)

133. R. Haggkvist, J.C.M. Janssen, All-even latin squares. Discrete Math. **157**, 199–206 (1996)

134. M. Hall, Jr., *The Theory of Groups* (Chelsea Publishing, New York, 1976)

135. T. Hawkins, The origins of the theory of group characters. Arch. History Exact Sci. **7**, 142–170 (1971)

136. T. Hawkins, Hypercomplex numbers, Lie groups, and the creation of group representation theory. Arch. History Exact Sci. **8**, 243–287 (1972)

137. T. Hawkins, New light on Frobenius' creation of the theory of group characters. Arch. History Exact Sci. **12**, 17–243 (1974)

138. T. Hawkins, Emergence of the theory of Lie groups, in *An Essay in the History of Mathematics 1869–1926*. Sources and Studies in the History of Mathematics and Physical Sciences (Springer, New York, 2000)

139. T. Hawkins, *The Mathematics of Frobenius in Context* (Springer, New York, 2013)

140. M.T. Heidemann, D.H. Johnson, C.S. Burrus, Gauss and the history of the fast Fourier transform. Arch. Hist. Exact Sci. **34**, 265–277 (1985)

141. H. Helling, Eine Kennzeichnung von Charakteren auf Gruppen und assoziativen Algebren. Commun. Algebra **1**, 491–501 (1974)

142. A.O.F. Hendrickson, Supercharacter theory constructions corresponding to Schur ring products. Commun. Algebra, **40**, 4420–4238 (2012)

143. H.-J, Hoehnke, *Konstruktive Methoden in der Theorie der Algebren. I*, vol. 2 (Monatsb. Deutsch. Akad. Wiss. Berlin, 1960), pp. 138–46

144. H.-J. Hoehnke, Uber Beziehungen zwischen Probleme von H. Brandt aus der Theorie der Algebren und den Automorphismen der Normenform. Math. Nachr. **34**, 229–255 (1967)

145. H.-J. Hoehnke, Spatial matrix systems and their theories. A contribution to bilinear algebra, in *Proceedings of the Second International Symposium, n-Ary Structures (Varna, 1983)* (Center for Applied Mathematics, Sofia, 1985), pp. 109–170

146. H.-J. Hoehnke, Einheit von Strukturtheorie der Algebren und Formentheorie. Preprint (unpublished)

147. H.-J. Hoehnke, K.W. Johnson, The 1-,2-, and 3-characters determine a group. Bull. Am. Math. Soc. **27**, 243–245 (1992)

148. H.-J. Hoehnke, K.W. Johnson, The 3-characters are sufficient for the group determinant, in *Proceedings of the Second International Conference on Algebra*. Contemporary Mathematics, vol. 184 (1995), pp. 193–206

149. H.-J. Hoehnke, K.W. Johnson, k-characters and group invariants. Commun. Algebra **26**, 1–27 (1998)

150. S.P. Humphries, Cogrowth of groups and the Dedekind-Frobenius group determinant. Math. Proc. Camb. Philos. Soc. **121**, 193–217 (1997)

151. S.P. Humphries, Weak Cayley table groups. J. Algebra **216**, 135–158 (1999)

152. S.P. Humphries, K.W. Johnson, Fusions of character tables and Schur rings of abelian groups. Commun. Algebra **36**, 1437–1460 (2008)

153. S.P. Humphries, K.W. Johnson, Fusions of character tables. II. p-groups. Commun. Algebra **37**, 4296–4315 (2009)

154. S.P. Humphries, E.L. (Turner) Rode, Weak Cayley tables and generalized centralizer rings of finite groups. Math. Proc. Camb. Philos. Soc. **153**, 281–318 (2012)

155. S.P. Humphries, E.L. Rode, A class of groups determined by their 3-S-rings. Rocky Mountain J. Math. **45**, 1–17 (2015)

156. S.P. Humphries, K.W. Johnson, B.L. Kerby, Fusions of character tables III: fusions of cyclic groups and a generalisation of a condition of Camina. Israel J. Math. **178**, 325–348 (2010)

157. S.P. Humphries, K.W. Johnson, A. Misseldine, Commutative S-rings of maximal dimension. Commun. Algebra **43**, 5298–5327 (2015)
158. B. Huppert, *Endliche Gruppen. I*. Die Grundlehren der Mathematischen Wissenschaften, vol. 134 (Springer, Berlin, 1967)
159. T. Hurley, Group rings and rings of matrices. Int. J. Pure Appl. Math. **31**, 319–335 (2006)
160. T. Hurley, Convolutional codes from units in matrix and group rings. Int. J. Pure Appl. Math. **50**, 431–463 (2009)
161. T. Hurley, I. McLoughlin, A group ring construction of the extended binary Golay code. IEEE Trans. Inform. Theory **54**, 4381–4383 (2008)
162. I.M. Isaacs, *Character Theory of Finite Groups* (Academic, New York, 1976)
163. N. Ito, *Lectures on Frobenius and Zassenhaus Groups* (University of Illinois at Chicago, 1969)
164. N. Jacobson, *The Theory of Rings* (American Mathematical Society, New York, 1943)
165. N. Jacobson, Some applications of Jordan forms to involutorial simple associative algebras. Adv. Math. **48**, 149–165 (1983)
166. K.W. Johnson, S-rings over loops, right mapping groups and transversals in permutation groups. Math. Proc. Camb. Philos. Soc. **89**, 433–443 (1981)
167. K.W. Johnson, Loop transversals and the centralizer ring of a permutation group. Math. Proc. Camb. Philos. Soc. **94**, 411–416 (1983)
168. K.W. Johnson, Latin square determinants, in *Algebraic, Extremal and Metric Combinatorics 1986*. London Mathematical Society Lecture Notes Series, vol. 131 (1988), pp. 146–154
169. K.W. Johnson, On the group determinant. Math. Proc. Camb. Philos. Soc. **109**, 299–311 (1991)
170. K.W. Johnson, Sharp Characters of quasigroups. Eur. J. Comb. **14**, 103–112 (1993)
171. K.W. Johnson, The Dedekind-Frobenius group determinant, new life in an old method, in *Proceedings, Groups St Andrews 97 in Bath, II*. London Mathematical Society Lecture Notes Series, vol. 261, (1999), pp. 417–428
172. K.W. Johnson, E. Poimenidou, A formal power series attached to a class function on a group and its application to the characterisation of characters. Math. Proc. Camb. Philos. Soc. **155**, 465–474 (2013)
173. K.W. Johnson, S.K. Sehgal, The 2-character table is not sufficient to determine a group. Proc. Am. Math. Soc. **119**, 1021–1027 (1993)
174. K.W. Johnson, S.K. Sehgal, The 2-characters of a group and the group determinant. Eur. J. Comb. **16**, 623–631 (1995)
175. K.W. Johnson, J.D.H. Smith, Characters of finite quasigroups. Eur. J. Comb. **5**, 43–50 (1984)
176. K.W. Johnson, J.D.H. Smith, Characters of finite quasigroups II. Ind uced characters. Eur. J. Comb. **7**, 131–137 (1986)
177. K.W. Johnson, J.D.H. Smith, A note on character induction in association schemes. Eur. J. Comb. **7**, 139 (1986)
178. K.W. Johnson, J.D.H. Smith, Characters of finite quasigroups. III. Quotients and fusion. Eur. J. Comb. **10**, 47–56 (1989)
179. K.W. Johnson, J.D.H. Smith, Characters of finite quasigroups V, Linear characters. Eur. J. Comb. **10**, 449–456 (1989)
180. K.W. Johnson, J.D.H. Smith, On the category of weak Cayley table morphisms between groups. Sel. Math. (N.S.) **13**, 57–67 (2007)
181. K.W. Johnson, J.D.H. Smith, Matched pairs, permutation representations, and the Bol property. Commun. Algebra **38**, 2903–2914 (2010)
182. K.W. Johnson, P. Vojtěchovský, Right division in groups, Dedekind-Frobenius group matrices, and Ward quasigroups. Abh. Math. Semin. Univ. Hambg. **75**, 121–136 (2005)
183. K.W. Johnson, S.Y. Song, J.D.H. Smith, Characters of finite quasigroups, VI. Critical examples and doubletons. Eur. J. Comb. **11**, 267–275 (1990)
184. K.W. Johnson, S. Mattarei, S.K. Sehgal, Weak Cayley tables. J. Lond. Math. Soc. **61**, 395–411 (2000)

185. K.W. Johnson, M. Munywoki, J.D.H. Smith, The upper triangular algebra loop of degree 4. Comment. Math. Univ. Carol. **55**, 457–470 (2014)

186. I.L. Kantor, A.S. Sodolodnikov, *Hypercomplex Numbers* (Springer, Berlin, 1989)

187. J. Karlof, The subclass algebra associated to a finite group and a subgroup. Trans. Am. Math. Soc. **207**, 329–341 (1975)

188. G. Karpilovsky, *Projective Representations of Finite Groups* (Marcel Dekker, New York, Basel, 1985)

189. M.G. Karpovsky, R.S. Stankovic, J. Astola, *Spectral Logic and Its Application for the Design of Digital Devices* (Wiley, London, 2008)

190. A. Kerber, *Applied Finite Group Actions*, 2nd edn. (Springer, Berlin, 1999)

191. M. Kervaire, On the parallelizability of the spheres. Proc. Nat. Acad. Sci. U.S.A. **44**, 280–283 (1958)

192. H.M. Khudaverdian, T.T. Voronov, A short proof of the Buchstaber-Rees theorem. Philos. Trans. R. Soc. London, Ser. A **369**, 1334–1345 (2011)

193. W. Kimmerle, On the characterization of finite groups by characters, in *The Atlas of Finite Groups: Ten Years on (Birmingham, 1995)*. London Mathematical Society Lecture Note Series, vol. 249 (Cambridge University Press, Cambridge, 1998), pp. 119–138

194. W. Kimmerle, K.W. Roggenkamp, Non-isomorphic groups with isomorphic spectral tables and Burnside matrices. Chin. Ann. Math. Ser. B **15**, 273–282 (1994)

195. W. Kimmerle, R. Sandling, Group-theoretic and group ring-theoretic determination of certain Sylow and Hall subgroups and the resolution of a question of R. Brauer. J. Algebra **171**, 329–346 (1995)

196. F. Klein, *Lectures on the Icosahedron and the Solution of Equations of the Fifth Degree* (Dover, New York, 2003)

197. W. Knapp, On Burnside's method. J. Algebra **175**, 644–660 (1995)

198. M.A. Knus, Quadratic forms, in *Clifford Algebras and Spinors*. Seminars in Mathematics, vol. 1 (Departamento de Matemática, University Campinas, Campinas, 1988)

199. D. Kosambi, U.V.R. Rao, The efficiency of randomization by card shuffling. J. R. Stat. Soc. A **128**, 223–233 (1958)

200. S. Lang, in *Cyclotomic Fields I and II. Combined*, 2nd edn. Graduate Texts in Mathematics, vol. 121 (Springer, New York, 1990)

201. C.R. Leedham-Green, S. McKay, *The Structure of Groups of Prime Power Order*. London Mathematical Society Monographs. Oxford Science Publications, vol. 27 (Oxford University Press, Oxford, 2002)

202. P.G. Lejeune Dirichlet, *Lectures on Number Theory* (American Mathematical Society, Providence, 1999)

203. K.H. Leung, S.L. Ma, The structure of Schur rings over cyclic groups. J. Pure Appl. Algebra **66**, 287–302 (1990)

204. K.H. Leung, S.H. Man, On Schur rings over cyclic groups. Israel J. Math. **106**, 251–267 (1998)

205. D.A. Levin, Y. Peres, E.L. Wilmer, *Markov Chains and Mixing Times* (American Mathematical Society, Providence, 2009)

206. T. Luczak, L. Pyber, On random generation of the symmetric group. Comb. Probab. Comput. **2**, 505–512 (1993)

207. I.G. MacDonald, *Symmetric Functions and Hall Polynomials*, 2nd edn. (Clarendon Press, Oxford, 1995)

208. G.W. Mackey, Harmonic analysis as the exploitation of symmetry—a historical survey. Bull. Am. Math. Soc. (N.S.) **3**, 543–698 (1980)

209. G.W. Mackey, *The Scope and History of Commutative and Noncommutative Harmonic Analysis*. History of Mathematics, vol. 5 (American Mathematical Society, Providence; London Mathematical Society, London, 1992)

210. W. Magnus, A. Karrass, D. Solitar, *Combinatorial Group Theory* (Dover, New York, 1976)

211. A. Mann, C.M. Scoppola, On p-groups of Frobenius type. Arch. Math. (Basel) **56**, 320–332 (1991)

212. R. Mansfield, A group determinant determines its group. Proc Am. Math. Soc **116**, 939–941 (1992)
213. J. McKay, D. Sibley, Brauer pairs with the same 2-characters. Preprint
214. D.K. Maslen, D.N. Rockmore, Separation of variables and the computation of Fourier transforms on finite groups. J. Am. Math. Soc. **10**, 169–214 (1997)
215. D.K. Maslen, D.N. Rockmore, The Cooley-Tukey FFT and group theory. Notices Am. Math. Soc. **48**, 1151–1160 (2001)
216. S. Mattarei, Character tables and metabelian groups. J. Lond. Math. Soc. **46**(2), 92–100 (1992)
217. S. Mattarei, Retrieving information about a group from its character table, Ph.D Thesis, University of Warwick, 1992
218. S. Mattarei, An example of p-groups with identical character tables and different derived lengths. Arch. Math. **62**, 12–20 (1994)
219. S. Mattarei, Retrieving information about a group from its character degrees or from its class sizes. Proc. Am. Math. Soc. **134**, 2189–2195 (2006)
220. K.E. Morrison, Spectral approximation of multiplication operators. N.Y. J. Math. **1**, 75–96 (1995)
221. M. Muzychuk, M. Klin, R. Pöschel, The isomorphism problem for circulant graphs via Schur ring theory, in *Codes and Association Schemes (Piscataway, 1999)*. DIMACS: Series in Discrete Mathematics and Theoretical Computer Science, vol. 56 (American Mathematical Society, Providence, 2001), pp. 241–264
222. G.P. Nagy, P. Vojtěchovský, LOOPS: computing with quasigroups and loops in GAP, version 2.2.0. http://www.math.du.edu/loops
223. P.M. Neumann, The context of Burnside's contribution to group theory, in *The Collected Papers of William Burnside*, ed. by P.M. Neumann, A.J.S. Mann, J.C. Thompson (Oxford University Press, Oxford, 2004), 15–37
224. M. Niemenmaa, T. Kepka, On connected transversals to abelian subgroups in finite groups. Bull. Lond. Math. Soc. **24**, 343–346 (1992)
225. H.J. Nussbaumer, *Fast Fourier Transform and Convolution Algorithms* (Springer, Berlin, 1981)
226. L. Nyssen, Pseudo-représentations. Math. Ann. **306**, 257–283 (1996)
227. A. Okounkov, A.M. Vershik, A new approach to the representation theory of the symmetric groups. Sel. Math. (N.S.) **2**, 581–605 (1996)
228. S. Okubo, *Introduction to Octonian and Other Non-associative Algebras in Physics* (Cambridge University Press, Cambridge, 1995)
229. D.V. Ouellette, Schur complements and statistics. Linear Algebra Appl. **36**, 187–295 (1981)
230. J.C. Parnami, M.K. Agrawal, A.R. Rajwade. On some trigonometric diophantine equations of the type $\sqrt{n} = c_1 \cos(\pi d_1) + \ldots + c_\lambda \cos(\pi d_\lambda)$. Acta Math. Acad. Sci. Hung. **37**, 423–432 (1981)
231. A. Pfister, Zur Darstellung von -1 als Summe von Quadraten in einem Körper. J. Lond. Math. Soc. **40**, 159–165 (1965)
232. A. Pfister, Multiplikative quadratische Formen. Arch. Math. **16**, 363–370 (1965)
233. H. Poincaré, Sur les nombres complexes. Comptes Rendues Acad. Sci. Paris **99**, 740–742 (1884), Oeuvres **5**, 77–79
234. H. Poincaré, Sur l'intégration des équations linéaires et les périodes des intégrales abéliennes. J. des Math. Pures Appl. **9**(5), 139–212 (1903), Oeuvres **3**, 106–166
235. H. Poincaré, *Calcul des Probabilités* (Gautier-Villars, Paris, 1912)
236. H. Poincaré, *Papers on Fuchsian functions*. J. Stillwell (translator) (Springer, New York, 1985)
237. R. Poschel, Untersuchungen von S-ringen, insbesondere im Gruppenring von p-Gruppen. Math. Nachr. **60**, 1–27 (1974)
238. C. Praeger, Quasiprimitivity: structure and combinatorial applications. Discret. Math. **264**, 211–224 (2003)
239. C. Procesi, The invariant theory of $n \times n$ matrices. Adv. Math. **19**, 306–381 (1976)

240. R. Reams, S. Waldron, Isometric tight frames. Electron. J. Linear Algebra **9**, 122–128 (2002)
241. D.St.P. Richards, Algebraic methods toward higher order probability inequalities II. Ann. Probab. **32**, 1509–1544 (2004)
242. E.L. Rode, On a generalized centralizer ring of a finite group which determines the group. Algebra Colloq. **1** 31–50 (2019)
243. F. Roesler, Darstellungstheorie von Schur-Algebren. Math. Zeitshrift **125**, 32–58 (1972)
244. M. Roitman, A complete set of invariants for finite groups. Adv. Math. **41**, 301–311 (1981)
245. M.H. Rosas, B.E. Sagan, Symmetric functions in noncommuting variables. Trans. Am. Math. Soc. **358**, 215–232 (2006)
246. G.-C. Rota, J. Shen, On the combinatorics of cumulants. J. Comb. Theory A **91**, 283–304 (2000)
247. R. Rouquier, Caractérisation des caractères et pseudo-caractères. J. Algebra **180**, 571–586 (1996)
248. L. Saloff-Coste, *Random Walks on Finite Groups, in Probability on Discrete Structures* Encyclopaedia of Mathematical Sciences, vol. 110 (Springer, Berlin, 2004), pp. 263–346
249. S. Sahi, Higher correlation inequalities. Combinatorica **28**, 209–227 (2008)
250. J.W. Sands, Base change for higher Stickelberger ideals. J. Number Theory **73**, 518–526 (1998)
251. I. Schur, Über die Darstellung der Endliche Gruppen durch gebrochenen lineare Substitutionen. J. reine angew. Math. **127**, 20–50 (1904), Ges. Abh. **I**, 86–116
252. I. Schur, *Neuer Begründung der Theorie der Gruppencharaktere* (Sitzungsber. Akad. Wiss, Berlin, 1905), pp. 406–432; Ges. Abh **I**, 143–169
253. I. Schur, Untersuchungen über die Darstellung der Endliche Gruppen durch gebrochenen lineare Substitutionen. J. Reine Angew. Math. **132**, 85–137 (1907); Ges. Abh. **I**, 198–250
254. I. Schur, Neuer Beweis eines Satzes von W. Burnside. Jahresber. Deutsch. Math.-Verein **17**, 171–176 (1908); Ges. Abh **I**, 266–271
255. I. Schur, Zur Theorie der einfach transitiven Permutationsgruppen. Sitzungsber. Preuss. Akad. Wiss. Phys-Math Klasse 598–623 (1933); Ges. Abh III, 266–291
256. W.R. Scott, Half homomorphisms of groups. Proc. Am. Math. Soc. **8**, 1141–1144 (1957)
257. W.R. Scott, *Group Theory* (Prentice Hall, Englewood Cliffs, 1964)
258. J.A. Sjogren, Connectivity and spectrum in a graph with a regular automorphism group of odd order. Internat. J. Algebra Comput. **4**, 529–560 (1994)
259. J.D.H. Smith, Centraliser rings of multiplication groups of quasigroups. Math. Proc. Camb. Philos. Soc. **79**, 427–431 (1976)
260. J.D.H. Smith, Induced class functions are conditional expectations. Eur. J. Comb. **10**, 293–296 (1989)
261. J.D.H. Smith, Combinatorial characters of quasigroups, in *Coding Theory and Design Theory, Part I: Coding Theory*, ed. by D. Ray-Chaudhuri (Springer, New York, 163–187, 1990)
262. J.D.H. Smith, A left loop on the 15-sphere. J. Algebra **176**, 128–138 (1995)
263. J.D.H. Smith, *An Introduction to Quasigroups and Their Representations* (Chapman and Hall, London, 2006)
264. J.D.H. Smith, A.B. Romanowska, *Post-Modern Algebra* (Wiley, New York, 1999)
265. T.P. Speed, Cumulants and partition lattices. Aust. J. Stat. **25**, 378–388 (1983)
266. A. Speiser, Gruppendeterminante und Körperdiskriminante. Math. Ann. **77**, 546–562 (1916)
267. R.S. Stanković, C. Moraga, J. Astola, *Fourier Analysis on Finite Groups with Applications in Signal Processing and System Design* (Wiley-IEEE Press, Hoboken, 2005)
268. D. Stanton, Orthogonal polynomials and Chevalley groups, in *Special Functions: Group Theoretical Aspects and Applications*. Mathematics and Its Application (Reidel, Dordrecht, 1984), pp. 87–128
269. R.P. Stanley, Invariants of finite groups and their relations to combinatorics. Bull. Am. Math. Soc. (New series) **I**, 475–511 (1979)

270. L. Takács, Harmonic analysis on Schur algebras and its applications in the theory of probability, in *Probability Theory and Harmonic Analysis*, ed. by J.-A. Chao, W.A. Woyczyński (Marcel Dekker, New York, 1986)

271. O. Tamaschke, *Schur-Ringe, (Vorlesungen an der Univ. Tübingen 1969)* (Bibliographisches Institut, Mannheim, 1970)

272. O. Taussky, A note on group matrices. Proc. Am. Math. Soc. **6**, 984–986 (1955)

273. O. Taussky, Matrices of rational integers. Bull. Am. Math. Soc. **66**, 327–345 (1960)

274. O. Taussky, History of sums of squares in algebra. American mathematical heritage: algebra and applied mathematics (El Paso, Tex., 1975/Arlington, Tex., 1976), pp. 73–90, Math. Ser., 13, Texas Tech Univ., Lubbock, Tex., 1981

275. R.L. Taylor, Galois representations associated to Siegel modular forms of low weight. Duke Math. J. **63**, 281–332 (1991)

276. A. Terras, *Fourier Analysis on Finite Groups and Applications*. London Mathematical Society Student Texts, vol. 43 (Cambridge University Press, Cambridge, 1999)

277. J. Tits, Free subgroups in linear groups. J. Algebra **20**, 250–270 (1972)

278. C.B. Thomas, Chern classes of representations. Bull. Lond. Math. Soc. **18**, 225–240 (1986)

279. R. Tolmieri, M. An, C. Lu, *Algorithms for Discrete Fourier Transforms and Convolutions* (Springer, New York, 1989)

280. E. Trachtenberg, Singular value decomposition of Frobenius matrices for approximate and multi-objective signal processing tasks, in *SVD and Signal Processing*, ed. by E. Deprettere (1988), pp. 331–345

281. E.L. Turner (Rode) k-S-rings, Thesis, Brigham Young University, 2012

282. B.L. van der Waerden, *Algebra I and II* (Springer, New York, 2003)

283. V.S. Varadarajan, *Supersymmetry for Mathematicians: An Introduction*. Courant Lecture Notes in Mathematics, vol. 11 (American Mathematical Society, Providence, 2004)

284. M.J. Vazirani, Extending Frobenius' higher characters. Sci. Math. Jpn. **58**, 169–182 (2003)

285. A.L. Vyshnevetskiy, E.M. Zhmud', Random walks on finite groups converging after finite number of steps. Algebra Discret. Math. **2**, 123–129 (2008)

286. S. Waldron, Tight frames, in *Finite Frames, Theory and Appications*, ed. by P.G. Casazza, G. Kutyniok (Birkhäuser, Basel, 2013), pp. 171–192

287. M. Ward, Postulates for the inverse operations in a group. Trans. Am. Math. Soc. **32**, 520–526 (1930)

288. W.C. Waterhouse, Composition of norm-type forms. J. Reine Angew. Math. **353**. 85–97 (1984)

289. H. Weber, Theorie der Abel'schen Zahlkörper I, section 3. Acta. Math. Bd **8**, 193–263 (1886)

290. H. Weber, Theorie der Abel'schen Zahlkörper IV, sections 2, 3. Acta. Math. Bd **9**, 105–130 (1887)

291. H. Weber, *Lehrbuch der Algebra*, vol. III (Chelsea Publishing Company, White River Junction, 1961)

292. B. Weisfeiler, *On Construction and Identification of Graphs*. Springer Lecture Notes in Mathematics, vol. 558 (Springer, Berlin, 1976)

293. H. Wielandt, *Finite Permutation Groups* (Academic, New York, 1964)

294. H. Wielandt, *Mathematical Works, Volume 1: Group Theory* (Walter de Gruyter, Berlin, New York, 1994)

295. E. Wigner, Restriction of irreducible representations of groups to a subgroup. Proc. R. Soc. Lond. Ser. A **322**, 181–189 (1971)

296. A. Wiles, On ordinary λ-adic representations associated to modular forms. Invent. Math. **94**, 529–573 (1988)

297. P.-H. Zieschang, *An Algebraic Approach to Association Schemes*. Springer Lecture Notes in Mathematics, vol. 1626 (Springer, Berlin, 1996)

298. D. Zipliess, Dividierte Potenzen, Determinanten und die Algebra der verallgemeinerten Spurpolynome. Dissertation, University of Düsseldorf, 1983

Index

© Springer Nature Switzerland AG 2019
K. W. Johnson, *Group Matrices, Group Determinants*
and Representation Theory, Lecture Notes in Mathematics 2233,
https://doi.org/10.1007/978-3-030-28300-1

Symbol Index

© Springer Nature Switzerland AG 2019

K. W. Johnson, *Group Matrices, Group Determinants and Representation Theory*, Lecture Notes in Mathematics 2233, https://doi.org/10.1007/978-3-030-28300-1

LECTURE NOTES IN MATHEMATICS   🐎 Springer

Editors in Chief: J.-M. Morel, B. Teissier;

Editorial Policy

1. Lecture Notes aim to report new developments in all areas of mathematics and their applications – quickly, informally and at a high level. Mathematical texts analysing new developments in modelling and numerical simulation are welcome.

   Manuscripts should be reasonably self-contained and rounded off. Thus they may, and often will, present not only results of the author but also related work by other people. They may be based on specialised lecture courses. Furthermore, the manuscripts should provide sufficient motivation, examples and applications. This clearly distinguishes Lecture Notes from journal articles or technical reports which normally are very concise. Articles intended for a journal but too long to be accepted by most journals, usually do not have this "lecture notes" character. For similar reasons it is unusual for doctoral theses to be accepted for the Lecture Notes series, though habilitation theses may be appropriate.

2. Besides monographs, multi-author manuscripts resulting from SUMMER SCHOOLS or similar INTENSIVE COURSES are welcome, provided their objective was held to present an active mathematical topic to an audience at the beginning or intermediate graduate level (a list of participants should be provided).

   The resulting manuscript should not be just a collection of course notes, but should require advance planning and coordination among the main lecturers. The subject matter should dictate the structure of the book. This structure should be motivated and explained in a scientific introduction, and the notation, references, index and formulation of results should be, if possible, unified by the editors. Each contribution should have an abstract and an introduction referring to the other contributions. In other words, more preparatory work must go into a multi-authored volume than simply assembling a disparate collection of papers, communicated at the event.

3. Manuscripts should be submitted either online at www.editorialmanager.com/lnm to Springer's mathematics editorial in Heidelberg, or electronically to one of the series editors. Authors should be aware that incomplete or insufficiently close-to-final manuscripts almost always result in longer refereeing times and nevertheless unclear referees' recommendations, making further refereeing of a final draft necessary. The strict minimum amount of material that will be considered should include a detailed outline describing the planned contents of each chapter, a bibliography and several sample chapters. Parallel submission of a manuscript to another publisher while under consideration for LNM is not acceptable and can lead to rejection.

4. In general, **monographs** will be sent out to at least 2 external referees for evaluation.

   A final decision to publish can be made only on the basis of the complete manuscript, however a refereeing process leading to a preliminary decision can be based on a pre-final or incomplete manuscript.

   Volume Editors of **multi-author works** are expected to arrange for the refereeing, to the usual scientific standards, of the individual contributions. If the resulting reports can be forwarded to the LNM Editorial Board, this is very helpful. If no reports are forwarded or if other questions remain unclear in respect of homogeneity etc, the series editors may wish to consult external referees for an overall evaluation of the volume.

5. Manuscripts should in general be submitted in English. Final manuscripts should contain at least 100 pages of mathematical text and should always include

   - a table of contents;
   - an informative introduction, with adequate motivation and perhaps some historical remarks: it should be accessible to a reader not intimately familiar with the topic treated;
   - a subject index: as a rule this is genuinely helpful for the reader.
   - For evaluation purposes, manuscripts should be submitted as pdf files.

6. Careful preparation of the manuscripts will help keep production time short besides ensuring satisfactory appearance of the finished book in print and online. After acceptance of the manuscript authors will be asked to prepare the final LaTeX source files (see LaTeX templates online: https://www.springer.com/gb/authors-editors/book-authors-editors/manuscriptpreparation/5636) plus the corresponding pdf- or zipped ps-file. The LaTeX source files are essential for producing the full-text online version of the book, see http://link.springer.com/bookseries/304 for the existing online volumes of LNM). The technical production of a Lecture Notes volume takes approximately 12 weeks. Additional instructions, if necessary, are available on request from lnm@springer.com.

7. Authors receive a total of 30 free copies of their volume and free access to their book on SpringerLink, but no royalties. They are entitled to a discount of 33.3 % on the price of Springer books purchased for their personal use, if ordering directly from Springer.

8. Commitment to publish is made by a *Publishing Agreement*; contributing authors of multiauthor books are requested to sign a *Consent to Publish form*. Springer-Verlag registers the copyright for each volume. Authors are free to reuse material contained in their LNM volumes in later publications: a brief written (or e-mail) request for formal permission is sufficient.

Addresses:
Professor Jean-Michel Morel, CMLA, École Normale Supérieure de Cachan, France
E-mail: moreljeanmichel@gmail.com

Professor Bernard Teissier, Equipe Géométrie et Dynamique,
Institut de Mathématiques de Jussieu – Paris Rive Gauche, Paris, France
E-mail: bernard.teissier@imj-prg.fr

Springer: Ute McCrory, Mathematics, Heidelberg, Germany,
E-mail: lnm@springer.com